建筑中國
ARCHITECTURE CHINESE
COMMERCLAL BUILDINGS
商·业·楼·盘

建筑中国编委会 主编

中国林业出版社

图书在版编目（CIP）数据

商业楼盘 /《建筑中国》丛书编委会 主编 . -- 北京 : 中国林业出版社 , 2013.9
（建筑中国）
ISBN 978-7-5038-7170-2
Ⅰ . ①商… Ⅱ . ①建… Ⅲ . ①商业建筑 - 建筑设计 - 作品集 - 中国 - 现代 Ⅳ . ① TU247
中国版本图书馆 CIP 数据核字 (2013) 第 199283 号

中国林业出版社 · 建筑与家居图书出版中心
责任编辑：李 顺　唐 杨
出版咨询：（010）83223051

出 版：中国林业出版社（100009 北京西城区德内大街刘海胡同 7 号）
网 站：http://lycb.forestry.gov.cn/
印 刷：北京卡乐富印刷有限公司
发 行：中国林业出版社
电 话：（010）83224477
版 次：2015年3月第1版
印 次：2015年3月第1次
开 本：889mm × 1194mm 1 ／ 8
印 张：65
字 数：200 千字
定 价：598 .00 元

Preface 前言

商业建筑不同于住宅写字楼，商业地产项目的服务内容和功能结构，已远远超出了常见的特定建筑类型范畴。对于建筑师而言，无疑是一个新的挑战。当代商业旗舰店的特点、创作变得更加自信，表意更加丰富、深刻，所表现出的思想观念和技术水平也更为先进。

商业建筑的外立面就是这种社会文化的媒介，它是商业文化的表征。基于“建筑　商业建筑表层与建筑实体等价”的新美学观念，给建筑带来更广阔、更深远的空间意象。在建筑的形态和意义上也给人以强烈的，视觉冲击和全新体验。以实现商业文化的强势传播。

新一代建筑师在建筑创作过程中，通过沉重的构件与粗糙的表面，来创造真实存在的气氛，极力地夸张建筑的物质性与真实感。使建筑自身呈现出新的逻辑关系及审美价值，也使人们身处不同的视觉角度，体验到不同的空间和时间的变化。

建筑的商业化已经促成了商业建筑外观设计手法的急剧变化，各种设计手法五花八门，层出不穷。经过建筑师们的努力创新，新的设计手法正在逐渐走向成熟，并产生了很多新的建筑语汇。更好地满足人们的物质和精神生活的需要。　本书多高层建筑，其社会价值和经济价值是显而易见。首先，其使人口高度集中，可利用建筑内部的竖向和横向交通缩短部门之间的联系距离，从而使公司和企业能够高效运作；其次建筑用地的面积大幅缩小，使得我们的大型企业在城市中心选址成为了可能；再次，更可以减少市政建设投资和缩短建筑工期。本书的案例皆从各个方面体现了其诸多优越性。

正是因为其诸多的优越性，我们更不能不考虑其存在的弊端，商业建筑随着层数的增加，其功能适用性、技术合理性和经济可行性，以及对其周围的环境影响都将发生了质的变化，这是我们设计师和建筑工程公司面临的巨大挑战。

本书中的案例，无一不是从城市规划、环境效益、人防设备、装修水平等问题上着力入手，逐一解决，都成为了当地的地标性建筑。

我们希望我国的设计师和建筑企业都能看到这本书，并从中获取自己想要的那部分内容，这也是我们策划出版本书的初衷。在编排和收集项目时，纰漏之处在所难免，我们恳切希望我们的设计师朋友不吝指正。

编者

2014年7月

Contents 目录

广州太古汇

建设地点：广州
建设单位：广州市设计院
项目名称：广州太古汇
项目类型：商业综合体

太古汇 — 太古地产位于广州的大型综合发展项目，由太古地产开发建设并运营管理。项目位于广州市天河中央商务区核心地段，总楼面面积约 35.8 万 m^2（不含文化中心），由一个大型购物商场、两座甲级办公楼、一个文化中心、广州首家文华东方酒店及酒店式服务住宅构成。该项目由广州市设计院设计，并由太古地产管理。传承集团多年国际化专业经验，太古地产悉心打造下的太古汇将成为集休闲娱乐、商贸活动、文化艺术于一体的综合商业体。

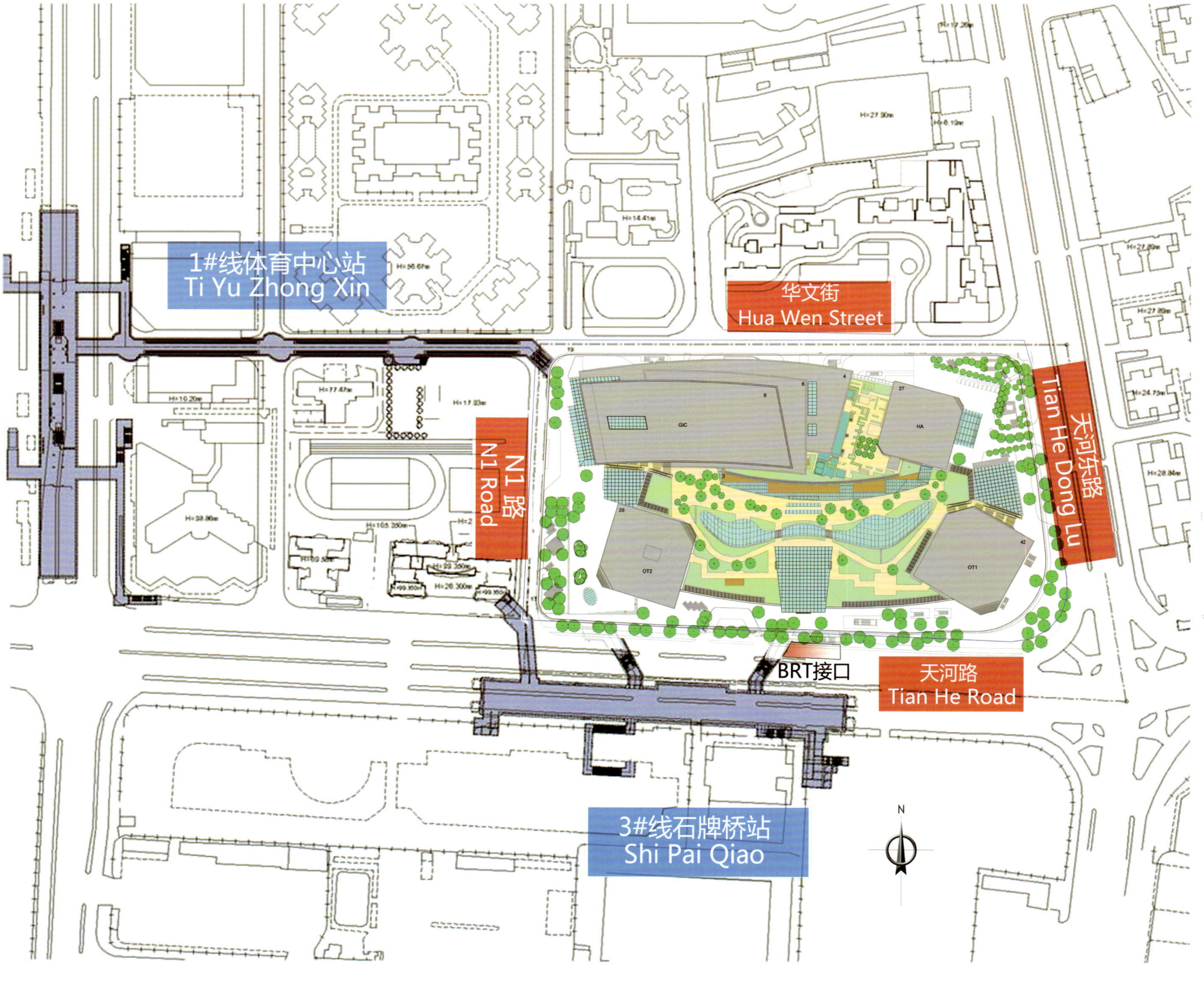

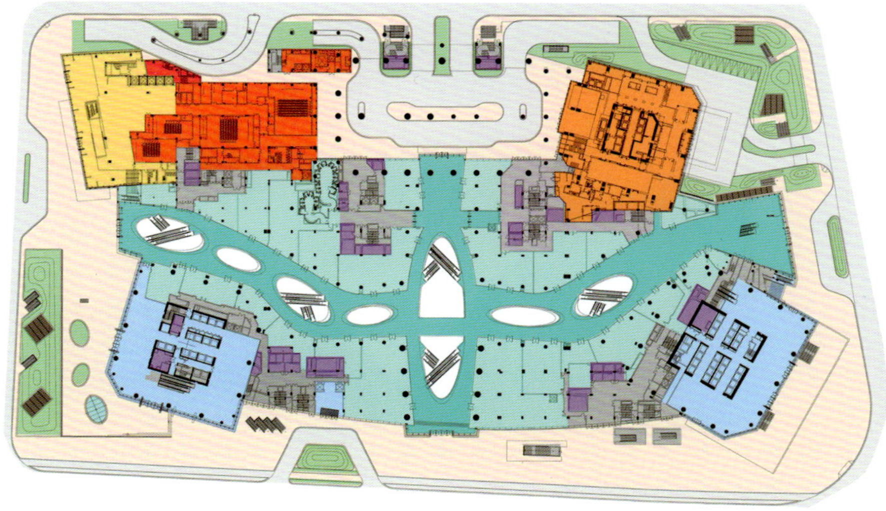

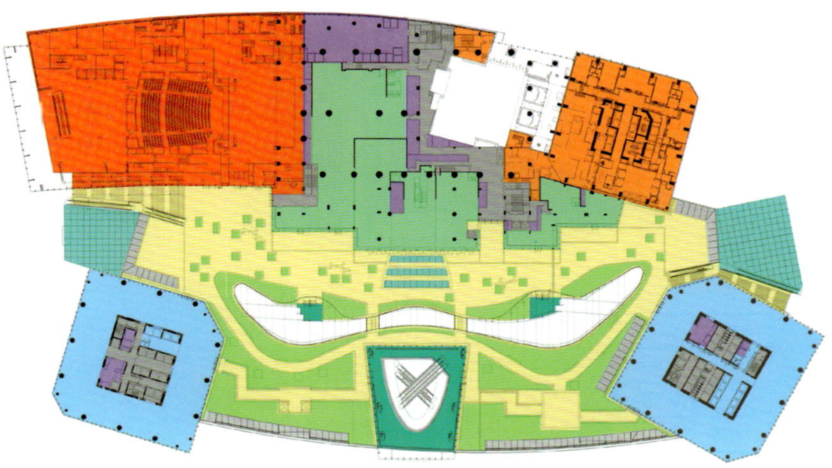

太古汇

PRADA

PRADA
PRADA
COACH
PRADA
CHANEL

Guangzhou shengfeng square

广州圣丰广场

建设地点：广东省广州市
建设单位：广东省建筑设计研究院
建筑面积：154663m^2
建筑层数：40 层

广州圣丰广场位于广州市广州大道 366 号，地处繁华的天河中心商业圈。项目主要由甲级写字楼、五星酒店、会议展览区和地下车库等部分组成。

建筑平面布局合理紧凑，交通组织安排有条不紊，最大限度的吸引商业人流，疏导办公人流。酒店、办公相对布置，充分利用每一分空间，同时又保留较大的舒适度。

立面造型，生动活泼。以流动的弧线作为造型的主要元素。写字楼以椭圆形平面为蓝本逐层收分，形成高耸挺拔的形象；酒店以"L"形平面和写字楼组合，形成了良好的外部空间，成为天河商务中心的一道靓丽风景线。

本工程为大底盘的超高层双塔楼建筑，其中北塔楼为 40 层超高层办公楼，建筑高度为 196m，采用框架－核心筒结构；南塔楼为 27 层豪华酒店，建筑高度为 103m，采用框架－剪力墙结构。部分塔楼及裙楼的竖向构件不连续，分别形成 25m 跨度钢骨梁承托 4 层、20m 跨度桁架承托 24 层及 19m 跨度跨层钢骨梁承托 20 层等多个大跨度高位转换结构；在两个塔楼中间的主入口大广场设置大跨度单层网壳结构雨棚，雨蓬外形近似半球面，直径约为 32m。

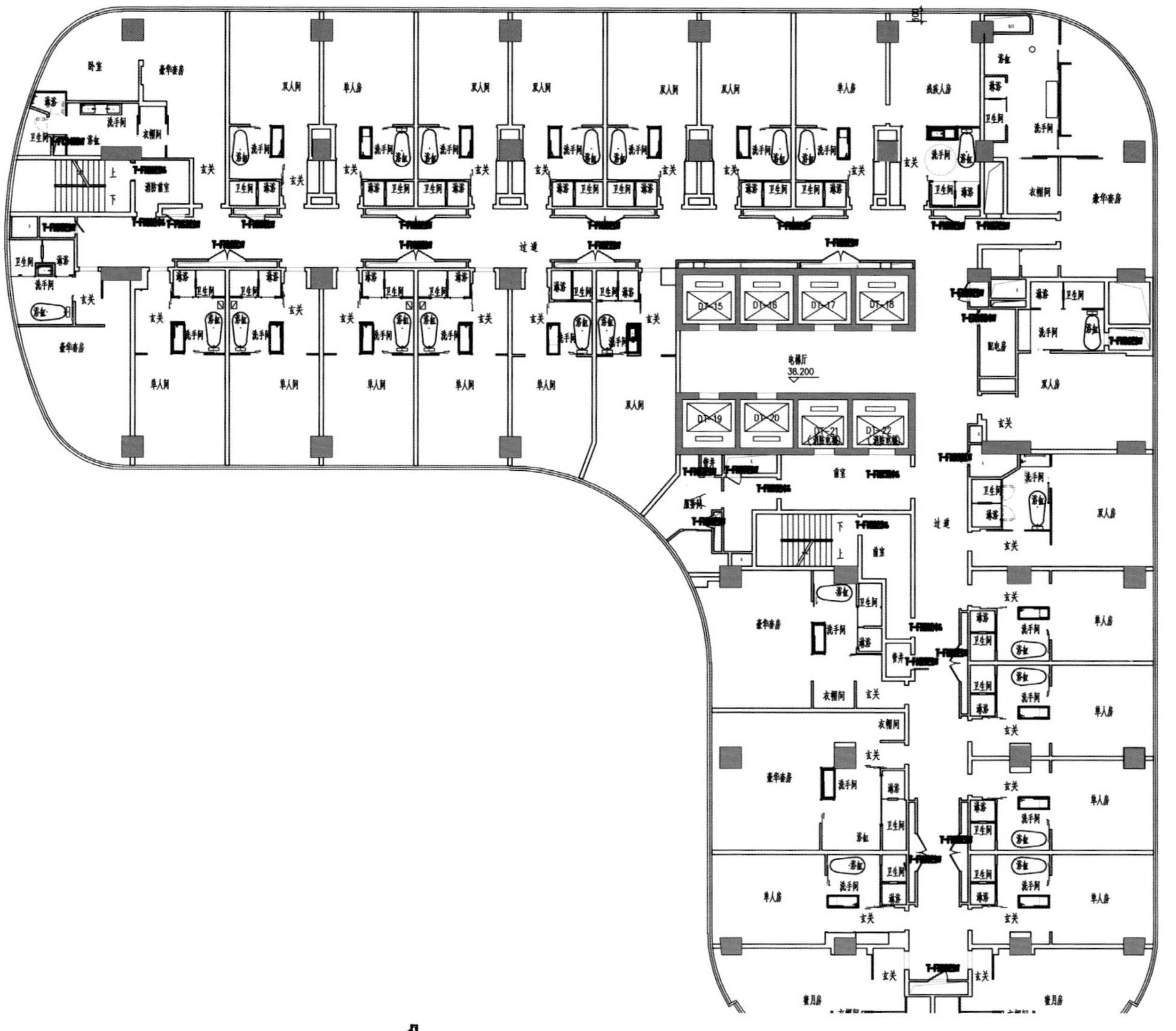

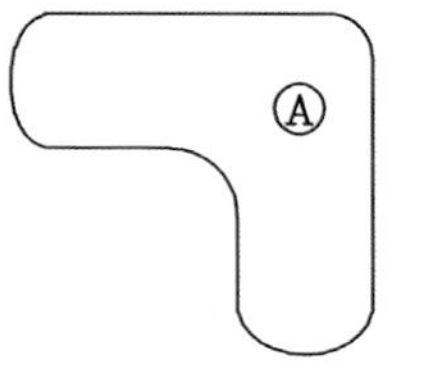

防火分区	面积(㎡)
A	1722

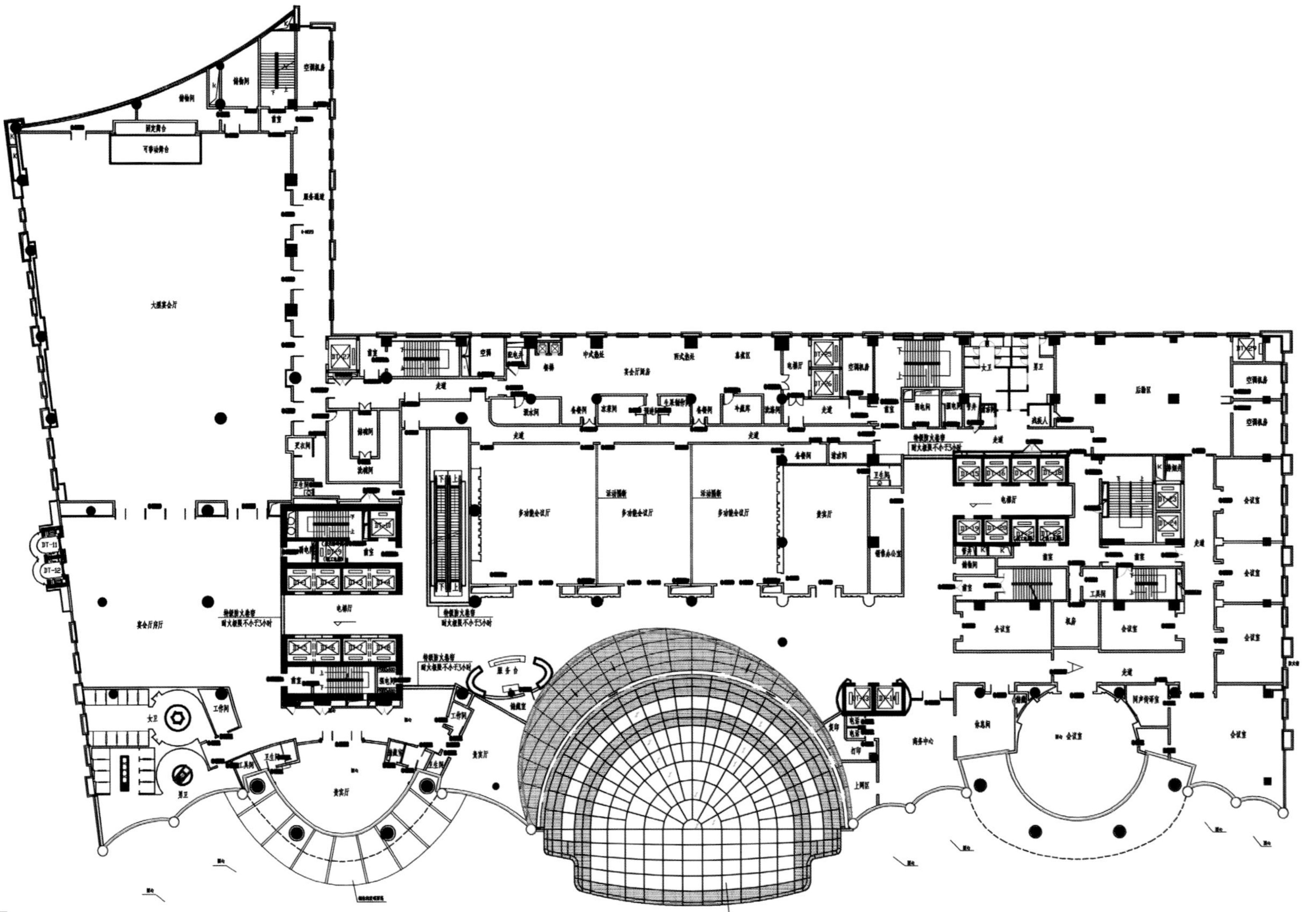

防火分区	面积(㎡)
3-1	1998
3-2	1989
3-3	1743

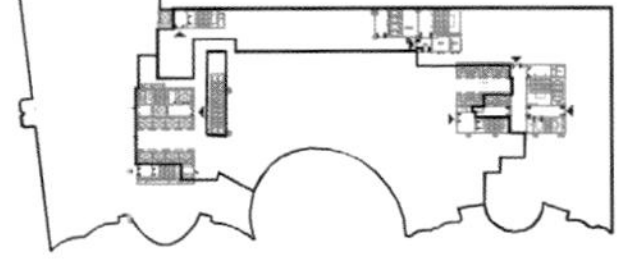

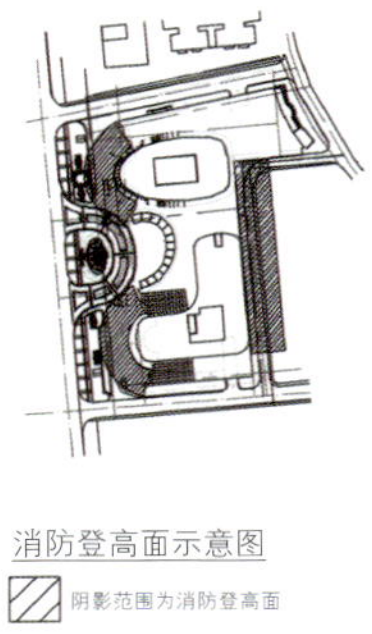

消防登高面示意图

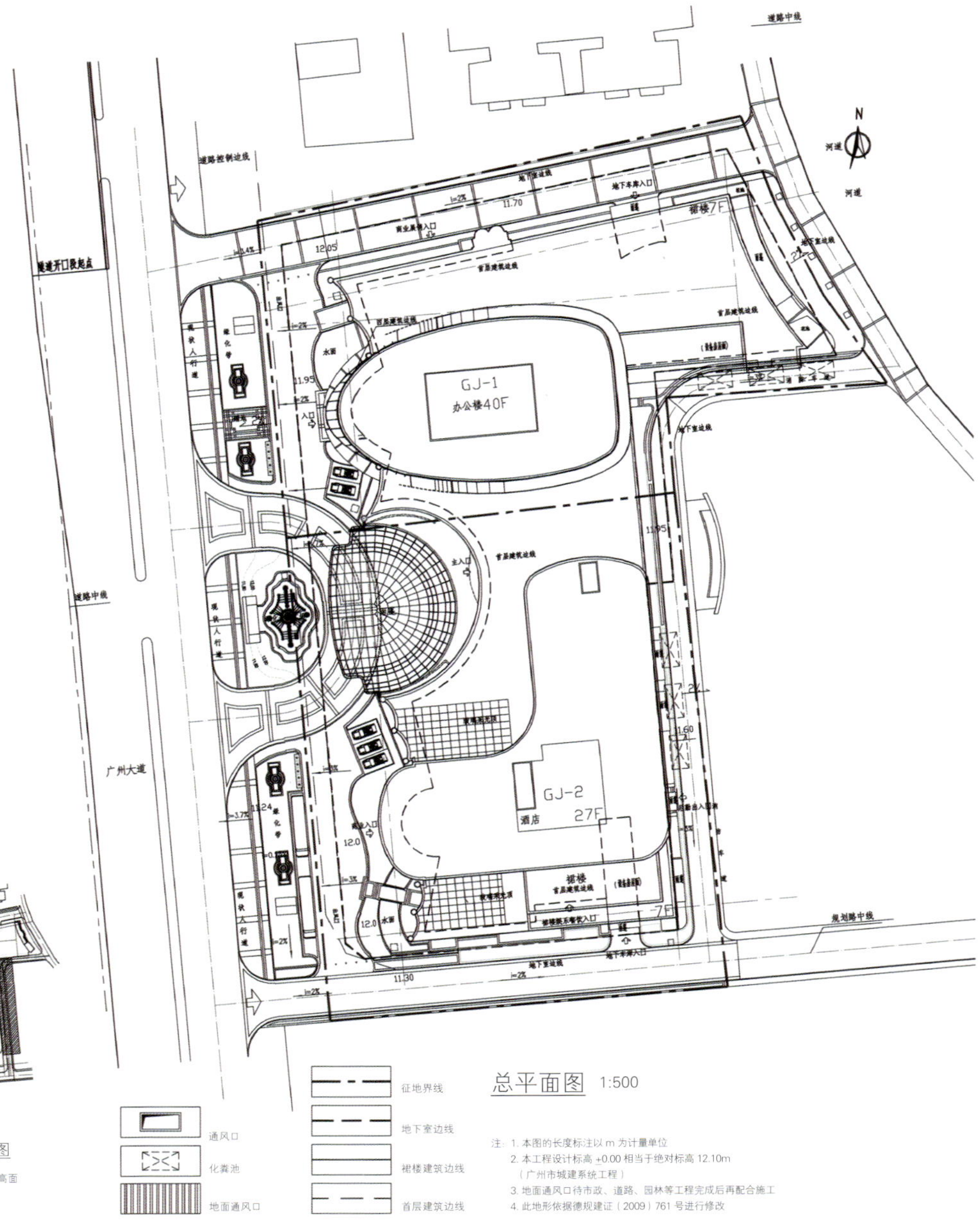

总平面图 1:500

注：1. 本图的长度标注以 m 为计量单位
2. 本工程设计标高 ±0.00 相当于绝对标高 12.10m（广州市城建系统工程）
3. 地面通风口待市政、道路、园林等工程完成后再配合施工
4. 此地形依据德规建证（2009）761 号进行修改

Guiyang international conference exhibition center

贵阳国际会议会展中心

建筑设计：华阳国际设计集团
项目地址：贵阳市金阳区
用地面积：总图设计 51.8hm²（总规划 120.6hm²）
建筑面积：88.8hm²
建筑高度：201m
容 积 率：1.76

项目概况

贵阳国际会议展览中心项目位于贵州省贵阳市金阳新区迎宾路以南，观山东路以北，长岭北路以西，紧临贵阳市市政中心，西侧拥有原生态景观的观山公园。由中天城投集团投资建设。做为一个以会展会议为核心的城市综合体开发项目，其占地51.8hm²，集会展中心、会议中心、总部办公、五星酒店、SOHO办公和公寓、商业、餐饮和文化娱乐等功能于一体。

规划布局

项目以“生态筑成－会通天下”、“山水会展－原乡国际”作为规划主题，以一心、两轴、四片作为总体规划结构，共分A、B、C、D四个区，共14个建筑子项。其中，会展中心约13万m²，会议中心约6.5万m²，总部办公约5.1万m²，五星酒店约7万m²，SOHO公寓和办公约18.5万m²，各种形态商业合计30万m²，地下车库13万m²。与该项目紧邻的南北用地，均规划设计为会展城居住小区。项目本身深刻体现了多种要素之间的统筹运行于良性互动。将基地内各类复合功能空间，按照城市街区的设计原理和城市活动特征，进行有效生动的城市空间组合。因此，“贵阳国际会议展览中心”项目是一座大规模的，以会展会议为依托的城市综合体建筑。它的建成不仅将成为贵阳对外经贸活动的重要窗口，同时它也将成为激活金阳新区城市生活的舞台。

会展中心↑　　会议中心↓

201 大厦

SOHO 区

风情商业街

Beijing international logistics port of flowers

北京国际花卉物流港

建筑设计：清华大学建筑设计研究院
项目地址：北京顺义区
建筑面积：166492m²

项目概况

第七届中国花卉博览会主场馆——北京国际花卉物流港工程由1、2、3号展馆、物流中心与交易中心组成，地上6层、地下1层，建筑最大高度为44m，总建筑面积为166492m²，建设规模较大。

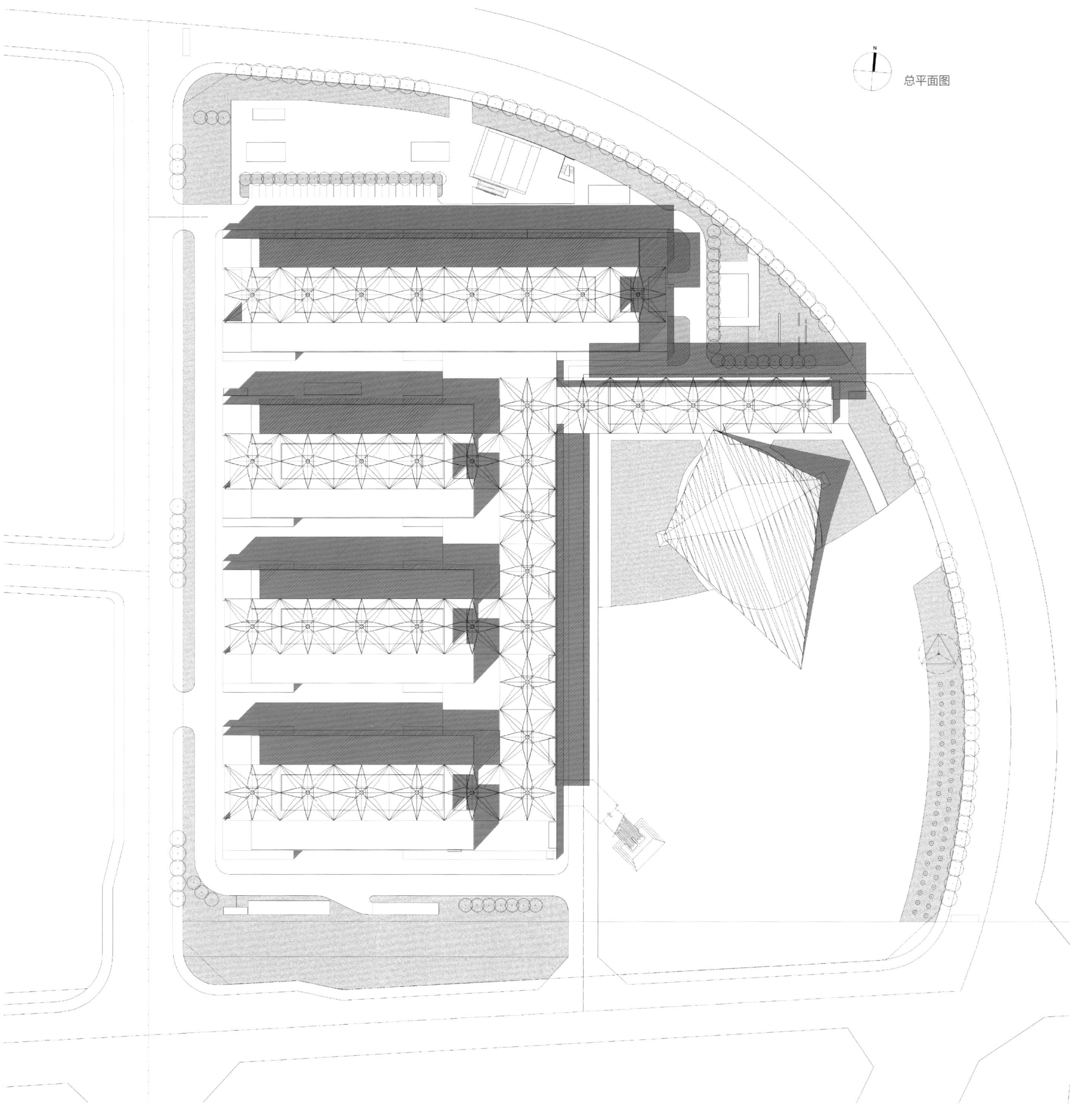

总平面图

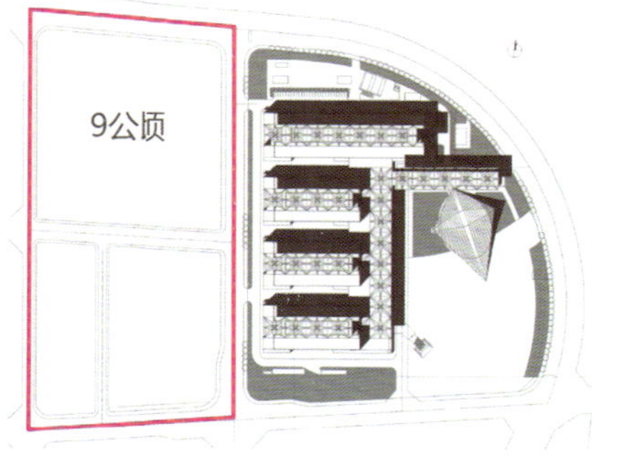

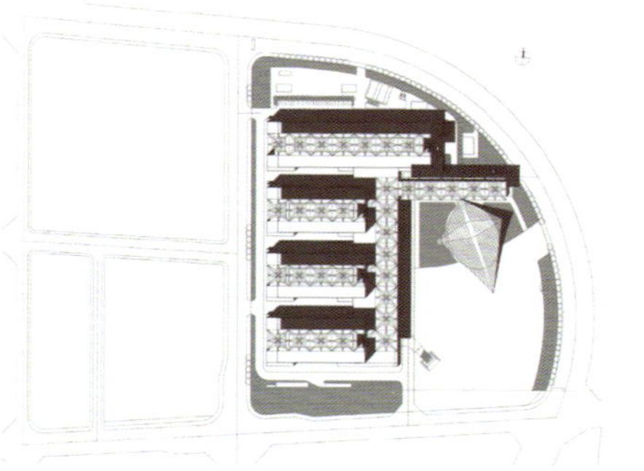

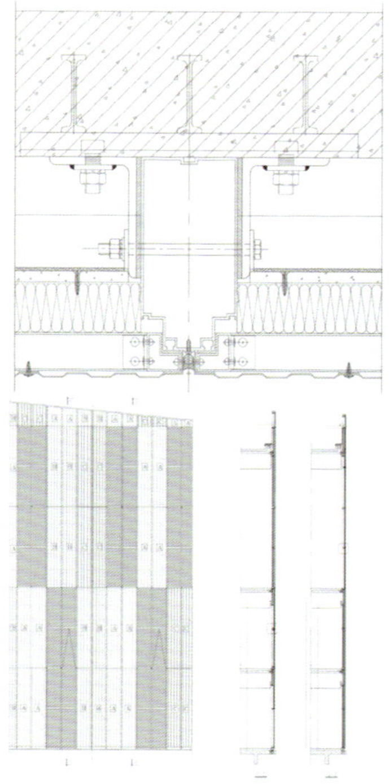

设计遵循可持续发展理念，综合考虑了节约土地、减少建筑能耗、收集利用场地内部雨水、节约建筑材料、缩短施工建设周期等五个要素，提出“五节”的设计策略，即：节地、节能、节水、节材、节时。

节地

采取紧凑化布局将场地分为东西两块，主体建筑集中布置在交通便利的东侧场地，西侧约9hm^{2}土地节约出来，会时作为售票、停车及展销使用，会后可进行二次开发建设。

节材

主体结构采取模数化钢结构体系，材料生产、施工及回收皆环保经济。建筑造型方正规矩模数化，节约材料使用。建筑外墙采用单元体幕墙体系，构造连接简单易于施工。

节时

在场地内满布12*12m网格矩阵，主体建筑及室外工程景观皆以此为基本建造模数。建筑主体采用钢结构体系，做到工厂批量生产。标准化构件在工地组装，实现模块化预加工体系与建造过程的并行化，有效缩短工期。

节水

通过花伞单元收集屋面雨水，过滤后可作为景观灌溉、马桶冲洗及冲洗外墙。全年约可回收雨水34000吨，占全年用水量的14%，非传统水源利用率达到20%以上。

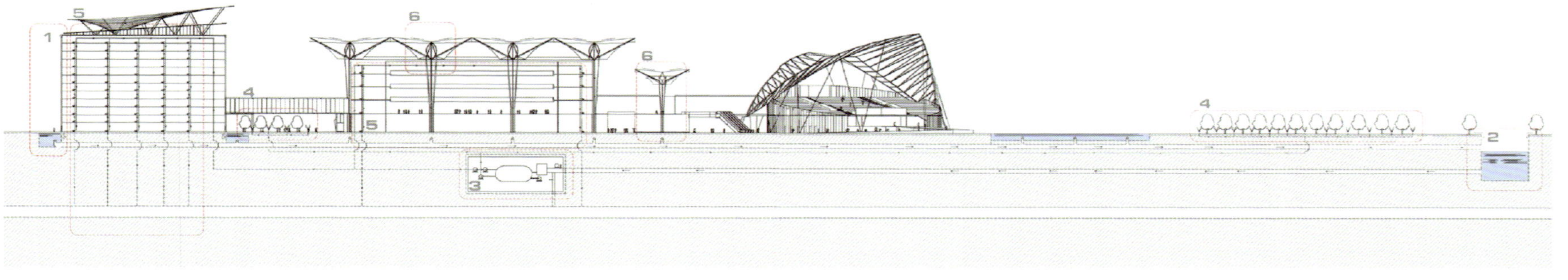

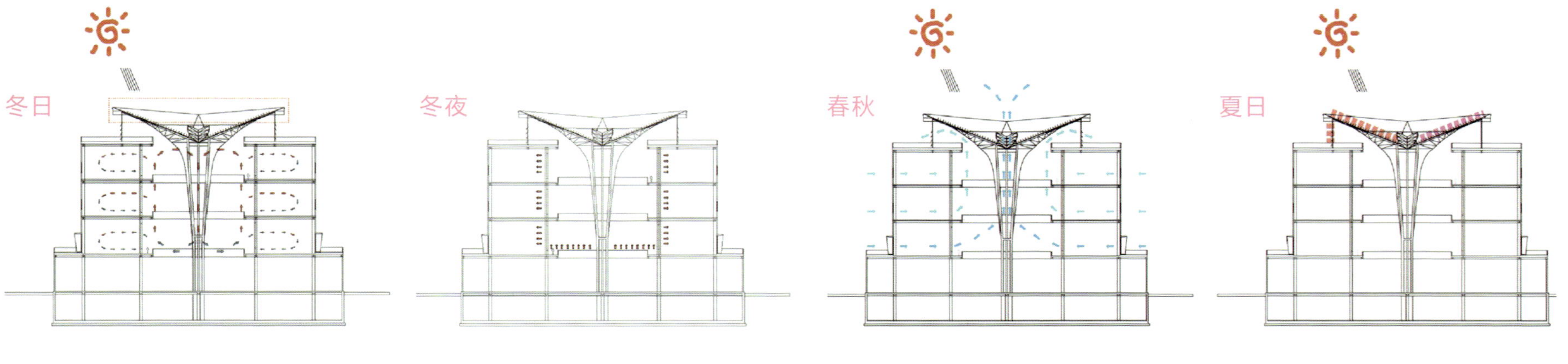

节能

展馆内部设置中庭空间，中间竖以花伞结构，中庭屋面(即花伞屋面)辅以PTFE膜，在强化室内景观的同时兼做屋顶遮阳、隔热及雨水收集构件。冬日关闭屋面天窗，利用温室效应及墙体蓄热，对新风进行预热并提高室内自然室温。经计算冬季平均温度上升3度，冬季节省新风加热能耗15%；冬夜蓄热墙体及楼面向室内释放热量，从而降低第二天的采暖负荷；过渡季利用花伞结构竖向钢管所挂设的空气导流板及花伞顶部所设玻璃烟囱，加强中庭烟囱效应，强化房间自然通风，并利用CFD模拟方法对热压通风相关构建进行形式优化，提高了热压通风中和面，增加了全楼的新风换气次数；过渡季及夏季夜间，利用屋面及外墙外窗的可开启性，进行夜间通风，以降低次日空调负荷。经流体动力学模拟计算，仅利用热压通风可对全楼产生3.5次的换气次数，经详细的空调、采暖及系统计算，利用热压通风可节省建筑整体能耗20.2%利用权衡计算法优化了中庭采光天窗的综合透光率，使得天窗的采暖能耗、空调能耗及照明能耗增益总和最小。经模拟软件计算，中庭天窗减少了12%的照明能耗。主展馆办公部分采用冷冻方式的温湿度独立控制策略，利用低温冷水机组处理潜热负荷，高温冷水机组处理显热负荷。有效地提高了综和制冷能效比（可节能5%左右），同时提高了室内的舒适度和洁净度。交易中心采用地源热泵机组。经详细的空调、采暖、照明能耗计算，建筑相对《公共建筑节能设计标准》要求的节能公建基础上，进一步节能21.5%。

花伞设计

花伞作为花博会整个建筑中最为突出的设计亮点，在建筑节能、组织建筑室内外空间、建筑识别形态方面发挥着巨大作用。花伞共计 36 把，以矩阵方式排列，将建筑室内外空间有机的组织在一起。36 把花伞分为 15 把室内伞与 21 把室外伞，每把花伞皆由花蕊、花冠、花茎三部分组成。花蕊最高点距地为 39.3m，花冠为 24m*24m，花径直径为 3m，由 8 根直径 406mm 钢管组成。搭建花伞的主要材料为无缝焊接钢管，表面涂以白色超薄型的防火涂料。每把花伞所集结的多种高科技手段以及所承载的节能功效，更是体现了“高水平、有特点”的办博主旨。

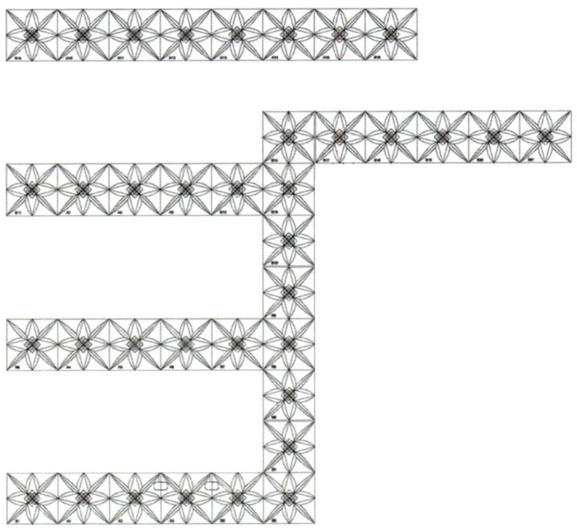

材料

室内花伞的“花冠”作为建筑屋面的维护结构，起到保温隔热、引进自然光线的作用。我们采用 PTFE 膜材，此材料遮阳系数达到 0.2，大大减少了夏季太阳辐射，降低空调负荷。PTFE 膜材的透光系数达到 12%，可实现柔和自然采光效果，降低白天照明能耗。同时 PTFE 膜材本身具有自洁功能，可减少后期维护成本。

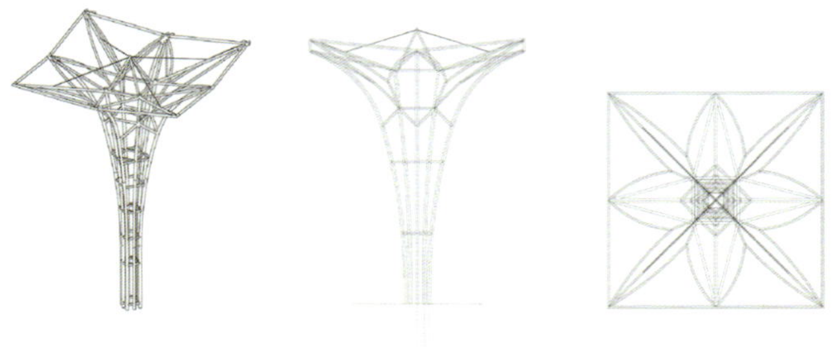

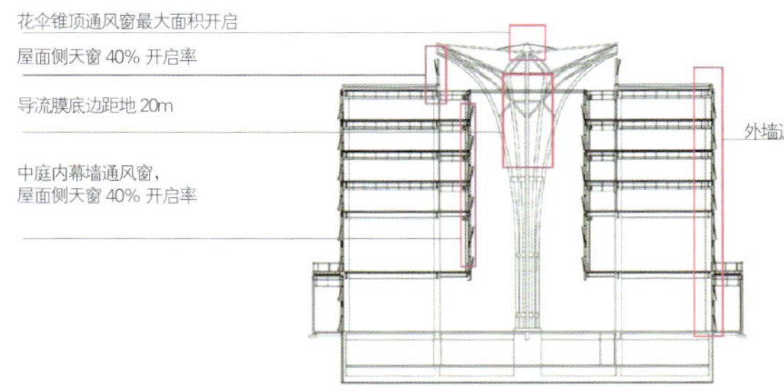

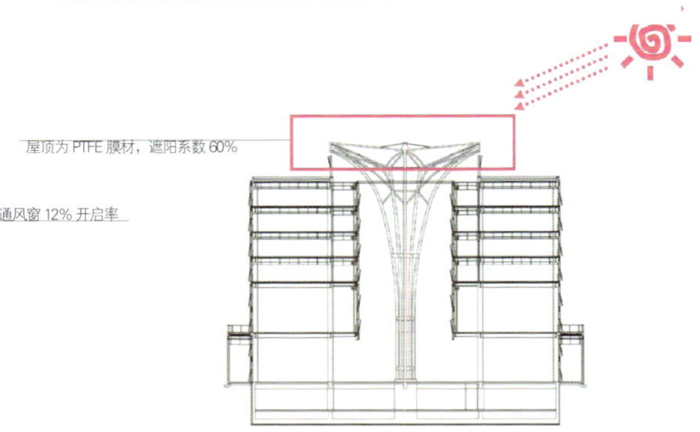

节能

构成花蕊的玻璃锥，上为电动窗，过渡季节及夏季一定时段开启，其下设置 10m 高的空气“导流膜”，形成“拔风烟囱”，配合中庭内部维护幕墙的特殊开窗率及开窗形式，有效组织展馆内部的冷热空气流动．从而为整个展馆降低空调能耗发挥了巨大作用。

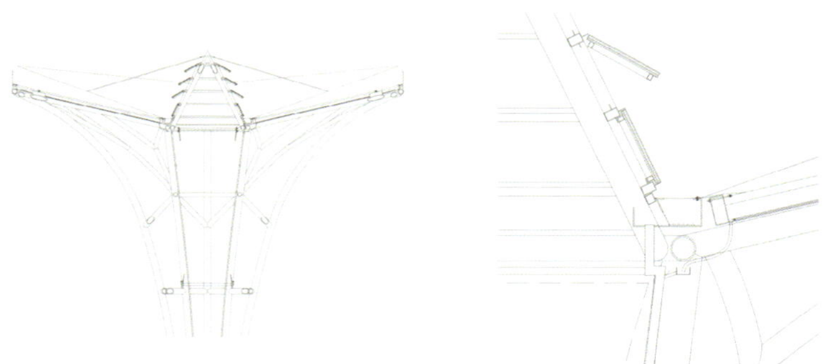

照明

考虑到花伞冬季可能积雪，我们在花伞上部设计了电伴热的融雪装置，冬季当室外温度及湿度同时达到临界值时，电伴热融雪装置可自动开启，从而将花伞上部的积雪融化。花伞同时也起到了固定景观照明灯具的“灯柱”作用，这些照明灯具即可以照亮花伞，表现花伞自身形态之美，同时也对展馆室内中庭的光环境发挥了积极作用。

节水

花冠的坡度是在考虑北京地区的平均降雨量及雨水流速后特别设计，同时通过屋面构造措施，可最大限度的收集屋面雨水。这些雨水经处理后可用于浇灌植被、冲洗建筑外墙及卫生间用水。花伞屋面采用了虹吸式排水，在汇水面积、雨水流速、管道安装坡度上都优于传统的重力排水，雨水管采用 HDPE 管材，可随着花伞主管的弧度一同弯曲，实现与花伞最好的造型拟合。

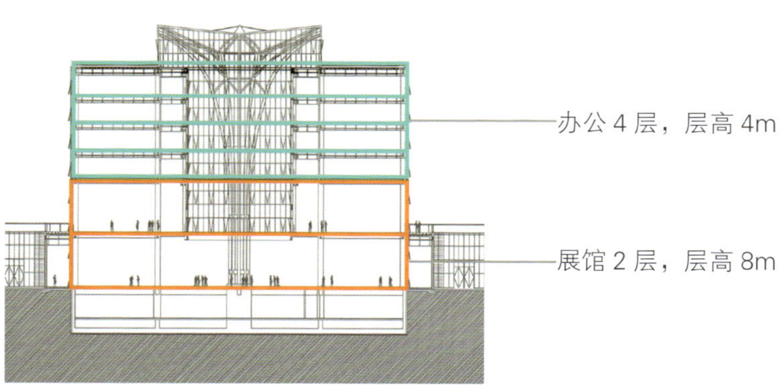

会后利用

方案提出从会后利用出发考虑会时设计的思路，从而为项目的可持续利用提供保证。展馆层高 8m，会后可仍为展览或是改为商业。展览辅助用房层高 4m，会后可改为办公。

城市公共空间

为了营造具有亲和力和活跃性的城市公共空间，结合交易中心的设计对标志性的需求，一方面，将东部椭圆形体量的屋顶设计成台阶状的斜面，并覆以叶状的轻钢结构屋顶，营造了可用于开幕式、招待会、发布会等大型公共集会需要的极富吸引力的城市空间。另一方面，包括物流中心和展厅在内的主体建筑则以花形单元结构来组织公共空间，包括室内的共享中庭和室外的城市平台。

Longjiang art exhibition center

龙江艺术展览中心

建筑设计：哈尔滨工业大学建筑设计研究院
项目地址：黑龙江哈尔滨
用地面积：24863.7m^2
建筑面积：9364m^2
容 积 率：0.18

项目概况

本工程为群力新区龙江艺术展览中心，位于群力新区，北侧为群力大道，南向为群力第五大道，东临景江东路，西接景江西路。本工程总用地面积 24863.7m^2，总建筑面积 9364m^2，主体建筑地上 2 层、地下 1 层，建筑总高度 15.5m。本工程采用框架结构，建筑设计使用年限为 50 年，抗震设防列度为 6 度，工程等级为一级。本工程主要功能为展览功能。

设计特色

本工程在形象上以“林海雪原”为设计理念，形象抽象于“林海”，一方面以“林”的万古长青与高大挺拔象征着龙江地区深厚的文化积淀与龙江人民淳朴热情的性格；另一方面以众树成林，聚之若“海”的磅礴气势体现着新世纪龙江人众志成城为建设新家园的坚定精神和高尚品质。

(1) 个性的总体布局

展览中心设计延续广场景观设计中代表金元文化的覆土建筑符号，与后方两覆土建筑形成具有标志性的城市空间，屋面则为覆土的种植屋面，使建筑与广场相互呼应顺畅过渡，浑然天成。展览中心被设计为一个高度集中的单一型体锥体，它在不同方向被切削，最终呈现出坚实、简洁、新颖、具有十足现代感、为城市提供可识别性的建筑形象。

(2) 绿色的景观建筑

为了积极贯彻落实国家发展绿色建筑的节能环保政策，响应哈尔滨市委、市政府提出的“创建三个适宜城市”和打造全新群力的指示精神，哈尔滨群力新区开发建设管理办公室决定建设弘扬龙江地域优秀文化的长期固定性场所——群力新区龙江艺术展览中心，将其打造成为哈尔滨城市人文新景观和哈尔滨群力新区的标志性的新亮点，并提出将其建设成为国家绿色建筑示范工程的设想。在整体环境规划中强调的是建筑与景观环境的关系，解决建筑与地貌、植被、水土、风向、日照与气候的关系。在单体建筑设计中，则主要是通过构造、技术手段创造舒适的室内环境，减少能耗，减少排放。

(3) 独特的展览空间

建筑为地上二层，地下一层。建筑功能布局为地下一层布置有设备用房，厨房，库房，咖啡厅，展厅；一层全部为展厅，并附设公厕；二层为拍卖厅和休息厅。建筑内部标高富有趣味性和独特性，分别通过四个逐一变化的地面标高，充分结合布展形式，将人的参观流线顺畅引导，让参观人员步移景异，逸趣横生。灵活多变的室内空间使龙江艺术展览中心成为城市广场上一颗璀璨的明珠。

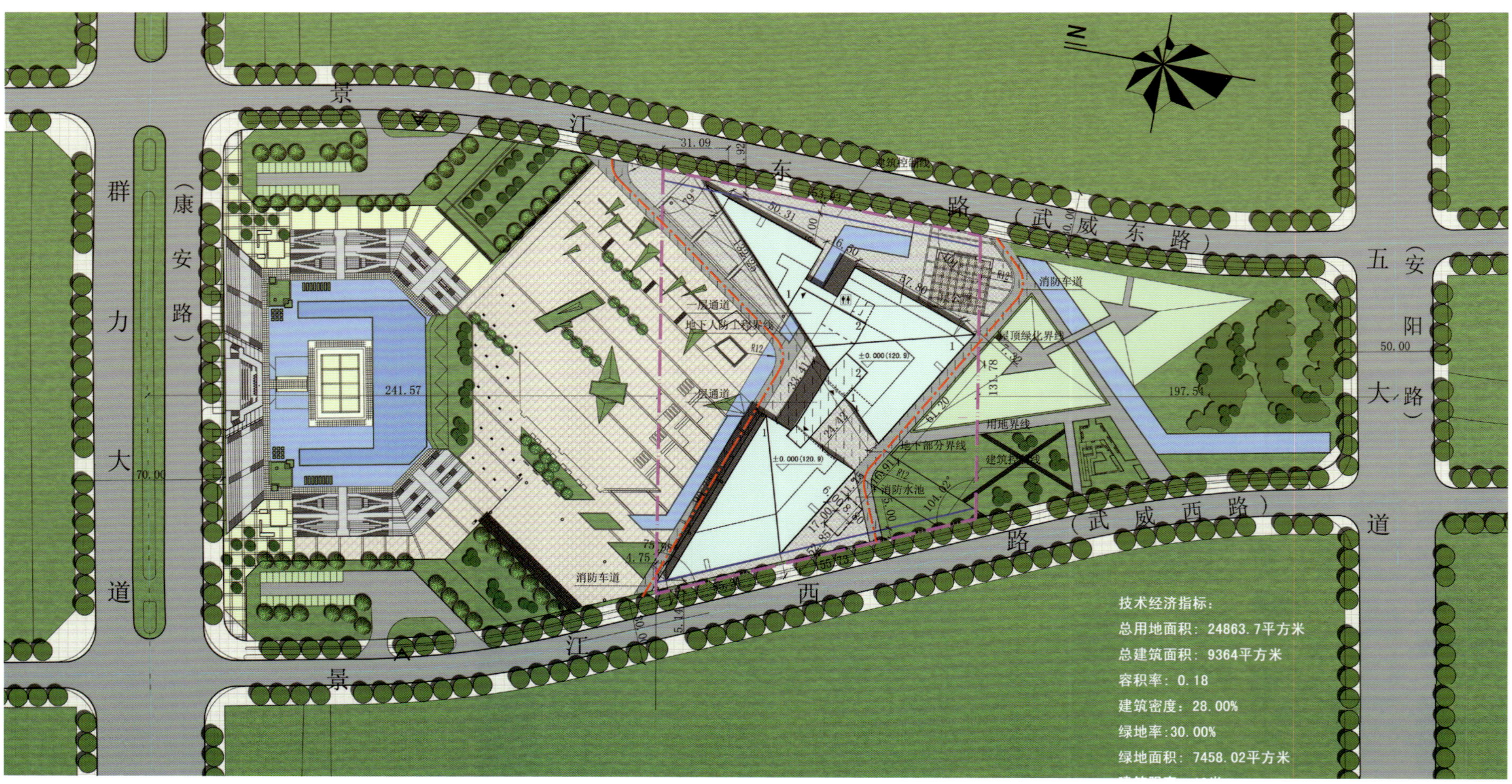

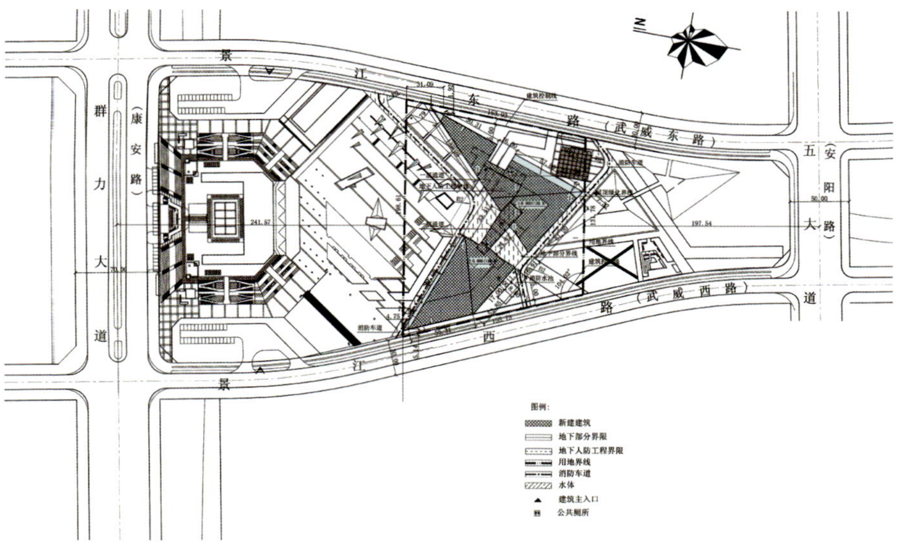

总平面位置图

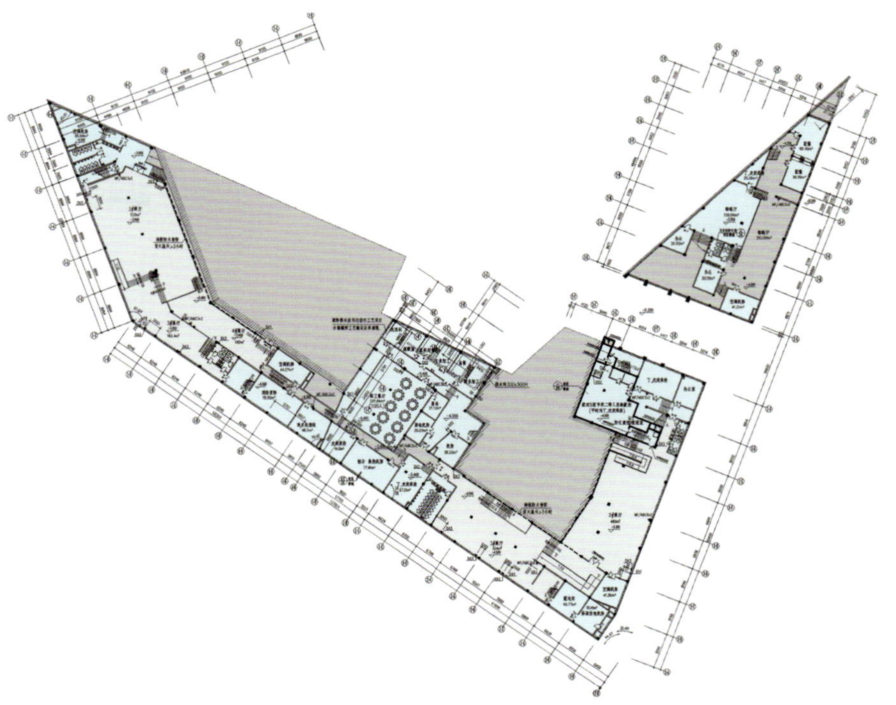

地下一层平面图

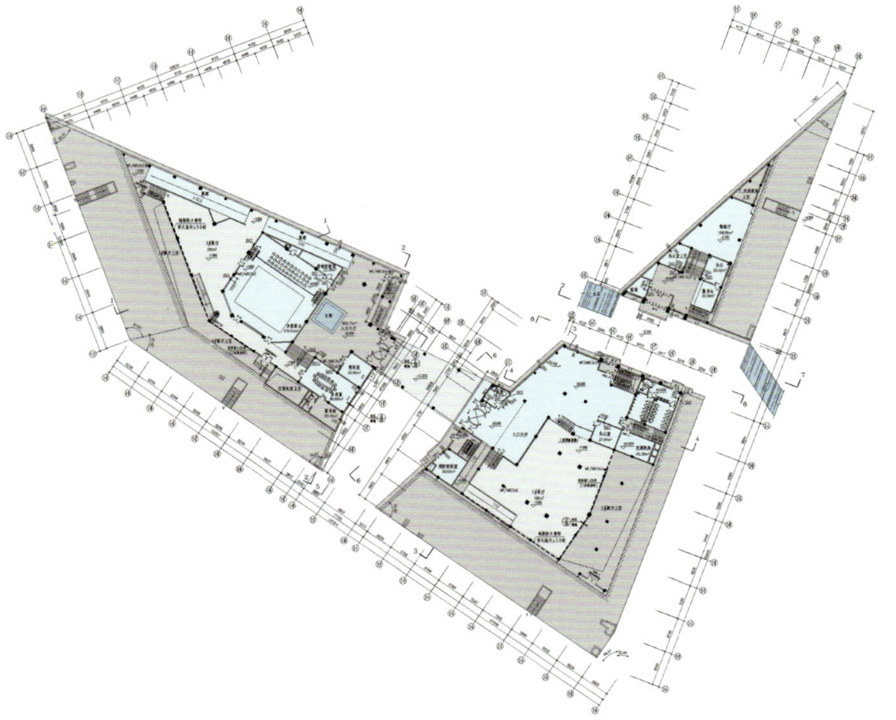

一层平面图

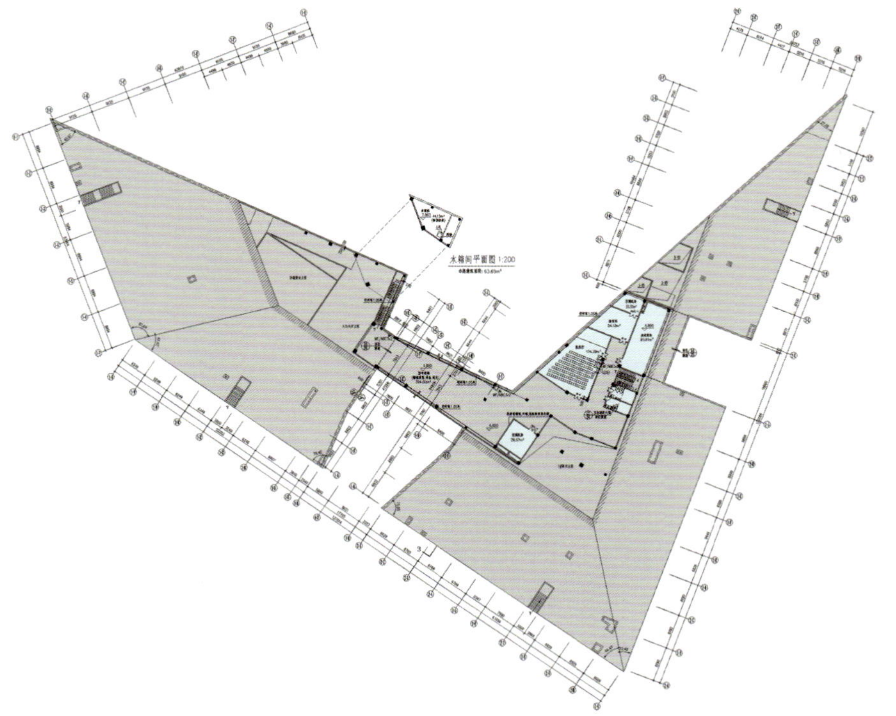

二层平面图

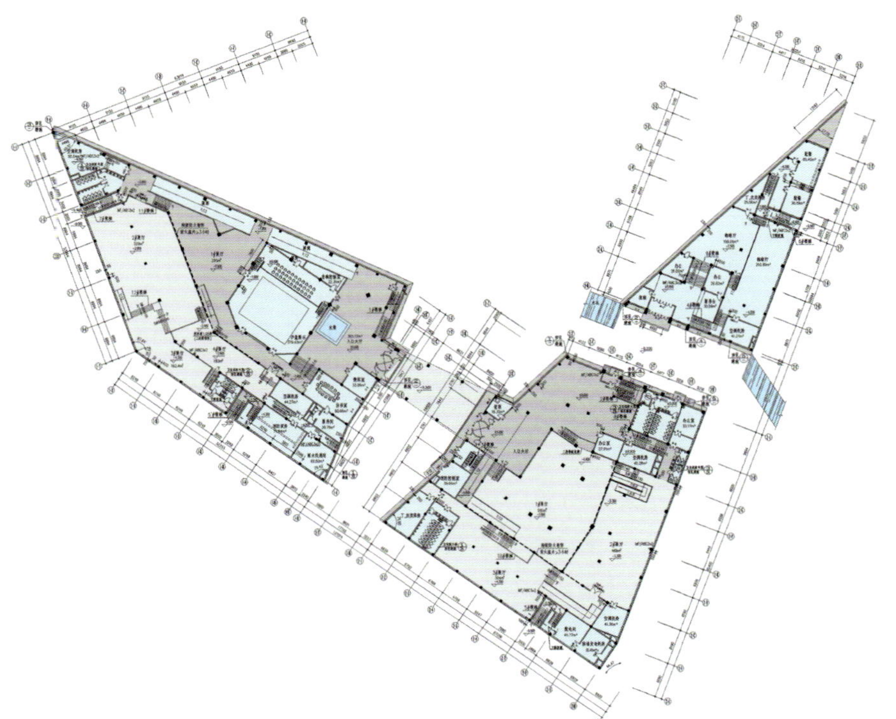

种植屋面及一层楼板下部空间平面组合图

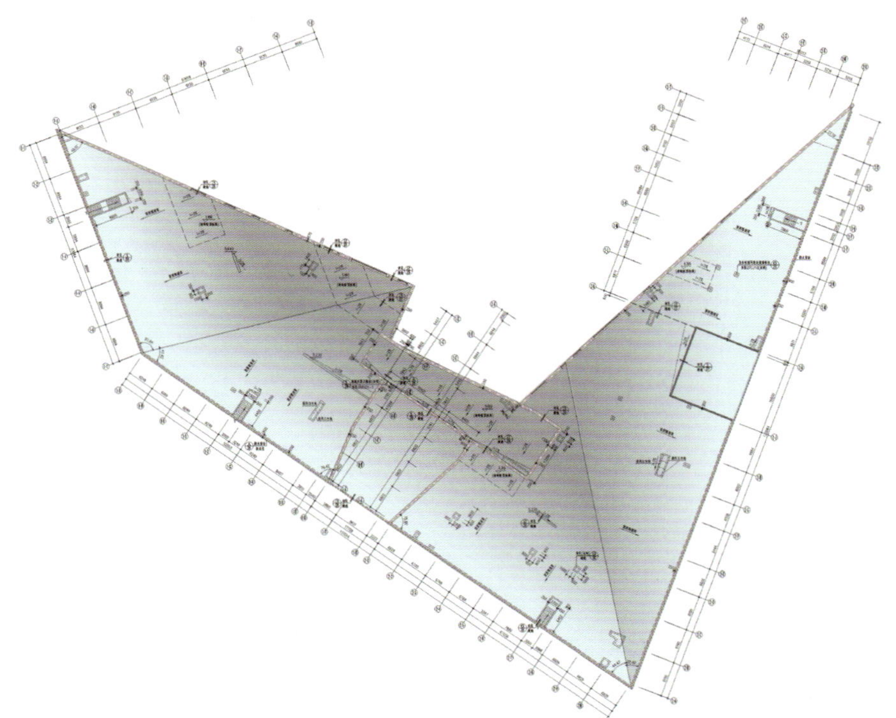

屋面排水平面图

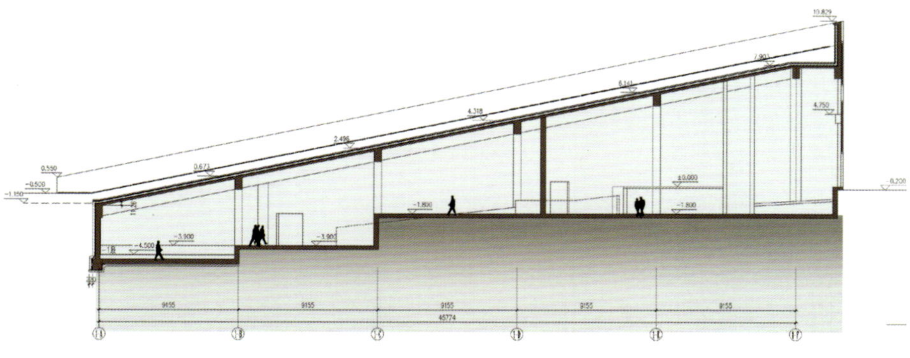

1-1 剖面图

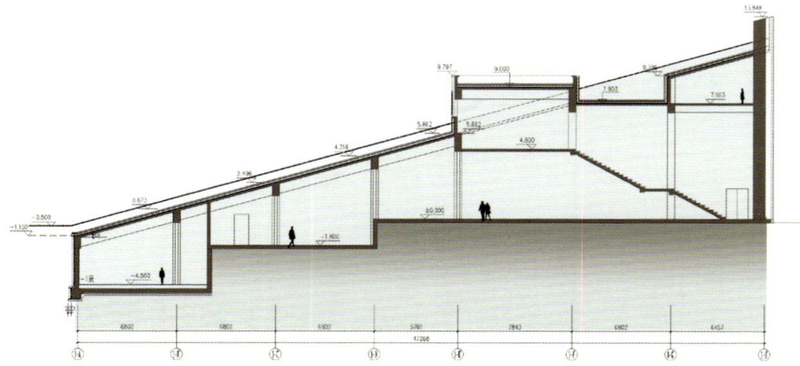

2-2 剖面图

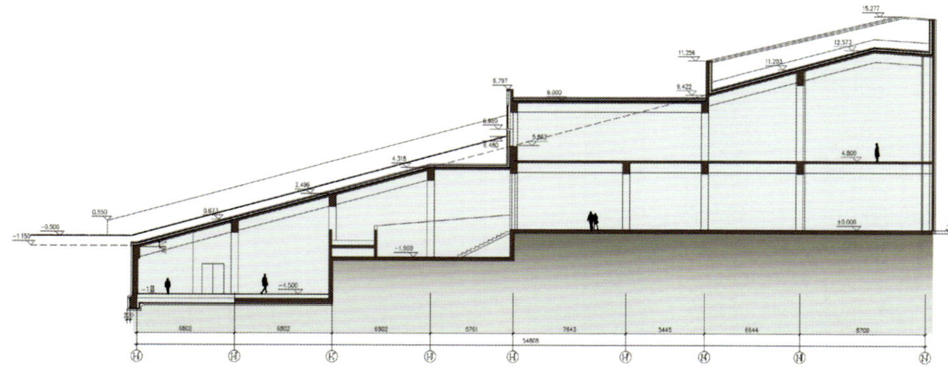

3-3 剖面图

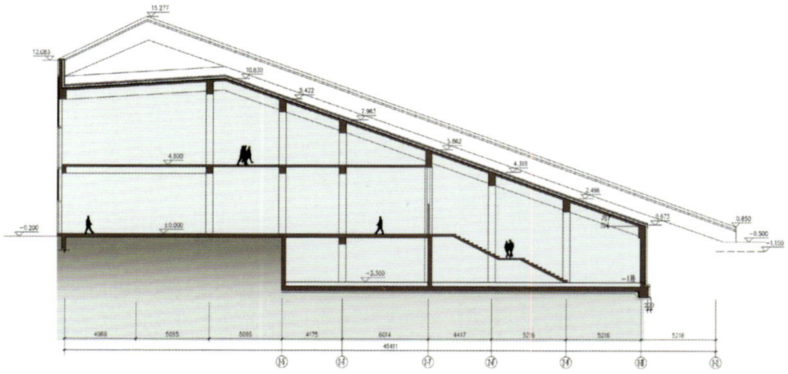

4-4 剖面图

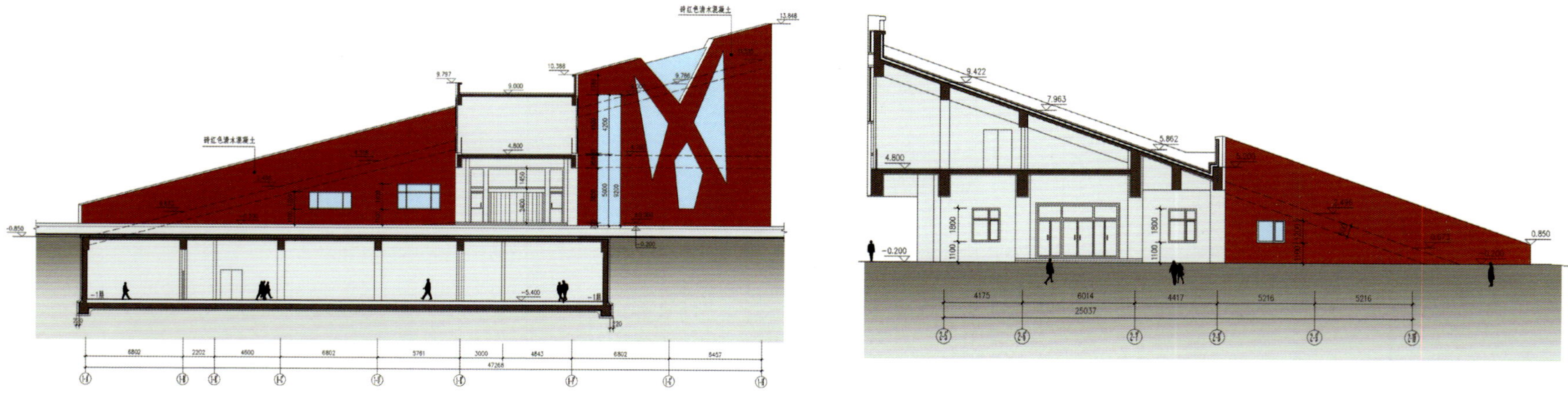

7-7 剖面图

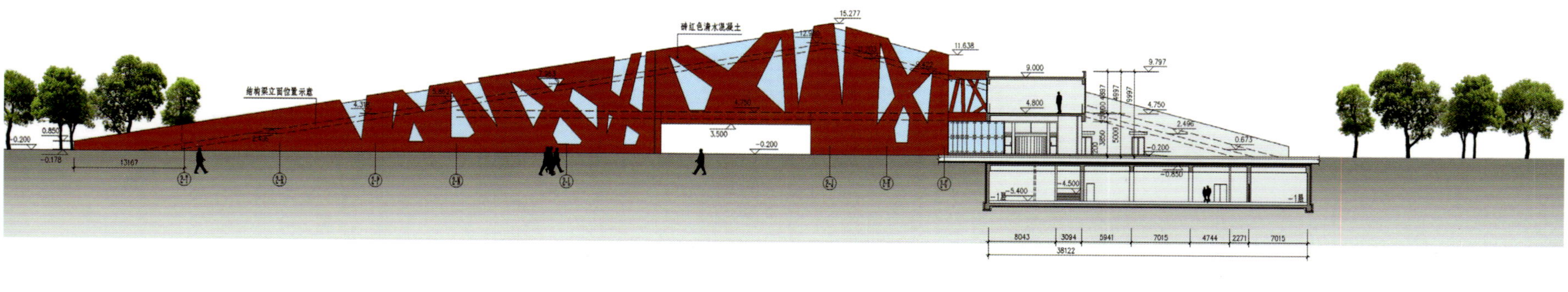

6-6 剖面图

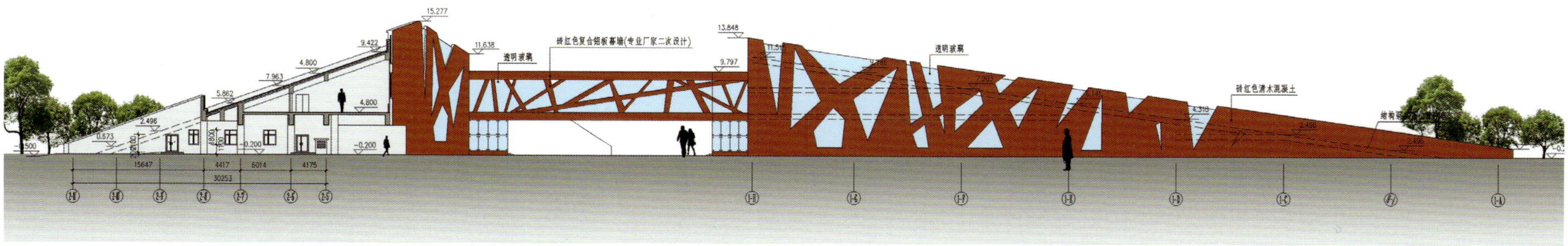

8-8 剖面图

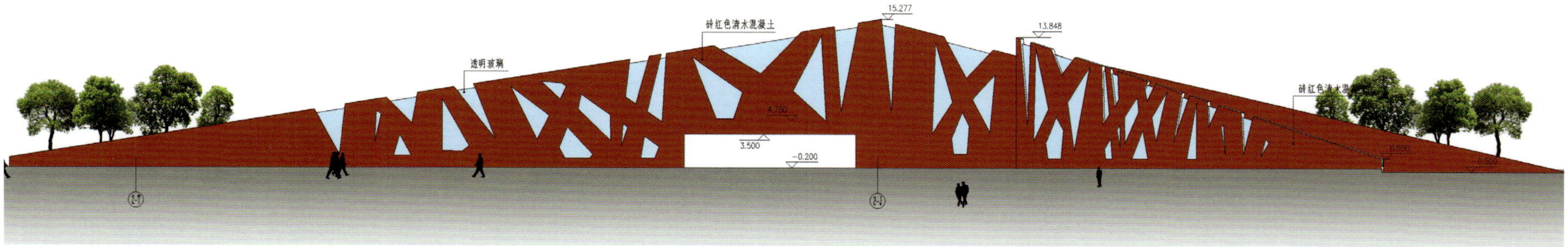

2-T—2-J 轴立面图

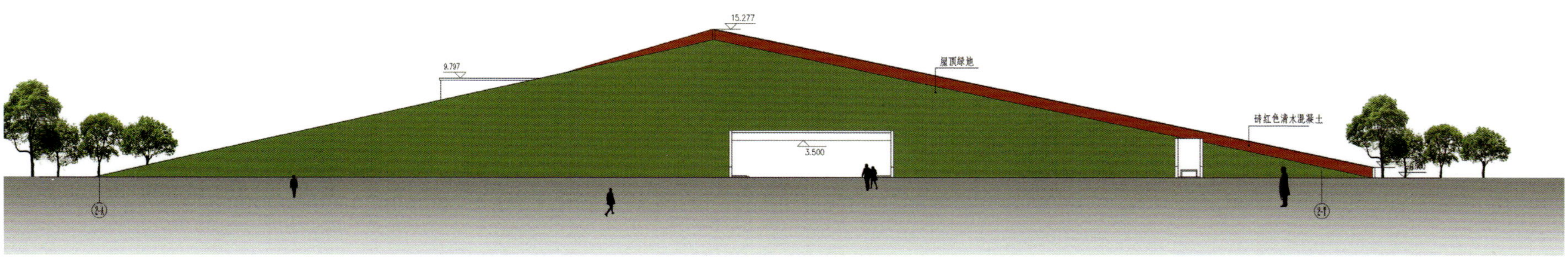

2-A—2-T 轴立面图

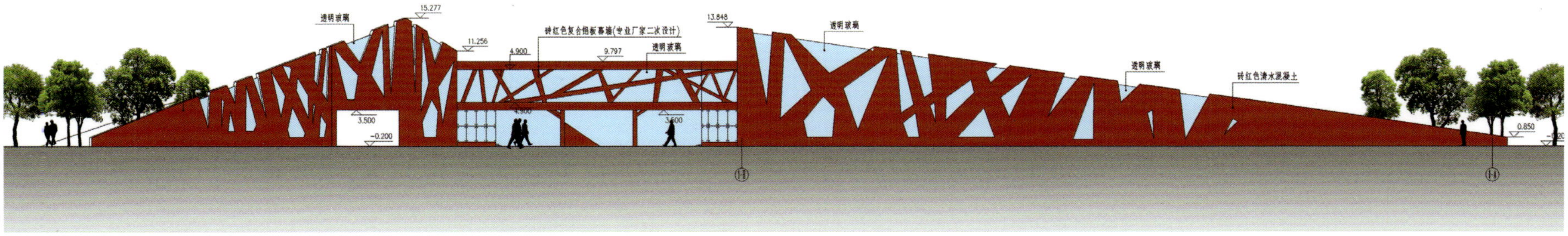

1-H—1-A 轴立面图

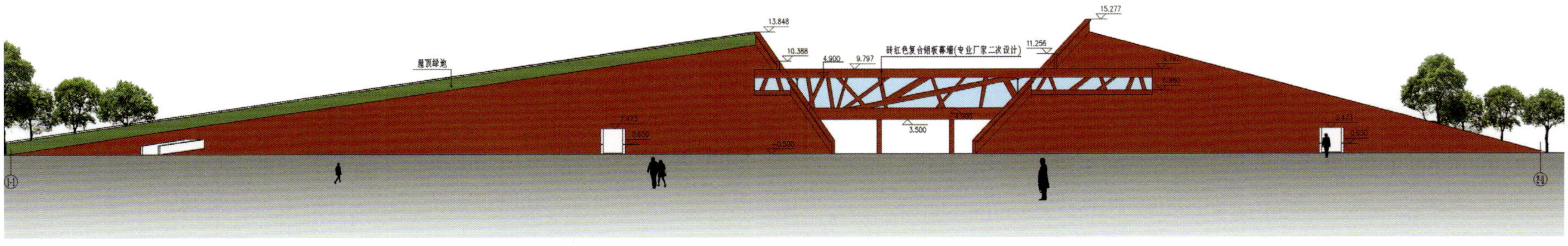

1-1—2-11 轴立面图

Guangzhou international sports center for performing arts

广州国际体育演艺中心

建筑设计：广州市设计院
项目地址：广东广州
用地面积：65430m²
建筑面积：130312.8m²

项目概况

广州国际体育演艺中心位于广州萝岗区中心区，是继北京五棵松体育馆、上海世博演艺中心之后全国第三个集体育、演艺活动为一体的大型综合场馆项目。本项目以环境的可持续发展作为设计目标，采用了如透水地面、雨水收集系统、新型节能材料等绿色建筑设计手段，通过了绿色建筑设计标识二星级的评定。

广州国际体育演艺中心
Guangzhou International Sports Arena
Harmonious Asia

GISA

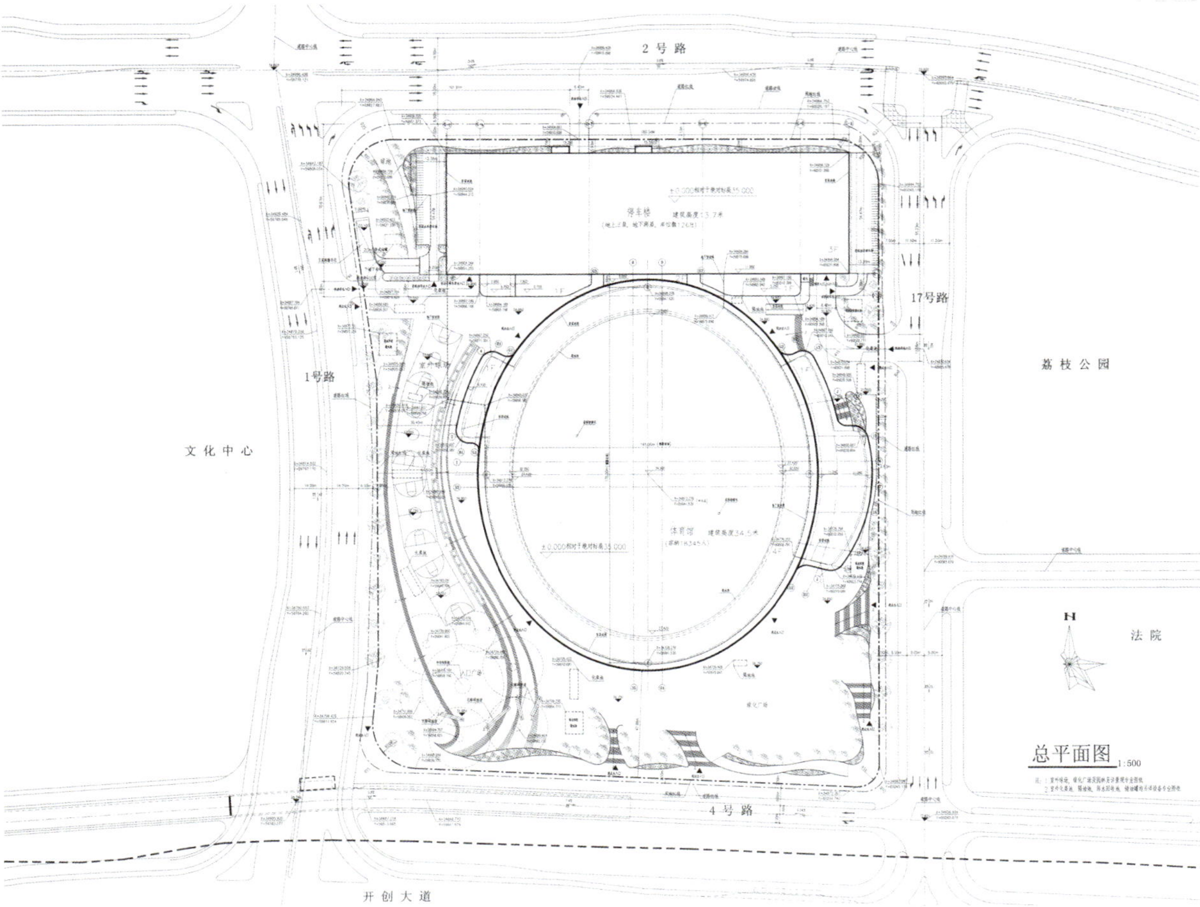

总平面图

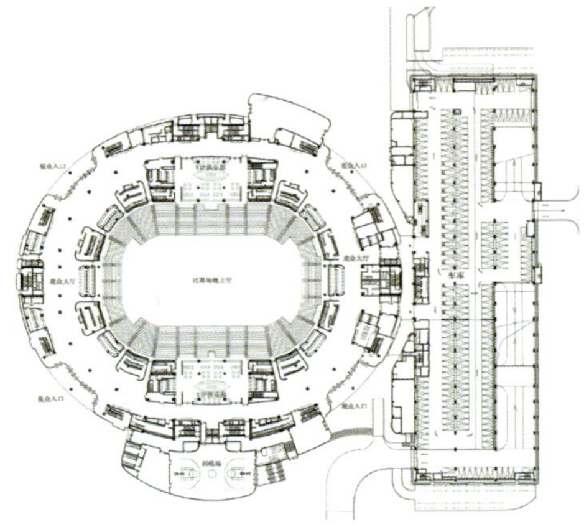

首层平面

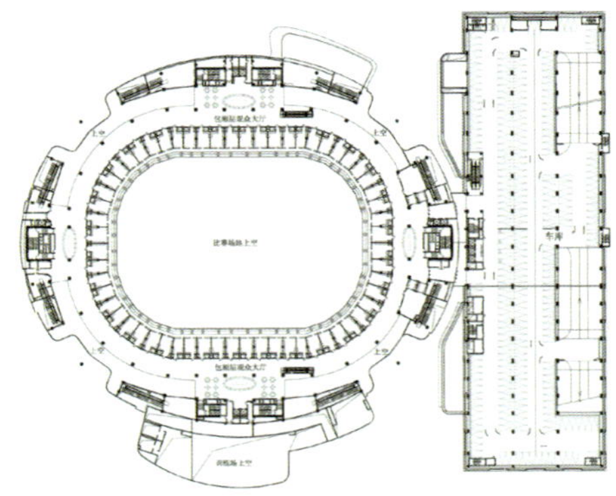

二层平面

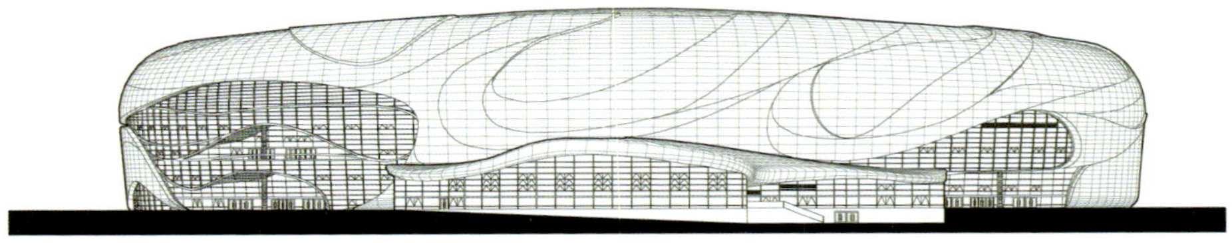

东立面

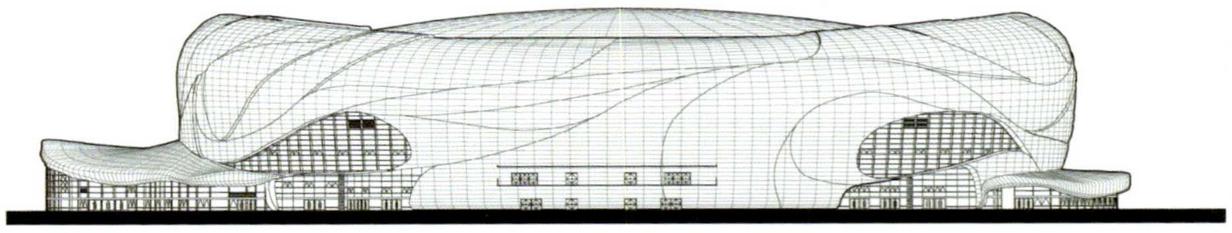

西立面

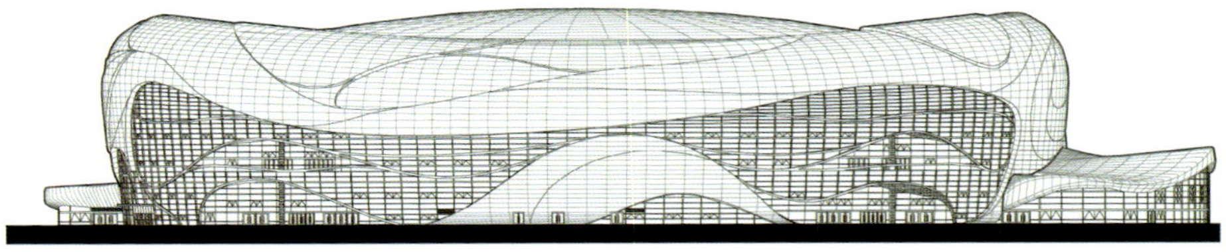

北立面

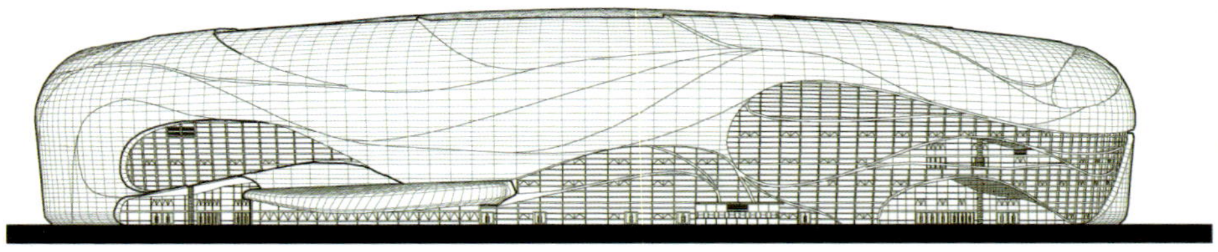

南立面

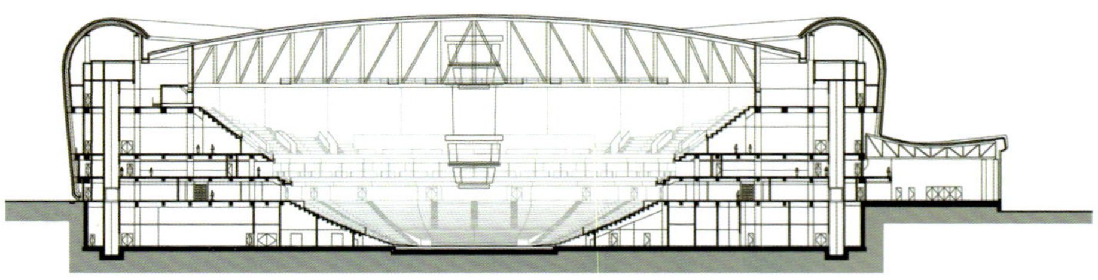

1–1 剖面

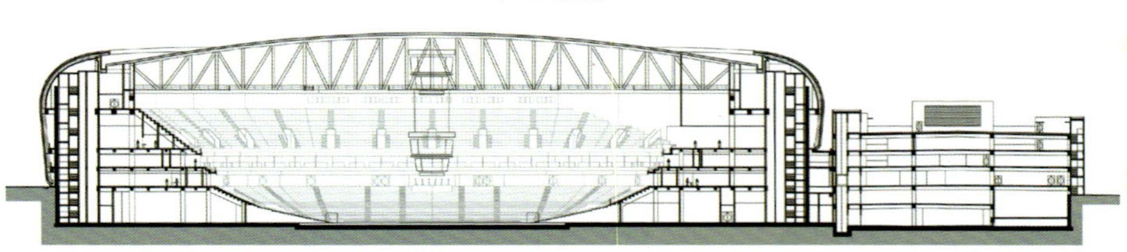

2–2 剖面

7-43 Elevation @ Restaurant Room

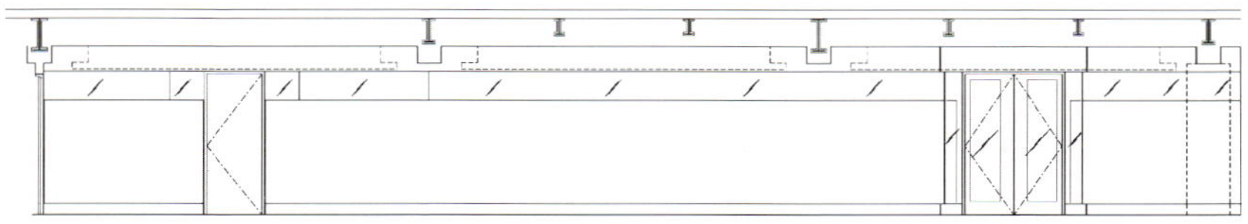

7-44 Elevation @ Restaurant Room

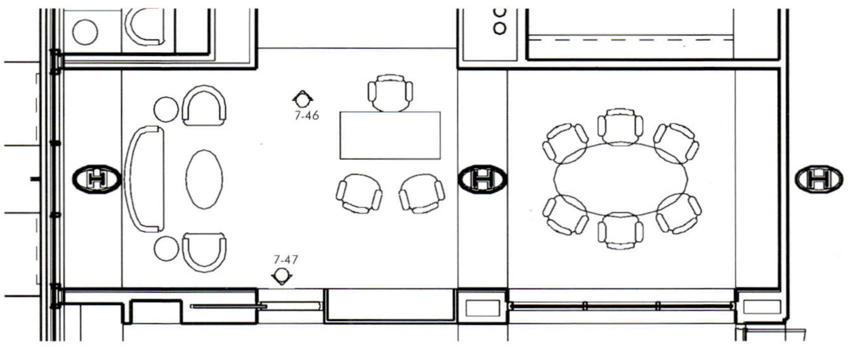

7-45 Murdoch Office Plan

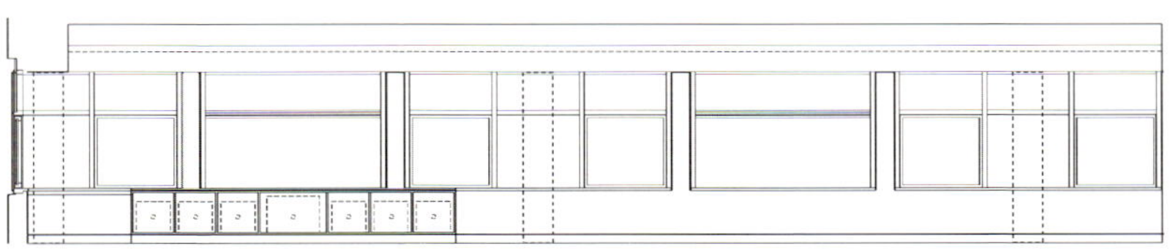

7-52 West Elevation @Murdoch Office

7-53 East Elevation @Murdoch Office

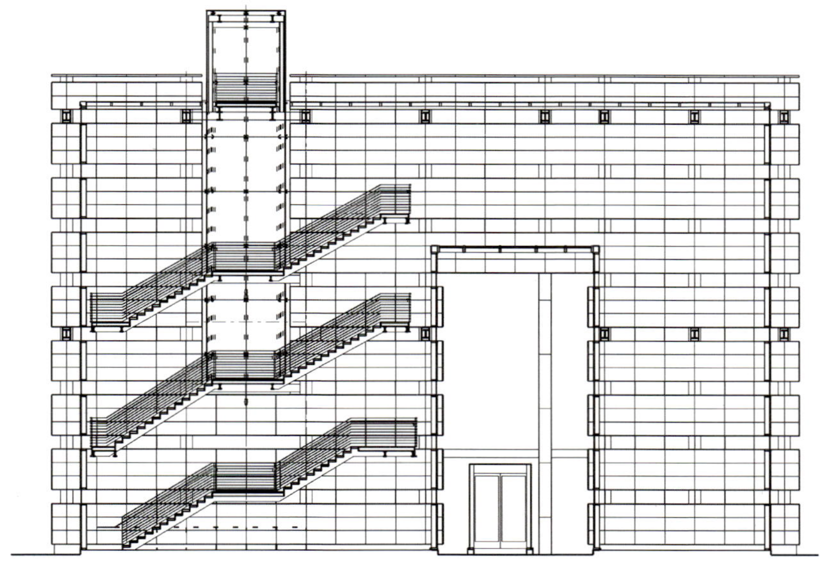

7-16 Developed Partition Elevation Of The Entry Court & Stone Wall

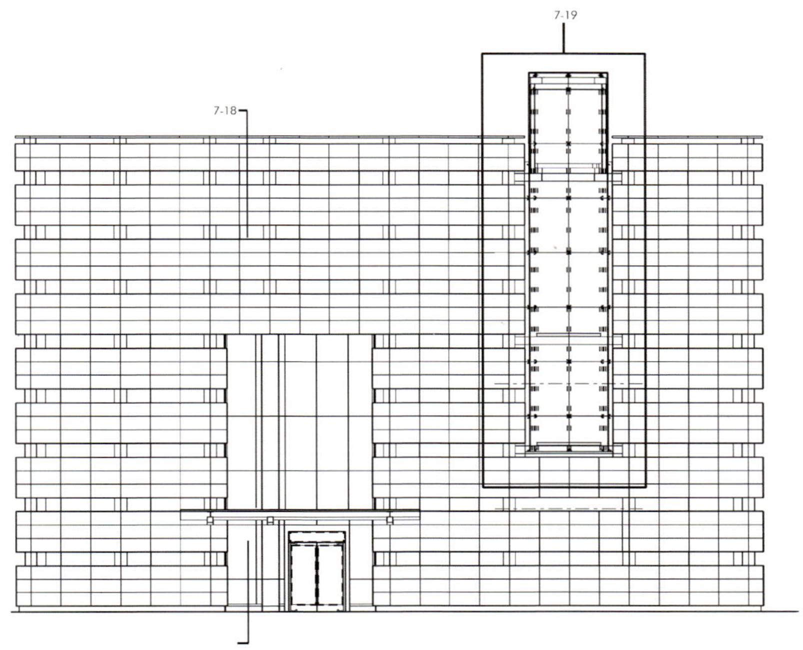

7-15 Developed Partition Elevation Of The Entry Court & Stone Wall

Mack-Cali Reality Corporation

海港财富中心广场

Location: Harborside Financial Center, Plaza 10, Jersey City, New Jersey, USA
Project size: 50,220m²
Services: Architecture, Landscape Architecture

位置：美国新泽西州泽西市广场10号Harborside金融中心
面积：50,220m²
服务内容：建筑设计、景观建筑设计

HLW's creation of a signature building for Mack-Cali Realty and its tenant, Charles Schwab, is the first of three projects to fulfill the architectural component of HLW's masterplan for this unique 5.5 acre site located along a distinctly urban Jersey City waterfront. The new facility will contain high-technology offices and retail space.

HLW successfully integrated the building into the context of Jersey City while creating a dramatic icon on the waterfront. The dual nature of the site is reflected in the facades of the building through a strong architectural dialogue. The south and east facades, which face a waterfront park and the river, use large areas of glass to maximize views, and give the building a modern feel. The north and west facades, which face the context of Jersey City, incorporate stone piers relating to the solidity of adjacent buildings. The building is crowned by an articulated canopy that adds excitement to the Jersey City skyline. Pedestrianscaled human elements such as an esplanade to welcome ferry passengers and a landscaped public park, help to create a sense of place.

HLW对一块位于独特的泽西城市水滨的22,258m²（5.5英亩）地段进行了整体建筑规划，总共包括3个项目，其为Mack-Cali投资公司及其业主设计的一座极具标志性的建筑便是其中之一。这座建筑将包括高科技办公空间和零售空间。

通过在水滨景观上建立一个宏伟的标志性建筑，HLM成功地将这座建筑与泽西城的地域特点融合在一起。通过颇有力度的建筑对话，这座建筑的外立面体现出该地段的双重性特点。南侧和西侧立面均朝向水滨公园和河流，大面积采用玻璃立面以获得最宽阔的视角，同时体现出该建筑的现代气息；北侧和西侧立面则朝向泽西城市区，利用石柱的厚重感与城市风貌和谐相融。该建筑的铰链式顶棚给泽西城的天际线带来几分跃动的激情。该项目还配有散步设施，如设有海滨散步区供来往渡船的乘客使用，而一个景观设计考究的公园则建立起一种地方感。

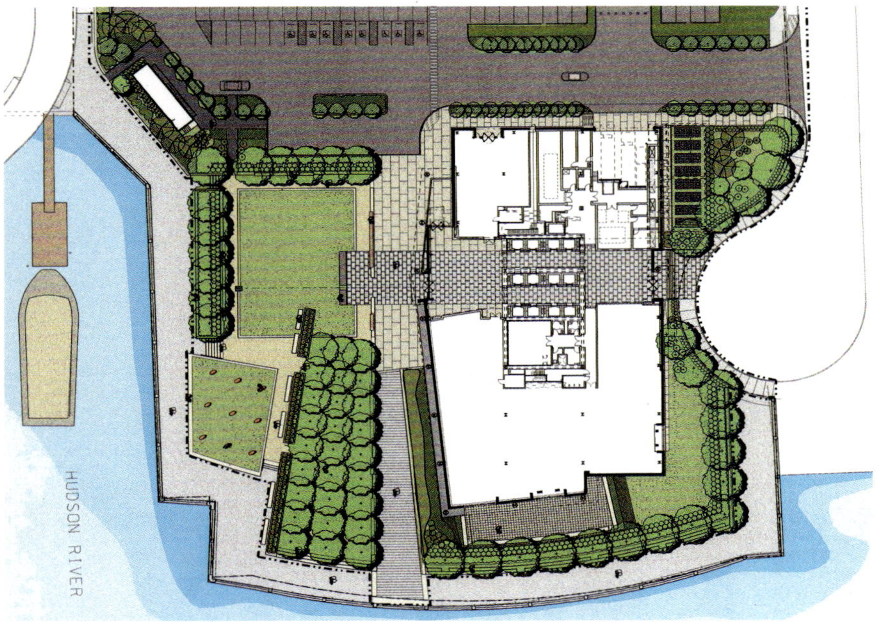
HUDSON RIVER

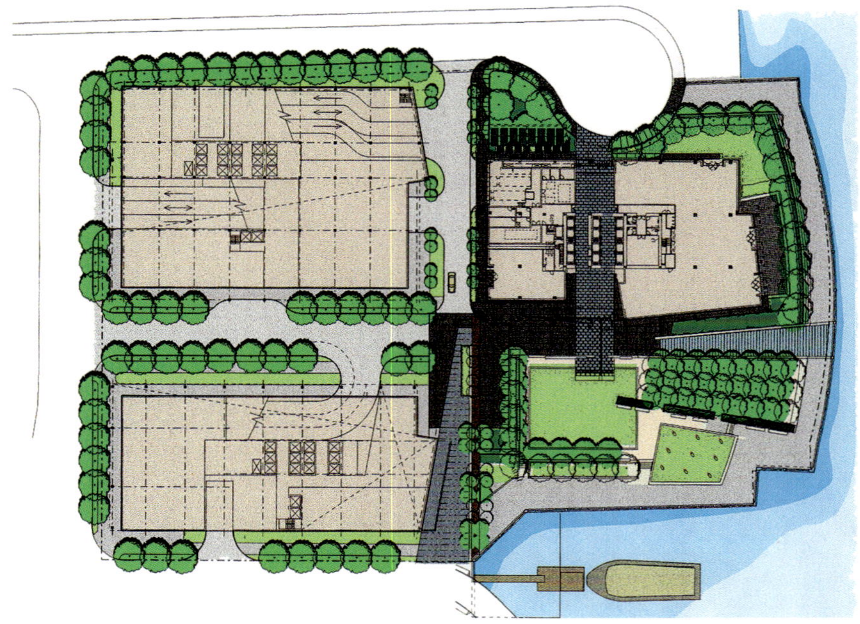

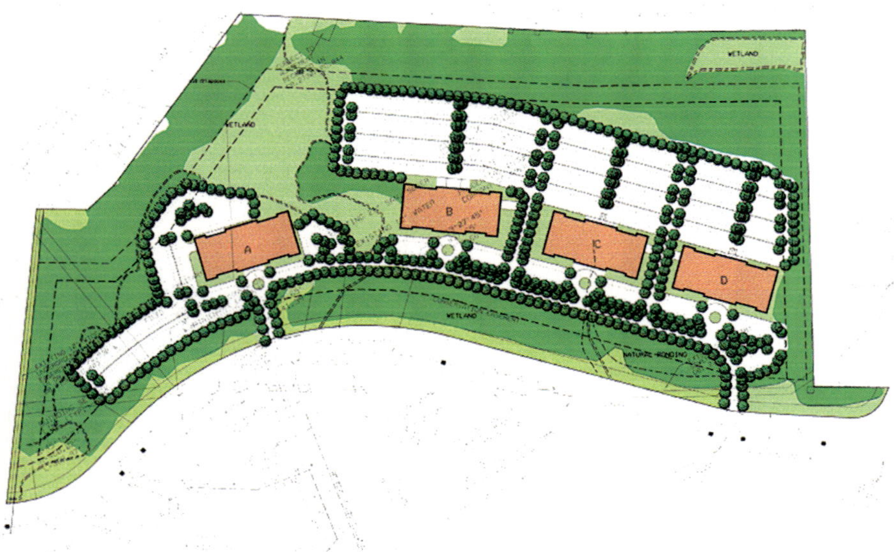
WETLAND
A
B
C
D

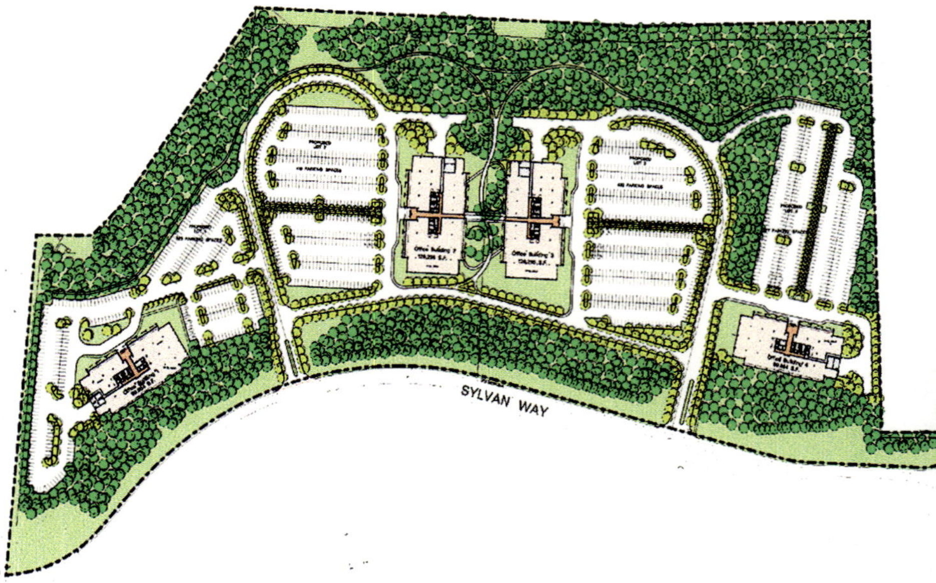
SYLVAN WAY

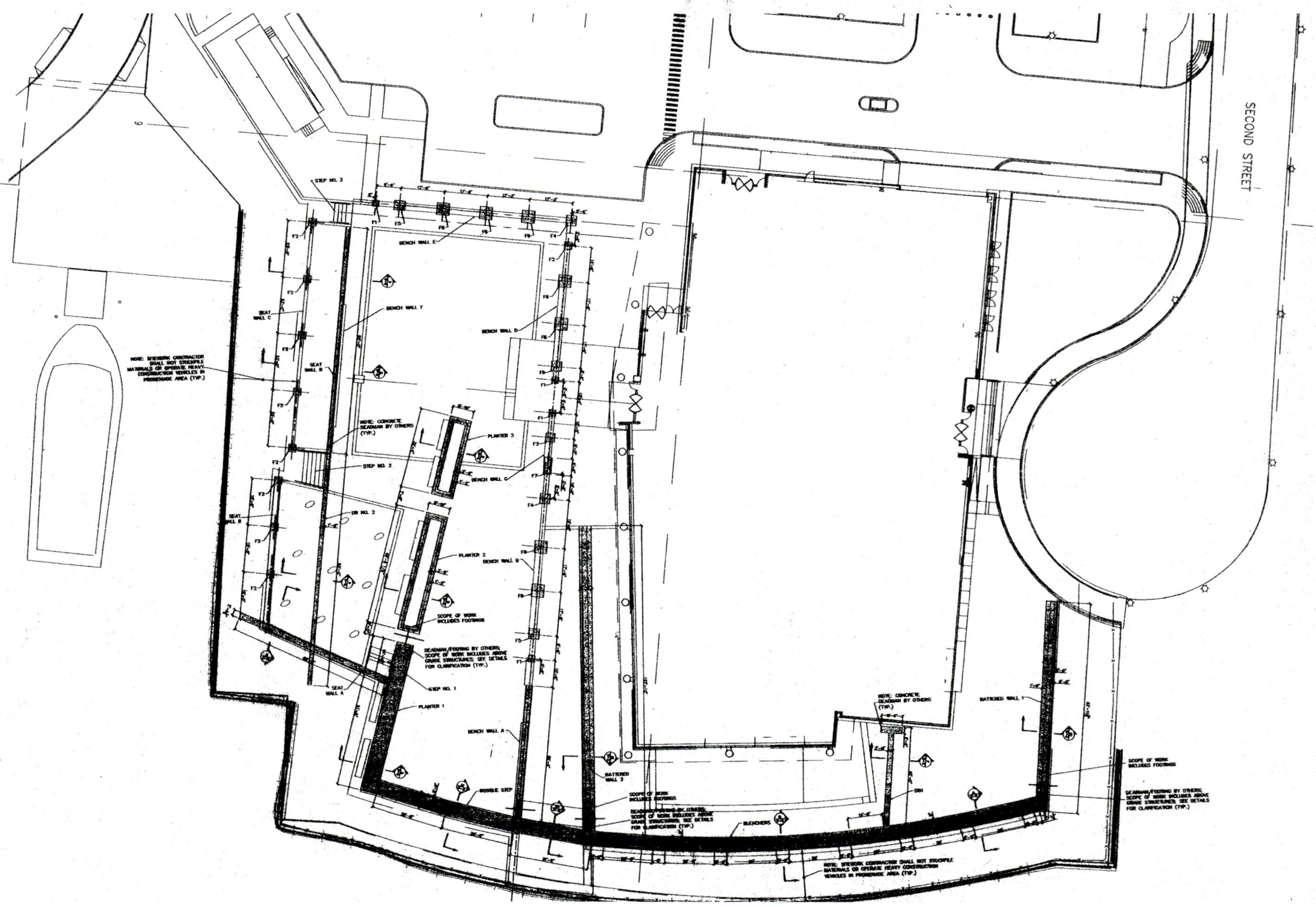
SECOND STREET

EXIT

Hong Kong Langham Place, The New Landmark of Mong Kok

旺角新地标——香港朗豪坊

Site area: 12,007m²
Building area: 167,000m²
Master plan and architecture: The Jerde Partnership
Client: The Great Eagle Holdings Ltd.
Location: Mong Kok, Hong Kong, China

总占地面积：12,007m²
总建筑面积：16.7万m²
总体规划和建筑设计：美国捷得国际建筑师事务所
业主：鹰君集团有限公司
项目地理位置：香港旺角

Located at the prime site of Mong KoK, the Langham Place is a large scale three-in-one comprehensive high end development project, the largest rebuilding project in downtown of Hong Kong, whose total building area is 167,000m². The Langham Place consists of the first five-star hotel (42 storeys) in Mong Kok, a grade A office building (59 storeys) and a 15-storey shopping mall with over 300 retailing shops and restaurants, performing the charm of the sleepless city by 24 hours.

The 59-storey grade A office building in Langham Place is 255m in height, the highest architecture in west Kowloon; with its unique design, it is the most high end grade A office building in Kowloon, providing world-class installation and the most matching facilities, such as the raised platform and broadband fiber optic network; visitors can enjoy the 360°full-sight of the context of Kowloon and Hong Kong. The building's façade uses blue low reflective glass wall, creating a circling smooth appearance — the main design concept of the integrated plan of Langham Place. A hemispheric structure is constructed on the top of the building,

朗豪坊坐落于旺角黄金地段，是九龙区内备受瞩目的大型三合一综合优质发展项目，是香港最大的市区重建项目，总建筑面积达16.7万m²。朗豪坊由旺角首间五星级酒店（42层）、一栋甲级写字楼（59层），及近300多间商店与食肆的15层高的大型购物商场所组成，全天候24小时展现旺角的不夜城魅力。

朗豪坊内高59层的甲级写字楼是西九龙区内的最高建筑物，255m高，设计独特，是九龙区内最高档的甲级写字楼，提供世界级设施及完善配套，包括升高式地台及光线宽频网络等，从大楼上可360°毫无遮挡地欣赏九龙及香港的优美景致。大厦以蓝色低反光玻璃帷幕覆面，圆浑的外观是整项朗豪坊计划的设计主线。大厦顶部建有半球状结构，并于每晚不断改变灯光的颜色。朗豪酒店则为旺角区唯一的五星级酒店，让旅客既能体验舒适豪华的酒店住宿，亦能尽享该区四通八达的交通设施。

至于朗豪坊商场，则开创本港先河，引入多项崭新的建筑及设计特色：包括全港最高、超过9层楼高、气势不凡的中庭楼底；两条全港最长

and constantly changes its color every night. Langham Place Hotel is the only five-star hotel in Mong KoK, providing the visitors with the comfortable and luxury accommodations and the all-direction transportation infrastructure.

The Langham Shopping Mall pioneers to introduce the many new architectural design features, including the 9-storey high ground floor hall (the highest in Hong Kong), two interior handrail elevators (the longest in Hong Kong) and the first "Digital Sky" in Hong Kong. The Langham Place is grand and spacious: the handrail elevator won't turn around each floor, instead the 82m long "stairway to heaven" leads the way directly from the ground floor to the top one; the multi-layered convoluted shopping corridor spirals downward from the top floor to the ground floor. The Langham Place is enthusiastic: crowing with visitors is preferred, and the ceiling is covered by a huge "Digital Sky", constantly playing animations of fireworks, swimming fish and snow flakes; looking through the glass wall, you would find the ancient street is moving slowly. The design of this project is as follows. Over 300 retailing shops and restaurants are orderly placed in different characteristic areas. The two storeys underground and the first three floors on the ground consist of a department store, some shops of high end brands and electronic products retailing shops. On the fourth floor is the gourmet plaza, connecting with the entrance of the "Plaza to Heaven". On the fifth to seventh floors are shops of fashion clothes, sports apparel and lifestyle taste. On the eighth to twelfth floors are the small speciality stores of teenager and young style.

The "Plaza to Heaven" is the converging site of the crowd flow, opened 24 hours a day. It is specially designed for all kinds of activities, such as unions with friends, senior citizens exercising shadow boxing and even casual business meeting. Two non-airy way "Stair to Heaven" elevators cross through the "Plaza to Heaven" from its central space. One of the elevators directly connects the fourth floor to the eighth, where the other leads to the twelfth floor; these two elevators lead people to a totally new shopping trip in the magnificent cliff carving architecture. Starting from the top floor of the shopping mall, the convoluted shopping corridor spirals downward from the twelfth floor to the eight floor, which abondons the floor directed layout of the conventional stores,

的室内扶手电梯；以及全港首创的数码天幕。朗豪坊是宏阔的：扶手电梯不再一层层打弯，长达82m的“通天梯”从底层直上最高层，蔚为壮观；而多层构架的回转式购物廊，再从最高层通过不断螺旋向下的斜坡，直达底层，别出心裁。朗豪坊是热忱的：它喜欢人头攒动的感觉，以致穹顶铺排了一张巨大的数码天幕，不断播放着烟火、游鱼、雪花的影像，热闹非凡，透过玻璃幕墙望出去，眼底的老街仿佛在踯躅缓行。具体设计如下：300多间商铺和餐厅整齐有致地分布于不同特色区域内。地库两层及一至三楼由一家百货公司以及其他高级品牌和电子产品零售店铺组成；四楼是美食广场，与通天广场的入口相连；五楼至七楼集合了大型时装、运动及生活品味商铺；八楼至十二楼则是走年轻路线的小型专卖店。

四楼通天广场是朗豪坊的人流交汇点，全日24小时开放，专门为不同类型的活动而设。例如三五知己畅聚、长者练习太极，甚至可以是轻松的商贸聚会。通天广场中央连接两条全港最长的非架空式“通天梯”，其中一条由四楼直达八楼，在此可以再搭乘另一通天梯到达十二楼。这两条通天梯，引领游人在美轮美奂的石刻建筑特色下开展耳目一新的购物之旅。

从商场最高层出发，“回转购物廊”由十二楼一直蜿蜒至八楼，打破传统商场按层划分的陈规，营造出店铺连绵不断的感觉，让游人可沿单一的走道轻松浏览不同商铺。旋转走廊的尽头设有六间戏院，15层商场的任何一层均可通往。八楼和七楼由“西班牙式楼梯”连接；连接六楼和五楼的“时装天桥”，可用作举办时装表演和其它活动。至于中间位置的扶手电梯，则可提供另一个游逛路线选择。

通天广场的弧形天顶悬挂着一道“数码天幕”，200多部投影机无间断播放多媒体影片和特别视觉效果，在商场以至街外的途人都可以观看得到，晚间的景致尤其美丽。在数码天幕之下，还有两层名为The Ozone的餐饮娱乐区，为这个特色商场提供了一个别致的聚脚点。

除了建筑上的匠心外，真正奠定朗豪坊作为一个中高档购物中心地位的，除了300家店铺外，UA戏院、西武百货和无印良品是举足轻重的3根支柱。UA戏院遍布港九，位于朗豪坊的这间最有噱头，因为里面的座位都

creating a constant flow of shops and enabling people visit various shops as walking along the corridor. At the end of the spiral corridor is 6 theatres, accessible for each floor of the 15-storey shopping mall. The eighth and the seventh floor are connected by a "Spanish stairway"; the "Fashion Bridge", connecting the sixth and the fifth floor, can be used to hold many activities like fashion shows. The handrail elevator in the middle is another option for shopping route.

A huge screen of "Digital Sky" highly hangs on the curved ceiling ofo the "Plaza to Heaven". Over 200 projectors constantly play multi-media pictures and special visual works, which could be enjoyed by not only visitors in the shapping mall but also passby outside on the street, most beautiful at night. Under the "Digital Sky" is a two-storey dining and entertainment area, called "The Ozone", another nice spot in this unique shopping mall.

Besides the exquisite architectural design, it is the UA theatres, Seibu Department Store and MUJI — the three significant supports, except for the 300 shops — that establish the core status for medium top grade shopping mall of Langham Place. UA theatres located at Langham Place is the most stunt one among all of them wide around Hong Kong with its seats worth of 1,500RMB per one. When pulled upward, the handrails between the two seats will become love seats; the width between two rows is 1.1m. Seibu Department Store, originated from Japan, is the major stores of high fashion and cosmetics, whose other two branch stores respectively are at Admiralty Pacific Place and Causeway Bay at Hong Kong Island. In terms of space design, Seibu Department Store at Langham Place adopts a relatively compact spatial layout, so as to create the same crowding atmosphere as the one on the streets of Mong Kok, hoping to make "MK citizens" feel a sense of intimacy. Although MUJI is simply a shop of lifesyle furniture, it is quite popular among people in Hong Kong. Its new shop in Langham Place combines 4 height interior space and the loft space, so as to make the shop more spacious.

The Langham Place will be a significant force to promote the vigorous development of Mong Kok.

是价值1,500元，中间扶手可以拉起变成情侣双座，前后座位间隔更达1.1m；源自日本的西武百货，是香港主要的高档时装、化妆品销售场所，其余两家分店位于港岛区的金钟太古广场和铜锣湾。在空间设计上，朗豪坊的西武百货采用了较为紧凑密集的布局，以此模仿了旺角街头那种人头攒动的感觉，希望让"MK人"（旺角人）有种亲近感；无印良品虽然只是一家生活风格家具店，但深得香港人的喜爱。位于朗豪坊的新店采用了4m高空间配合阁楼的设计，整个店铺感觉更开阔。

朗豪坊将成为推动生机勃勃的旺角地区继续复兴的一个亮点。

The Greater Tacoma Convention and Trade Center

美国塔科马市会展中心

Site area: 18,600m²
Building Area: 22,700m²
Architect: MulvannyG2 Architecture
Location: Tacoma, WA, USA

用地面积：18,600m²
建筑面积：22,700m²
设计单位：美国MG2建筑设计公司
项目地点：美国华盛顿州塔科马市

Located at 1500 Broadway, the Greater Tacoma Convention & Trade Center (GTCTC) is one of Washington State's largest meeting and convention facilities. The GTCTC offers an impressive selection of usable event space including the 50,000 square-foot column-free exhibition hall, 25,000 square feet of ballroom / meeting space and more than 42,000 square feet of additional event space.

How did we design a 277,000-square-foot public space without destroying the scale of the neighborhood? We combine classicism with the contemporary and make clever use of a few optical illusions. At street level, this building's detail meets human scale. Then, as the eye moves up, the building seems to melt into the sky, thanks to reflective materials and the building's vertical proportions. A signature civic building that blends seamlessly with its surroundings—and a perfect capstone to a decade of downtown revitalization.

坐落于塔科马市百老汇大街1500号的大塔克马市会展中心是美国华盛顿州最大的会展设施。该会展中心提供了各类可供选择的集会空间， 包括50,000平方英尺（4,645m²）的无柱展厅，25,000平方英尺（2,322.5m²）的宴会/会议厅以及额外42,000平方英尺（3,902m²）的活动空间。

我们是如何设计这个巨大的277,000平方英尺（25,734m²）的公共空间而不破坏整个社区的尺度的呢？首先，我们将古典与现代有机地结合，灵活地运用了视觉虚幻。在街平面上，建筑物的细部尺度与人体尺度一致。但当视线上移，由于使用了反光材料，加之建筑物合理的数相比例，使得整个建筑物如同融化消失在蓝天之中。这幢标志性的市政建筑与其周围的环境有机地融合在一起， 为十多年来的城市复兴添加了美妙的一笔。

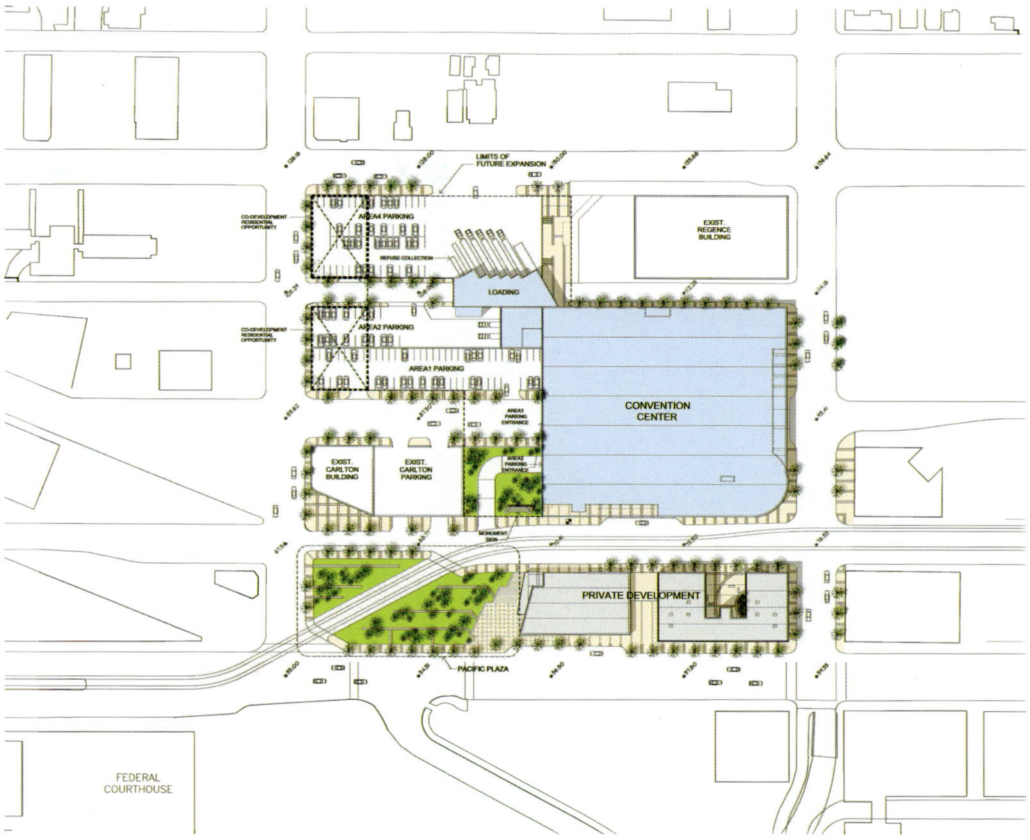
LIMITS OF FUTURE EXPANSION
EXIST. REGENCE BUILDING
LOADING
CONVENTION CENTER
EXIST. CARLTON BUILDING
EXIST. CARLTON PARKING
PRIVATE DEVELOPMENT
PACIFIC PLAZA
FEDERAL COURTHOUSE

Shenzhen International Chamber Centre

深圳国际商会中心

Location: CBD of Shenzhen, China
Site area: 8,192m^2
Building area: 134,081.38m^2
Volume Fraction: 11.0
Client: Shenzhen Rongchao Real Estate Development Co., Ltd.

建设地点:深圳市中心商务区（CBD）
总用地面积:8,192m^2
总建筑面积:134,081.38 m^2
建筑容积率:11.0
开发商：深圳市荣超房地产有限公司

Shenzhen International Chamber Centre is located at the south area of CBD of Shenzhen city, neighboring Shenzhen Convention & Exhibition Centre. This 216 meters' high tower, the tallest one in the South CBD of Shenzhen, is a simple sculptured building with four sides theme taken from the sail which symbolized its landmark position, precisely following the design concept of the CBD by Obermeyer Engineering Consulting Ltd. The lift core is crisscross shaped. The structure of reinforced concrete core inside frame outside left flexibility for zoning in office. Despite of its strict cost restrain, the landmark image and practical solution has made it most expensive development to sell in the CBD.

深圳国际商会中心位于深圳中心商务区（CBD）南区，毗邻深圳国际会展中心。建筑高度216m，是深圳中心区目前已建成的最高建筑。建筑造型设计严格遵循德国欧博迈亚公司所做的中心区城市设计，通过简洁的体块、面的处理形成一气呵成的建筑意象，塔楼四个面的造型则源自对风帆的意境概括，从而构成了本建筑在CBD中的地标地位；核心筒为流线合理、紧凑的十字形平面；结构采用内筒外框钢筋混凝土大小柱体系，十分利于办公区平面的划分。整个建筑的造价控制十分严格，而标志性的形象加上实用的布局使本建筑物的售价达到本区域的首高。

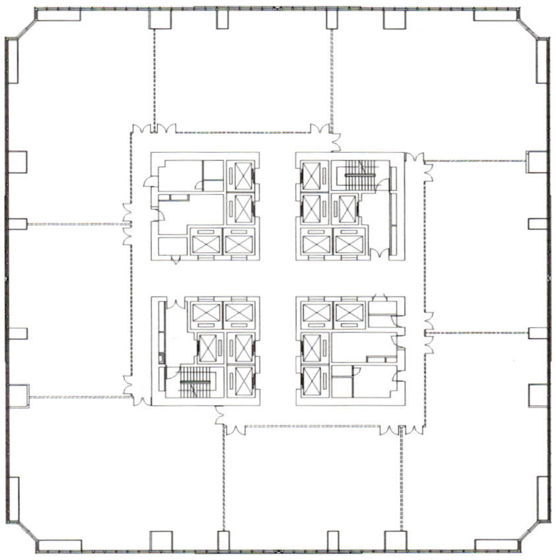

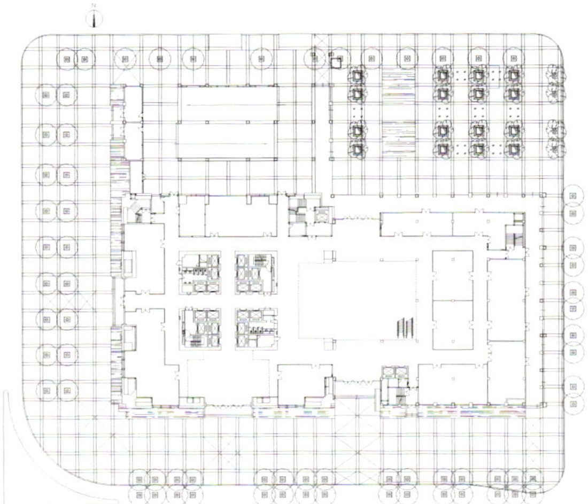

1层平面图 1:150

0 30m

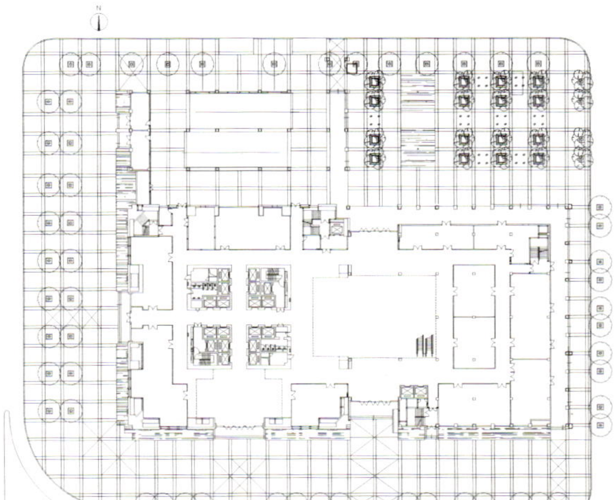

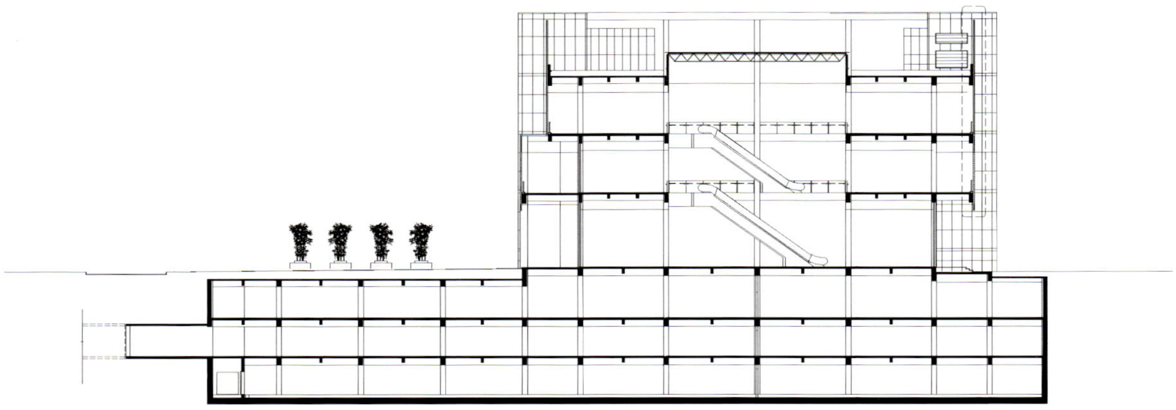

2-2剖面图 1:150

0 20m

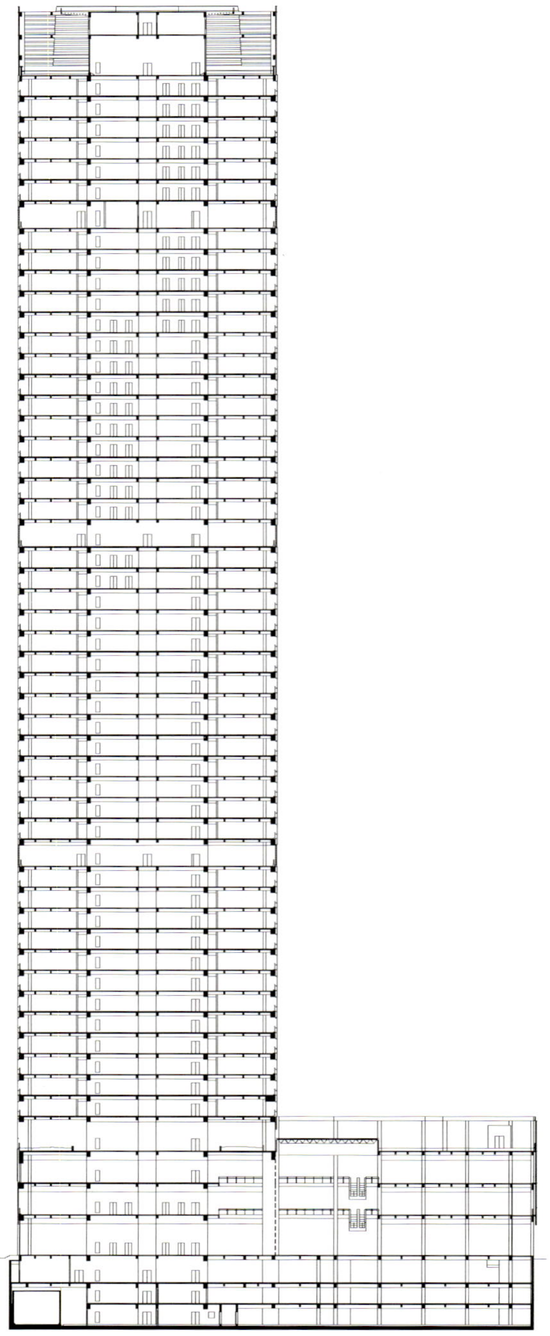

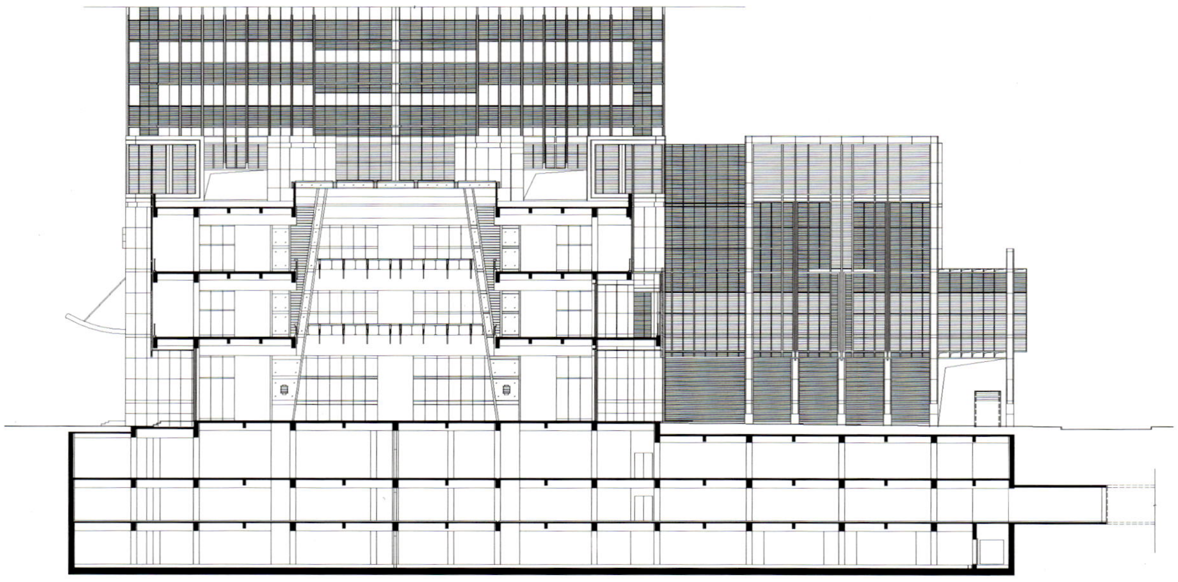

3-3剖面图 1:150

0 20m

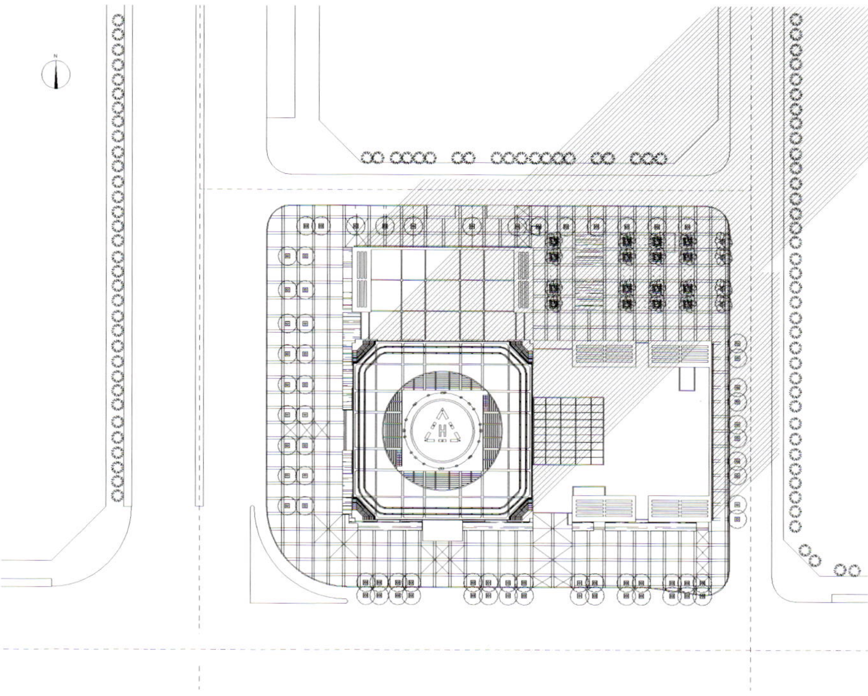

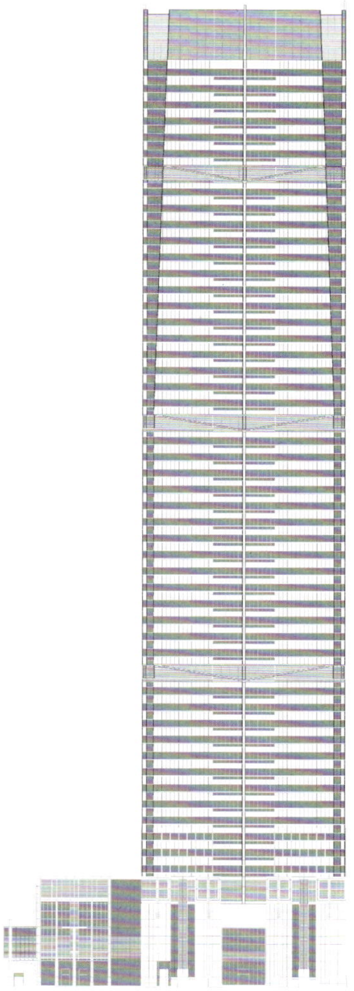

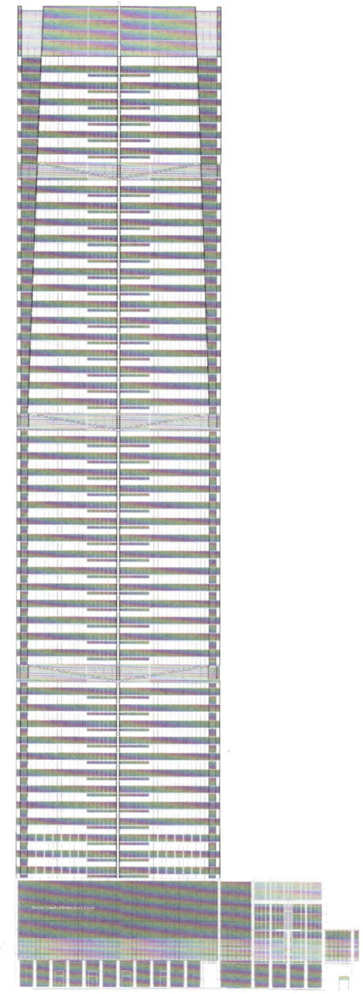

CDO Plaza

设计师广场

Client: Beijing Xinfeng Xinde Real Estate Development Co., Ltd.
Site Plan: Sako Architects
Architect: Beijing New Era Architectural Design Co., Ltd.
Building area: 103,218m²
Living density: 30%
Green area: 5,385m²
Green coverage: 30%
Building height: 80m

开发单位名称：北京鑫丰信德房地产开发有限公司
规划设计单位名称：SAKO建筑设计工社
建筑设计单位名称：北京新纪元建筑工程设计有限公司
建筑规模：103218m²
住宅密度：30%
绿地面积：5385m²
绿化率：30%
建筑高度：80m

This project well combines residential, commercial function and business entertainment, and supplements these functions with each other by utilizing the existing site features; it also keeps each functional area both relatively independent and easily interactive by interpenetrating the landscape. The site is designed to be a full open district. Along side the 4th Ring Road is a large shopping mall topped with twin towers, attracting the mass crowd by its huge volume. The bases of the other two towers are respectively independent, facing each other from north to south at the two sides of the shopping mall. All the buildings turn to the southeast by 45°, so as to acquire the best natural light and to keep the community accessible to all the way around.

The architectures follow the style of Columbia deconstruction. They apply the design logic and methods including contrast, staggering and shaking, creating world-class human residential architectural works, large scale public architecture in Beijing and the new landmark of the west 4th Ring Road.

本规划结合原有地形、地貌特点，充分发挥各功能区之间互补作用，将居住、商业及商务娱乐功能合理组合，并采用景观渗透的方式使各空间互动交流，使其既相对独立，又保持通畅的联系。地块规划为完全开放街区。沿四环路一侧布置大型商场基座及上盖双子座建筑，通过其庞大体量成为人群聚集中心。其余两栋楼座相对独立，南北遥望，呼应于商场之端翼。所有楼座向东南转角45°，以追求最佳采光，同时保证从社区四周通往中心商场的流线顺畅。

建筑单体沿袭哥伦比亚解构主义建筑风格，运用对比、易位、摇摆等设计逻辑和手法，竭力打造世界人居建筑作品，媲美北京大型公共建筑，塑造西四环新地标。

楼座规划对比北京南北轴向，优雅旋转45°角，使日照变得均匀，同时又和位于北京中轴的中心地区有着一个紧密的呼应。由于是开放式社区，因此形成的交通流线形成放射状从社区内部伸向四个方向，暗示着和外界灵活有机的联系，以及四通八达的意味。

The basement is planned to form a 45°angel with the north-south axis of Beijing city, acquiring even lighting as well as interacting with the central area on the middle axis of Beijing. As an open community, the project forms a radial routes spreading from the community to all the directions, indicating the flexible and organic relation with the outside and the auspicious connotation of "extending in all directions".

In terms of structural unit, all kinds of architectures consist of individual small structural units. Residential building can be considered only in the way of individual units, but most architectural projects won't represent only in the residential form. The CDO Plaza is designed according to personal feelings, considered as neither an individual unit nor an integrated mass. Weaving together the two colors, black and white, relieves the stress caused by high density buildings, so as to create a pleasant atmosphere. It is uniquely cool to weave black and white together, inspiring people to imagine a lot of things — chess board, flag of formula race and keyboard; and it is so astonishing to realize the complicated design by ingenious methods. One of the design requirements is that no column is visible from any angel; to meet this requirement, many new methods are applied, for example, the staggered arrangement and the roof covering which can hold visitors. Each cuboid of two layers is staggered with the other by 2m, creating a smooth architectural rhythm and a sense of solid; it also forms balconies contributed by the additional space. The staggering of the layers shapes up the building which is usually upright without any. The rhythm created by the two colors, black and white, relieves the stress caused by high density buildings, so as to create a pleasant atmosphere. When you look at the building formed by the staggered cuboid units of every two floors, you will reach the top of the 80m building unconsciously without recognizing it's such a tall one. The balconies formed by staggering become the roof coverings which people can visit; the roof coverings, located at the two sides of each cuboid, look like bows, and when the residents come to the balconies, they will feel like on the board. The 6-storey huge shopping mall faces the southwest of the 4th Ring Road; the raised structures projecting like tenons and the design of going backward provide

不论什么建筑从体块上来讲，都是由一个一个小的体块组成的，住宅项目可以仅从单体考虑，但大多数建筑都不会单纯的表现住宅的形式。设计师广场，在设计上既不作为单体考虑，也不做为一个整体，而从人的主观感受进行设计。采用两种颜色的交错来减轻高层建筑给人的压迫感，从而营造出一种舒适的感觉。这也是设计中运用黑白两色的初衷。这种黑白交错的形式本身就很酷，会产生很多联想（围棋、方程式的小旗，琴键），而且将非常复杂的东西用巧妙的方法表达出来，这就让人很震惊。在建筑的设计要求是从任何一个角度都看不出来一根柱子，也运用了许多新的方法，比如设计方式采用交错，以及加入可上人屋面的元素等。

每两层易位2m的长方体叠加，使建筑节奏优美舒畅，立体感足，并产生超空间错落阳台。易位，使通常那种直立缺少变化的楼宇有了姿势，黑和白两色形成的节奏，减弱了楼宇的高度给人和环境带来的压迫感，从而形成快乐的场。当视线追踪观察那些以两层为一个单位，呈长方体叠加，并各自错开2m的楼宇时，不知不觉地会爬到80m的高度，然而在视觉上它好像并没有那么高。由于易位产生出来的平台部分，自然成为可上人屋面，还因为可上人屋面位于长方体的两端，又像是船头，将来住在这里的住户，来到这里就好像来到甲板。6层高且巨大的商场部分面朝西南四环方向，那些在单一的面上出现一些像榫头一样挑出来的部分和退后处理，给这个立面增加了锐利的造型。那种随四时移动而形成的光影，显然为整个建筑群体增加了立体感。

随着长方体建筑体块稍稍错开之后，便出现有模数关系的1m²左右的正方形窗户。然而这些尺寸相同的窗，由于材质的原因，却有着不同的功能和表情。窗类型分为：固定窗、镜面窗、外旋窗和百叶窗。其中，“镜面窗”，活跃和丰富了建筑的立面，产生出极其美好的反射。它特有的效果，在大量的组窗中产生节奏，使我们的视线活跃起来，就像鱼身上的鳞光或者浪尖上跳跃的反光一样美妙。而立面上那些好像无规律却有着平衡感的窗，和那些室外机百叶大小一致，它们被交错布置在 一起，模糊了各户的分隔，从而强调了长方体的水平特征。

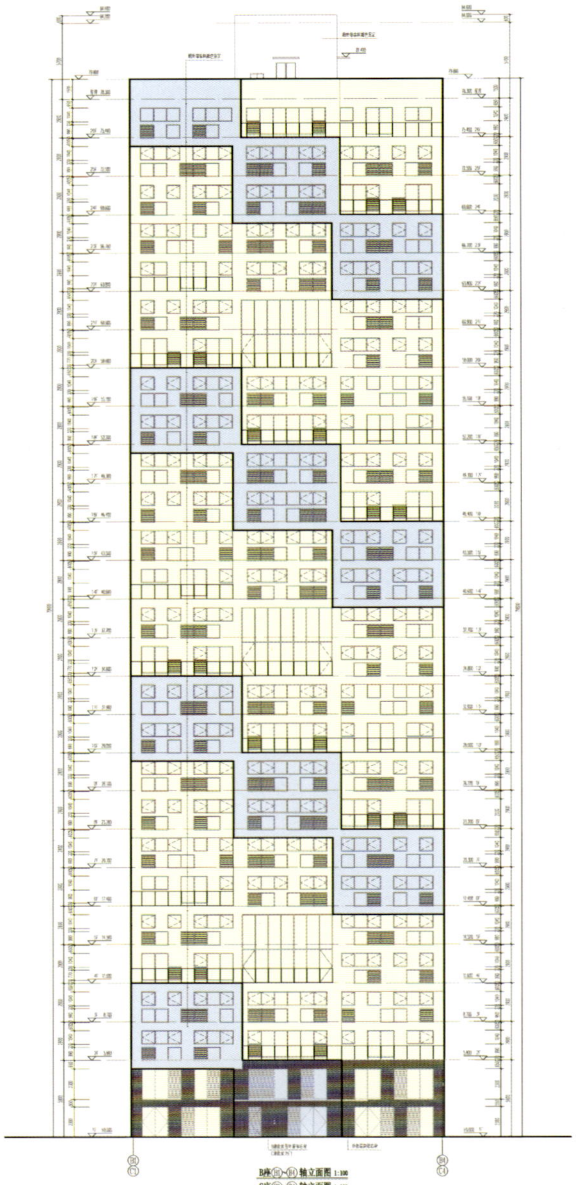

B座(B1)~(B4)轴立面图 1:100
C座(C1)~(C4)轴立面图 1:100

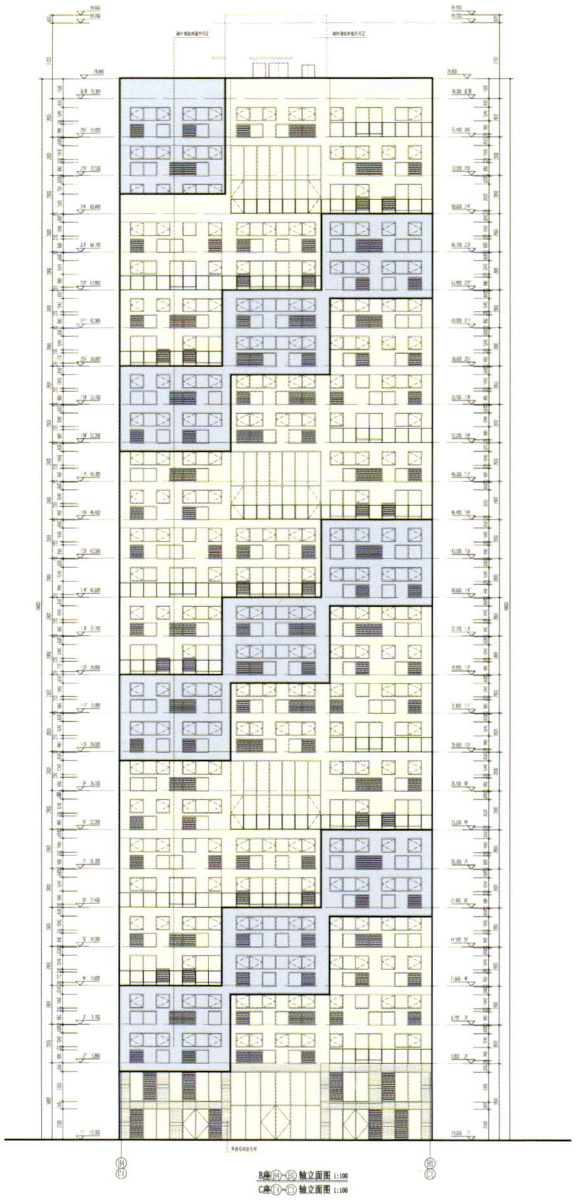

B座(B4)~(B1)轴立面图 1:100
C座(C4)~(C1)轴立面图 1:100

经济技术指标

建设用地面积				17949m²	建筑占地面积	5384m²
总建筑面积				99206m²	建筑密度	30.0%
	地上住宅及商业建筑面积			68244m²	容积率	3.8
	其中				机动车停车位(地下)	298辆
	1	A、D裙房商场(1-6层)	面积	20000m²	绿化率	30%
			高度	29.90m	总绿化面积	5385m²
	2	B、C首层商铺	面积	800m²	其中 实土绿化面积	3692m²
			高度	5.85m	覆土绿化面积	1693m²
	3	A座7-23层	面积	9575m²	总户数	1113户
			高度	79.8m	总人口	3117人
	4	B座2-26层	面积	14128m²		
			高度	79.8m		
	5	C座2-26层	面积	14128m²		
			高度	79.8m		
	6	D座7-23层	面积	9575m²		
			高度	79.8m		

注：此表为C座中2-8层非居住用房面积（位置详见图中虚线部分）

8层：30m²
7层：52m²
6层：105m²
5层：140m²
4层：216m²
3层：314m²
2层：403m²
合计：1260m²

其中	人防出口建筑面积		38m²
	地下工程建筑面积		30962m²
	地下三层	人防物资库(平时车库)建筑面积	7499m²
		设备用房建筑面积	1218m²
		物业用房建筑面积	1783m²
	地下二层	汽车库建筑面积	9514m²
		设备用房建筑面积	986m²
	地下一层	库房建筑面积	8440m²
		设备用房建筑面积	1522m²

说明：1. 本项目建成后对周边现状住宅的日照影响能够满足相关法规及设计规范中对日照的要求.

2. 本项目是中关村丰台科技园区项目之一。科技园区已根据京政发(2002)22号文件的要求.完成了相应配套设施的建设。

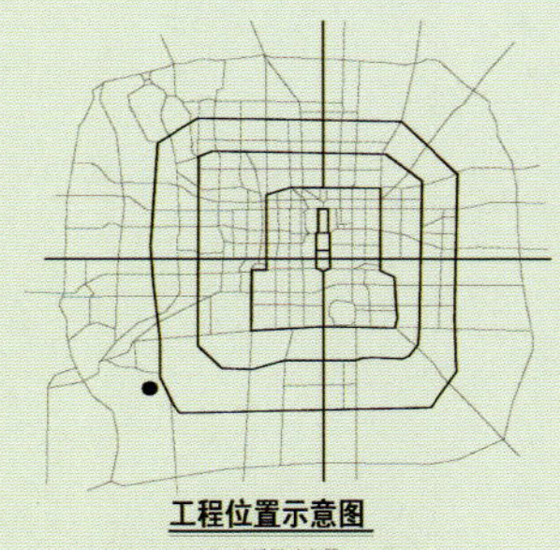

工程位置示意图

● 建设用地位置

总平面图 1:500

图中所注单位为米

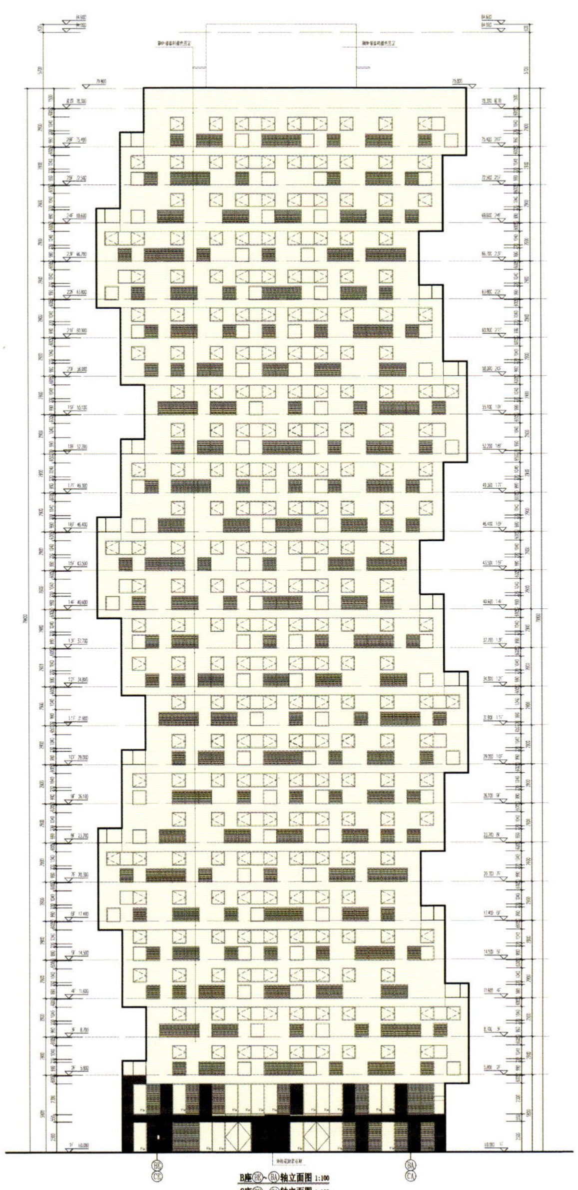

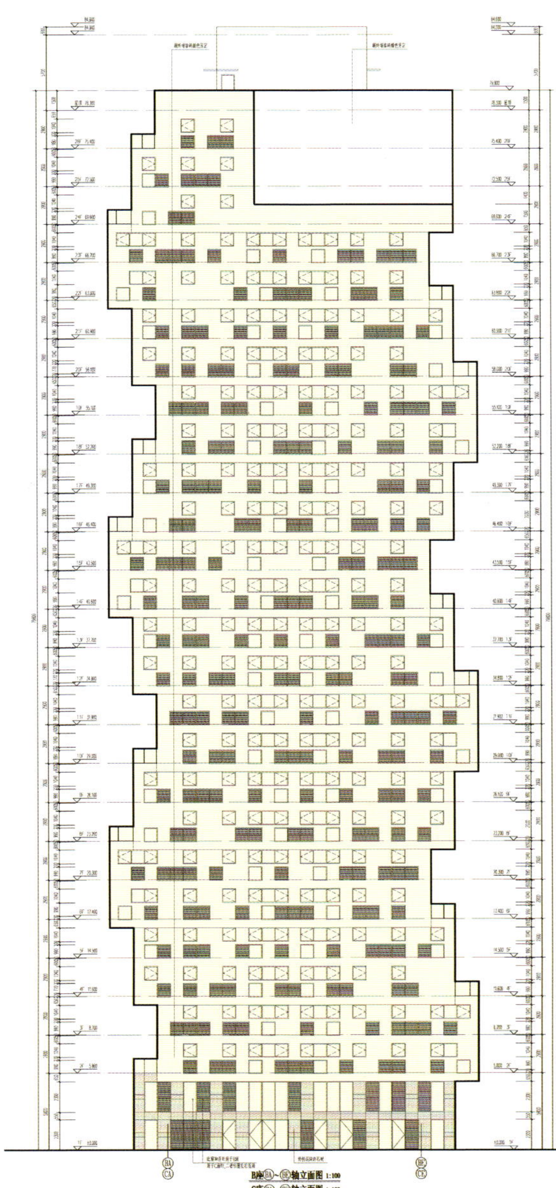

the façade with a sharp image. The light changing with the time sets up a sense of solid for the whole architecture complex.

When the cuboids of architectural units are staggered, the 1m X 1m square windows are formed with a modular dimention. Although these windows are in the same size, they have different functions and appearance because of their materials. The widow varies in form, including fix window, mirror window, pivoted window and window shutter. The mirror windows vitalize and enrich the façade, and they are as shining as fish scales or reflection on the top of wave.

Both black and white are exteme colors, and have no boundaries. White is to blend with other colors, chaning the color added to it into another one; black is to absorb any other colors, nothing changed when it is blended with other colors. It shortens the sensible distance between man and objects to use the stark contrast bewteen black and white. Both black and white are eternal colors, and they are always existing and being accepted. A maganificant visual effect is formed by compositing the mysterious, inclusive black color with the intangible, integrative white color, creating a new kind of black.

黑与白都是很极端的颜色，而且没有边界。白色是融合的颜色，会给加入的颜色带来一种变化；黑色可以吸收所有的颜色，任何颜色加到里面都不会有影响。利用黑色和白色的对比拉近人对物的感官距离。黑与白是永恒的色彩，永远都会存在和被接受。神秘的、包容的黑色与空灵的、融合的白色搭配形成很酷的视觉感受，创造一种新的黑色。

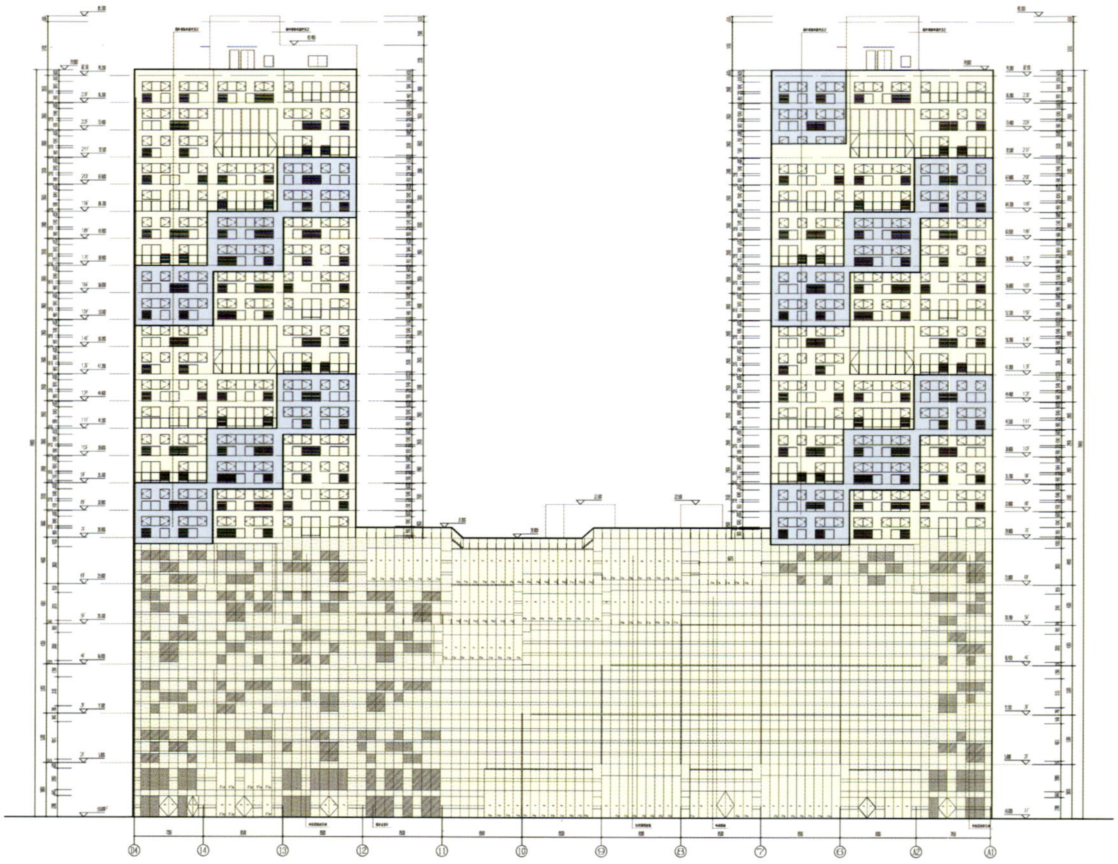

商业部分Ⓓ4~Ⓐ1立面图 1:150

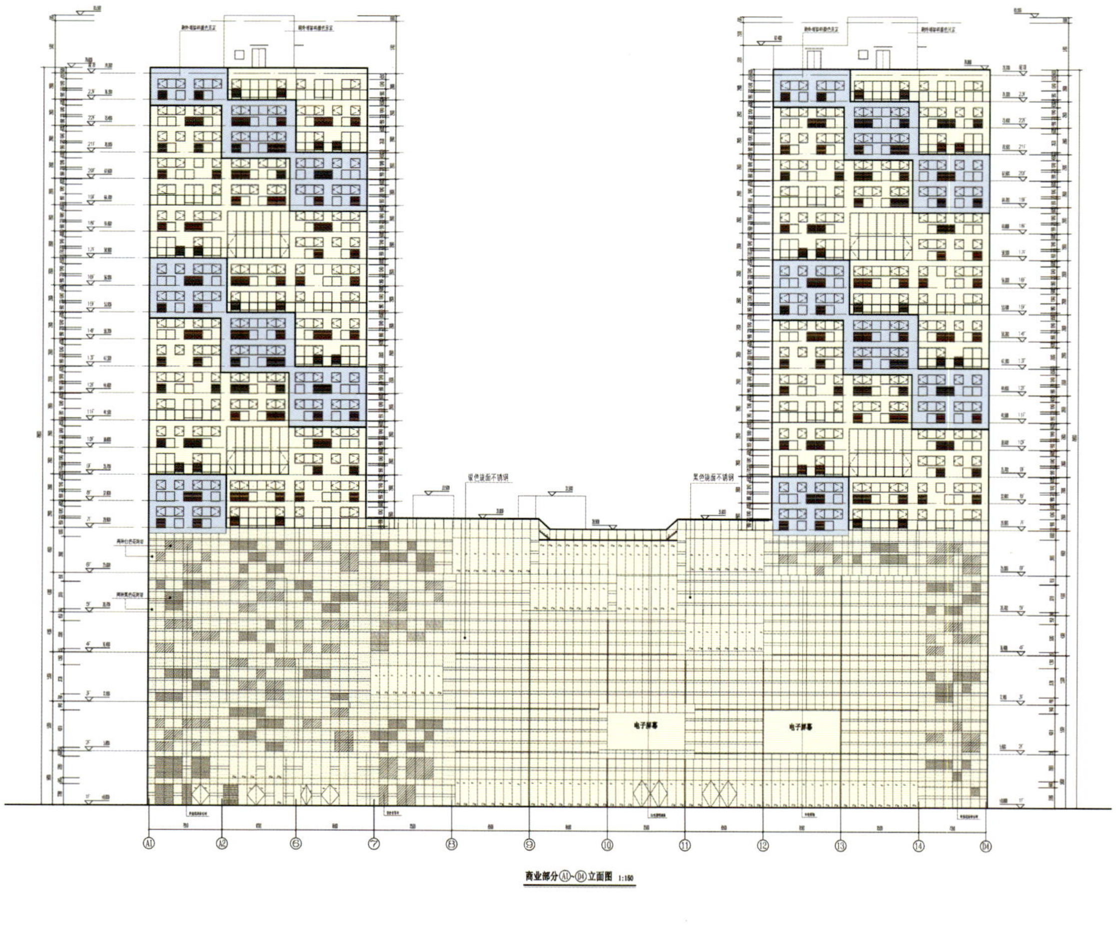

商业部分Ⓐ1~Ⓓ4立面图 1:150

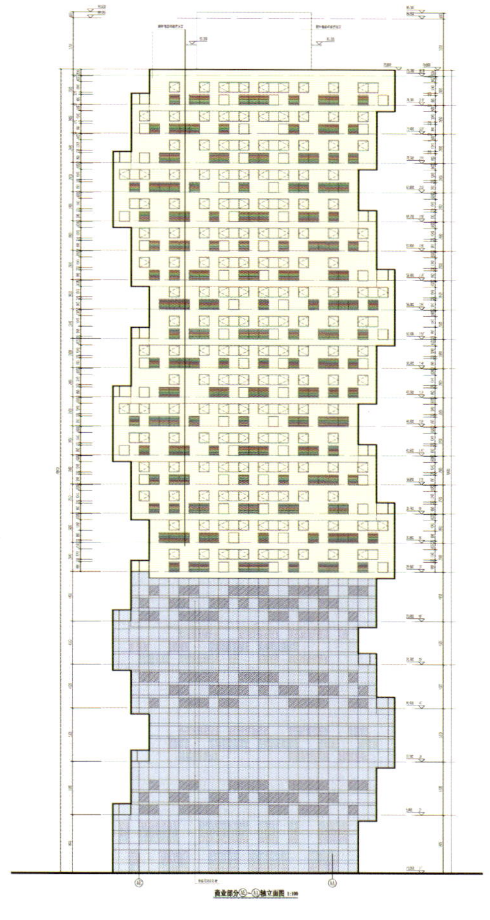
商业部分
轴立面图 1:100

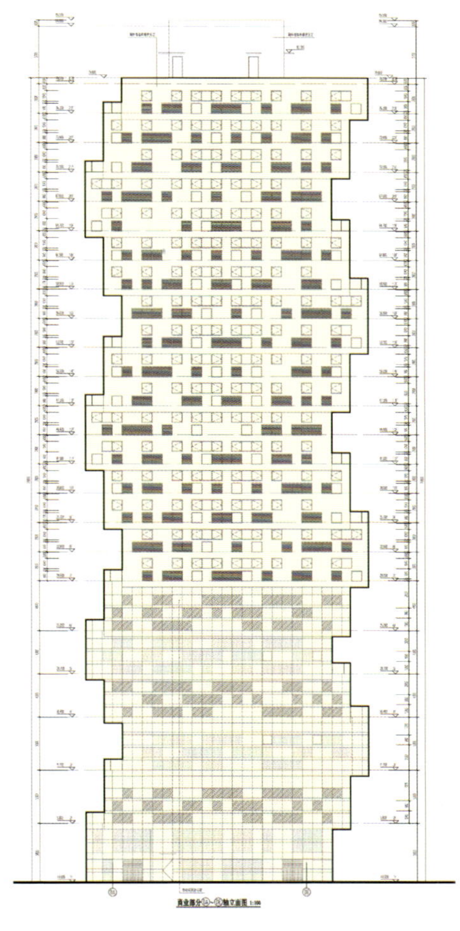
商业部分
轴立面图 1:100

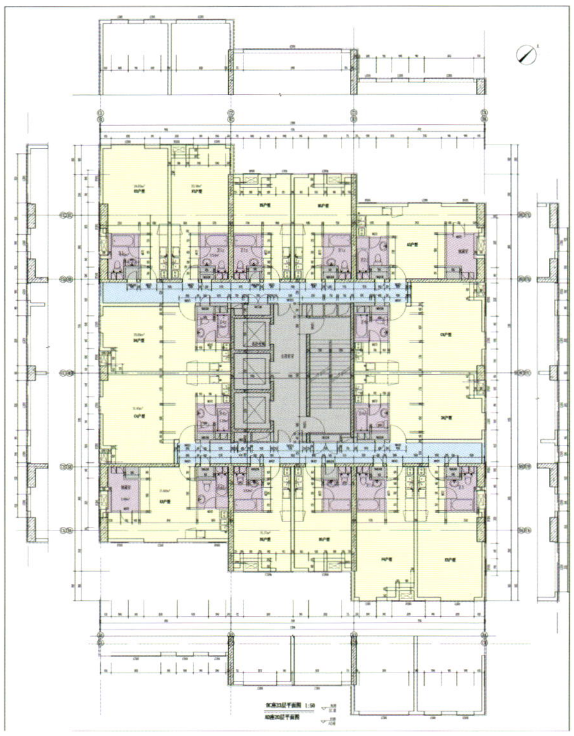

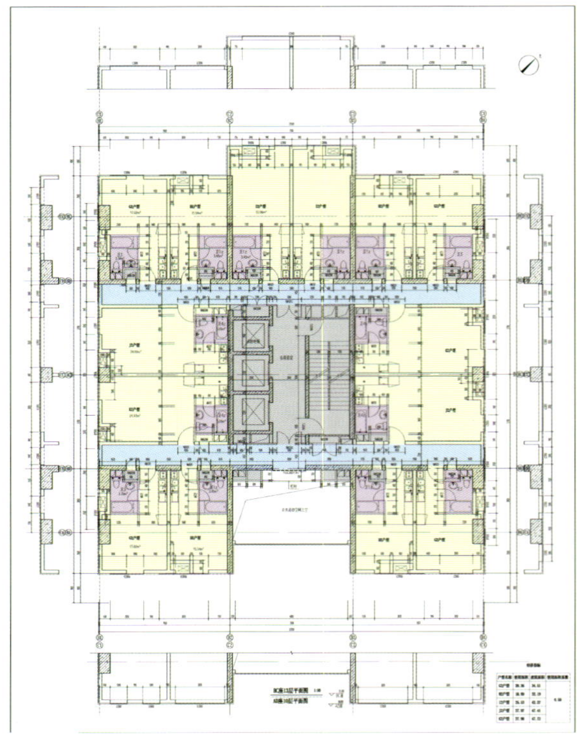

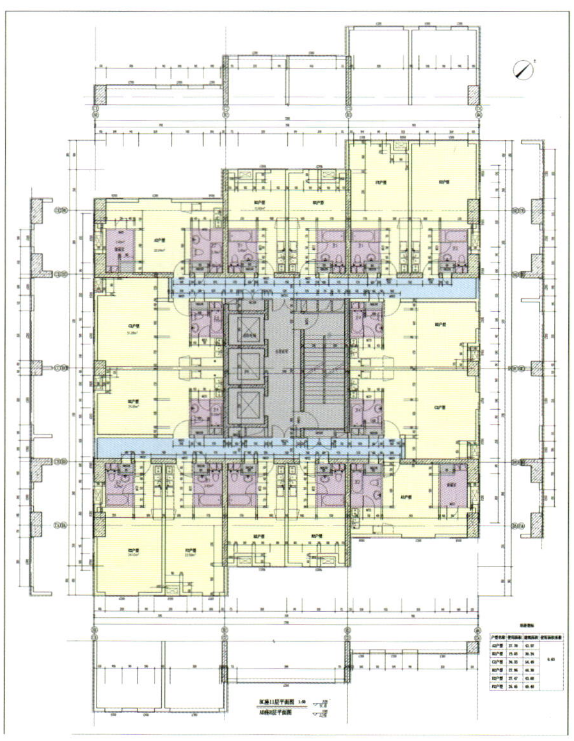

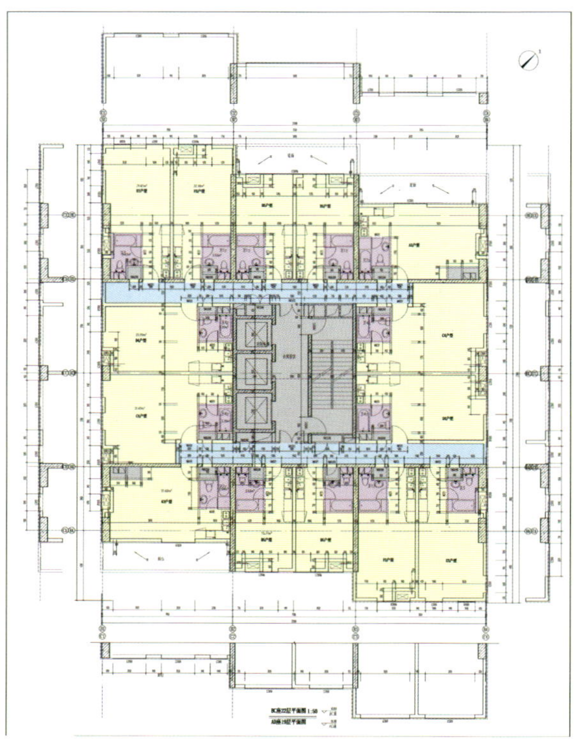

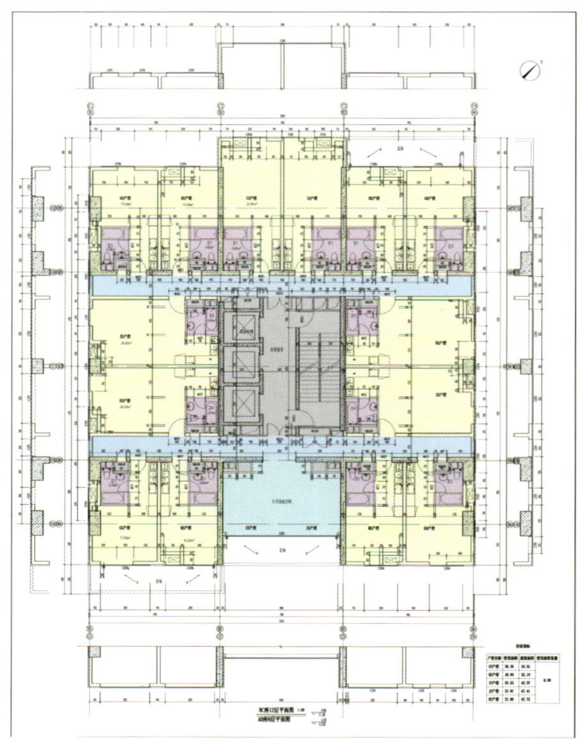

Beijing Fortune Plaza

北京财富中心

Developer: Xiangjiang International Group Co., Ltd.
Architure: IPPR International Engineering Institute
Co-architect: Kaida & Botao Architects Co., Ltd.
Design team: Fei Lin, Feng Fengfei, Chen Ziming, He Shan
Site area: 9,230m²
Building area: 28,082m²

开发商：香江国际集团
建筑设计：中元国际工程设计研究院
合作单位：凯达柏涛建筑师有限公司（香港）
建筑师：费麟、冯腾飞、赵永勃、陈自明、何山
占地面积：9,230m²
总建筑面积：28,082m²

Beijing Fortune Plaza is located at the middle section of east 3rd Ring Road in Beijing; it neighbours Jin Guang Center by north, Kerry Center by south; and it is facing with the new headquarter of CCTV, partitioned by the 3rd Ring Road. Beijing Fortune Plaza a comprehensive internationalized business community, housing grade A office building, high-end residential, five-star hotel, conference center, cultural & art center and recreational center. The preject of Beijing Fortune Plaza Phase I was firstly launched, which is multi-functioned with office and residential as well as equipted with many commercial and recreational facilities.

Taking full consideration of the existing situation of the new era, new Beijing and new market place, the design of this project demonstrates the special funtions suitable for the special site and the special environment. In term of the site planning, located at the core area of the CBD of Beijing, this project should meet the requirements of the master plan of CBD; it should maintain a special

北京财富中心地处北京东三环路中段，北邻京广中心，南邻嘉里中心，东侧隔三环路，与中央电视台总部新址相望。北京财富中心是一个集甲级写字楼、高档公寓、五星级酒店、会议中心、文化艺术心、休闲娱乐中心等功能于一体的综合性国际化商务社区。财富中心一期工程为其率先启动部分，有办公、居住等多种功能形态，同时配备多样的商业和文娱设施。

工程设计充分考虑新世纪、新北京、新市场的现状，用心注解了特殊地段和特殊环境的特殊功能。从规划角度看，项目处于北京商务核心区的核心，必须符合CBD总体规划的要求，有一个特殊的群体组合和环境设计，不能是几幢单体建筑的简单堆砌。整体项目应创造一种氛围，给人一种场所感，形成适合人们工作和生活的环境，使人们在其中能够感受到较浓的商业氛围，办公可提高公司的外在形象，居住可身处超高层而不远离地面生活。在处理好复杂的功能布局和现代化外型的同时，工程的设计更加重视城市设计和建筑的空间感受，重视室内外空间的人性化和生态设计、环保因素，尽心创造一种以人为导向的北京CBD应有的一种生活方式。

As night falls, the path among the buildings is sparkling with street lights, representing the florishing scences on the ground floor of the commercial center. 夜色降临，楼宇街道灯光熠熠生辉，首层的商业楼仍呈现出一篇繁华兴旺的景象

The standard floor of the office building features an uneven butterfly-shape façade. The varied façade and the regular traffic core form the varied office spaces, generating areas with different depths in each floor plan.

办公楼标准层外轮廓呈不规则的蝶型，变化的外轮廓与规则的交通核之间形成变化的办公空间，使楼层平面内具有不同进深的区域

The silver apartment building appears elegant and graceful, indicating its characteristic sophistication and maganificence.

银白的公寓楼优雅得体，显示出其特有涵养和大气

The apartment building is equipt with wide floor-to-ceiling windows.

有着宽大落地窗的公寓楼

mass composition and environment design, not simply several independent buildings piled together. The whole project should create such an atmosphere to make people feel a sense of place, and generate such a proper environment for work and life to make people feel a strong atmosphere of business; the office space would promote the exterior impression of the companies, and the residential would make people experience a life close to the ground at the height of skyscript. Besides the complicated functional arrangement and the mordernized façade, the design of this project highlights the space experience of the city planning and architecture, the humanized and ecological design, and the green elements, contributing to create a human-oriented lifestyle which CBD of Beijing should feature.

The complete space layout is composed by the 150m office tower by the 3rd Ring Road and the 40-storey apartment tower at the west angle of the southen site. The ground floor between the two towers is divided by the street, and the plate-buildings and the towers are engaged with each other by the linear curved connection between the 9-storey apartment buildings and the 8-storey plate-buildings of offices; they interconnect with each other, generating various space arrangement, and making the architectural complex the first landmark on the 3rd Ring Road.

The office tower is 40-storey on the ground, above the third floor is office space and on the ninth and tenth floor is two-storey high rooftop garden, featuring the pursuit of the ecological space. The graphic design of the standard floor of the office tower combines the traffic core and the changing surrrounding space. The traffic core is almost a square, equipted with lifts, toilets and machine room. Inside the core are four sets of passenger lifts, 16 lifts in total, operated respectively for high zone and low zone, and two fire and freight lifts and two smoke preventing staircases. The fire lifts and the smoke preventing staircases share one lobby room. The traffic core is arranged highly densed and minimums the service area, so as to increase the efficiency of the standard floor.

The office tower is 40-storey on the ground; above the third floor is office space. The standard floor of the appartment includes 12 living units. Ensuring the living quality, the standard floor of the appartment is divided into two towers — the northern one and the southern one — each of which has a core tube. At the connection of the two core tubes, each three floors have a roof garden, decorated by green plants. Each of the three floors are connected by the artium, a green space is created at the core tube and the elevator hall, which is full of sunlight. The green space transfers the traffic space into a space for communication and relaxation, enabling the residents of the high-level appartment to enjoy the natural environment at the smallest distance.

The standard floor plans of the two towers are dynamicly butterfly-shaped, forming the living units of different sizes (including large, medium and small sizes), and the appartments are various in their types, from one-roomed to five-roomed, meeting the multi-level demands. The wide floor-to-ceiling windows also provide the users with an open view.

The Fortune Plaza aims to create a multi-mode business space, in order to adapt to changing market demands. The phase one of the project is main for

临三环路的150m高的办公塔楼和南部基地西角的40层公寓塔楼共同组成了一个相对完整的空间布局，两楼之间底层被道路分割，以折线型的服务式公寓（9层）和办公板楼（8层）连接，板楼与塔楼咬合在一起，相互交错的形态构成了丰富多变的空间布局，使整个建筑群成为三环路上的第一个标志。

办公塔楼地上40层，3层以上为办公用房，在9层、10层设置两层楼高的空中花园，体现了对生态空间的品位追求。办公楼标准层平面设计将规则的交通核与变化的周边空间相结合。交通核基本成正方形，内设置电梯、卫生间、设备间。核内还设四组客梯，共16部，分高低区运行；设两部消防电梯兼货梯、两部防烟楼梯。消防电梯与防烟楼梯共用消防前室。交通核布置高度集中，尽量压缩辅助面积，以提高标准层使用率。

办公楼标准层外轮廓呈不规则的蝶型，变化的外轮廓与规则的交通核之间形成变化的办公空间，使楼层平面内具有不同进深的区域，适合不同客户安排不同功能的办公空间。标准层面积达2400m²，通过防火卷帘划分为两个防火分区。每个防火分区具有一部防烟疏散楼梯，并可借用相邻防火分区的另一部疏散楼梯。办公楼开设上悬外窗，有效开窗面积达标准层地面面积5%，可进行自然排烟，并具有良好的自然通风条件。

公寓塔楼地上40层，3层以上为公寓。公寓标准层有12个居住单元。为保证居住品质，公寓标准层平面分为南北两个塔楼，分设两个核心筒。在两个核心筒连接部位每三层设置一个空中花园，布置绿色植物。此三层通过中庭相通，在核心筒、电梯厅部位构成了一个充满阳光和绿化的空间，将交通空间转变为交往空间、休闲空间，使超高层公寓的居住者能最近距离享受到自然气息。

Rendering of Phase II　二期效果图

small-scaled business spaces — brand shops. The phase two is mainly large-scaled space — shopping mall. The phase one has many independent areas; meanwhile they are connected with the phase two. The various business spaces provide the users with multiple business modes.

The whole project uses glass walls, including the appartment building, the office tower and the annex. The streamline design of the appartment building brings a sense of lightness and elegance; the dynamic façade of its two towers is new and fresh, totally different from the conventional tall buildings. At the aspect of shape design, the tower body is divided into two parts, which shortens the width of the building, making the building more upstanding; the two parts of the tower body differ in their heights, enhancing the dynamic feature of the building.

The main building uses insulating glass for the glass wall. The use of low-reflective coated glass solves the problem of light pollution caused by glass wall of large area, and enables the wall to obtain a good heat engineering performance. The decorating stiles of the glass wall is visible from outside and their vertical lines are echoing with the volumn of the tall tower, which makes the orderly vertical decorating stiles and the simple layout of the glass wall generating a sense of skin texture. The glass wall of the annex combines the pointed fixed glass wall with the aluminum plates. The design of the façade is saking for rhythmic changes in simple layout, and aims to give the complex a harmony and integration by the texture contrast of glass and aluminum and steel plates.

This project provides the city with 30m wide green belt along the east of the 3rd Circle Road, generating a green screen between the project and the bustling road. Due to lack of land, the phase one of the project builds the office building and ground floor partly on the stilts along the north of the east-west street of the business center, forming the green belt along the street. Combining with the space design of the entrance to the business center, this green belt forms a pleasant

两塔楼的标准层平面为具有强烈动感的类似蝶形的平面，形成大小不等的居住单元（大、中、小不同的户型），从一居室到五居室，以多样性的户型适应多层次的需求。公寓楼宽阔的落地窗也为居住者提供了开阔的视野。

财富中心工程力图提供多种商业空间模式以适应多变的市场要求。一期工程商业空间以小型商业空间——精品店的形式为主。二期商业以大空间——百货商场的形式为主。一期商业既有相对独立的区域，又相互连通，并与二期商业相连。多种商业空间为业主提供了多种商业选择模式。

公寓楼、办公楼及裙房全部采用玻璃幕墙。公寓楼流线型的轮廓给人轻盈优雅的感觉，两塔楼具有活泼的外形，使人耳目一新，完全不同于传统的高层建筑。在体形处理上，设计将塔体"一分为二"，减少了建筑的宽度，使建筑显得更加挺拔，两翼的高度也有错落高低，增加了建筑的动感。

主楼部分的幕墙玻璃采用中空玻璃。低反射镀膜玻璃的使用解决了大面积玻璃幕墙的光污染问题，并使幕墙体系具有良好的热工性能。玻璃幕墙将装饰竖梃外露，使竖向线条与塔楼高耸的体量形成呼应，这样一来，规律排列的竖向装饰梃与幕墙玻璃简洁的排列方式产生了立体的肌理感。裙房部分的玻璃幕墙采用点式玻璃幕墙与铝板相结合的形式。整个立面设计在简单排列中寻求韵律变化，力求通过玻璃与铝板、钢材的质感对比，赋予建筑群和谐统一的美感。

工程在靠近三环路的东侧用地为城市保留30m宽绿化带，使工程与繁忙的三环路之间形成绿色屏障。一期用地较为紧张，为此在沿商务中心东西街的北侧，将办公楼和公寓首层局部做架空处理，形成沿街绿带。此绿带结合商业入口空间布置，形成良好的沿街景观，为进入商场人流提供了缓冲空间和室外休息空间。在办公楼的南侧布置块状绿化，形成相对独立的景观绿化。财富中心总体用地靠近三环路的中心地带一侧结合绿化带布置集中绿地，既为建筑群提供了绿色景观，同时也向城市开敞，将城市人流引入财富中心建筑群体，为市民提供了绿色休闲空间，也为大规模的建筑群体提供了可呼吸的绿肺。室外绿化空间的设计拉近了超高层建筑的巨大体量与人体尺度的距离。

The hall of the office building is elegant in colors, and comfortably spacious. 色彩淡雅、宽敞舒适的办公楼大厅

The rendering of the elevator lobby 办公楼首层电梯厅效果图

Floor plan of the appartment building　公寓楼层平面图

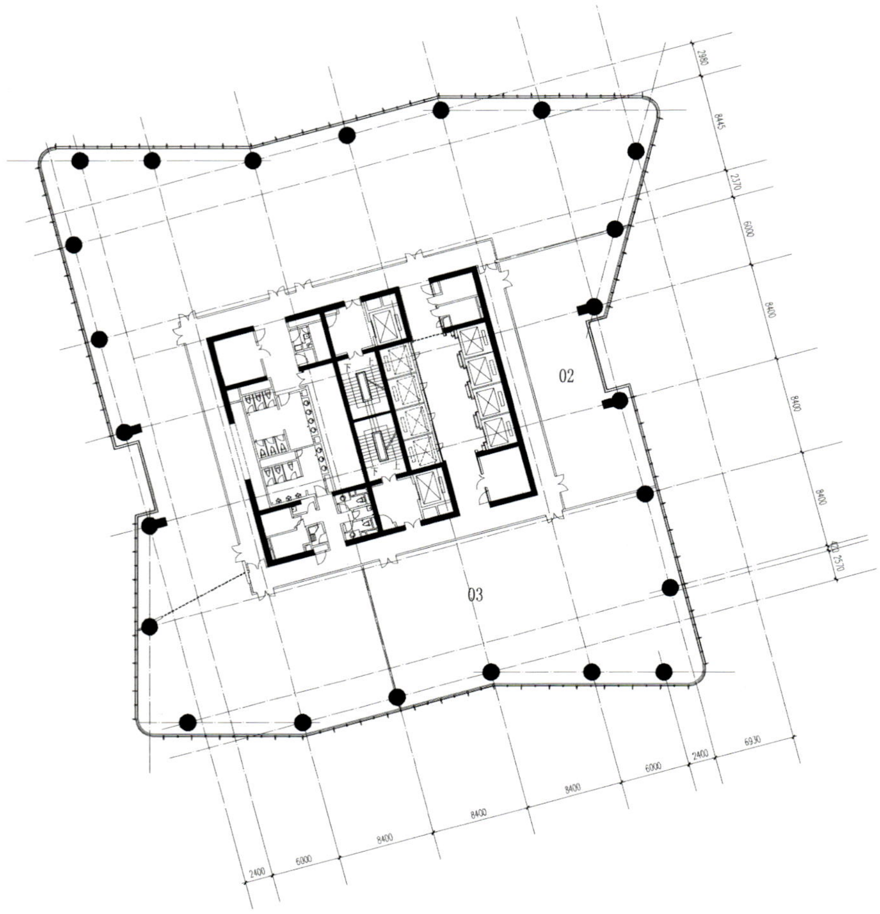

Plan of the 26th floor of the office building　办公楼二十六层平面图

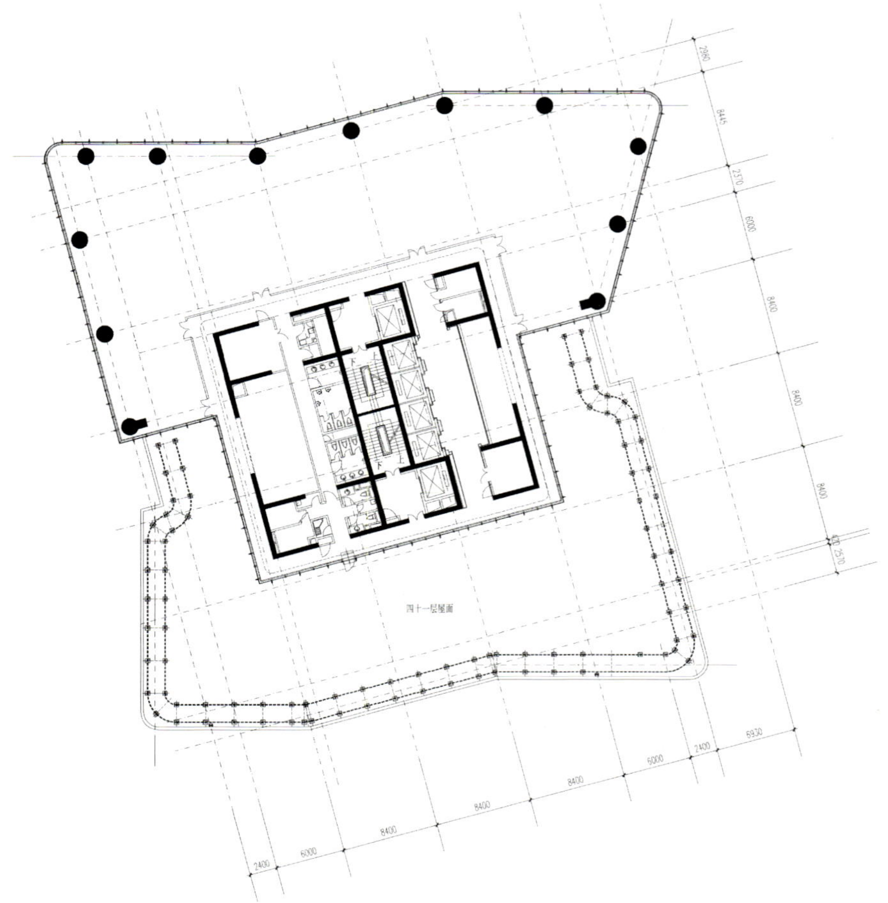

Plan of the floors from 37th to 40th of the office building　办公楼三十七至四十层平面图

Top view of the Fortune Plaza　财富中心顶视图

The pyramid decorator in front of the appartment hall features the craft design, and integrates with the check-shaped glass roof and the façade of the building.

公寓大堂前的锥形体诠释了工艺设计的特色，也与格子状的玻璃顶和建筑立面融为一体

The streamline design of the Fortune Plaza creates for the architecture an appearance of transparency, hamony and sophistication. For the sake of the positive interaction between community space and the mordern urban environment, the project creates a sense of place with the most vigor and coherence.

财富中心简洁的流线型设计创造了通透、文和、隽永的建筑外观，追求社区空间与现代城市空间的积极互动，通过开放、流动、立体化的设计组织，创造最具活力和聚合力的场所精神

The Fortune Plaza aims to create a multi-mode business space, in order to adapt to changing market demands. The rendering on the left shows part of the commercial district.

财富中心工程力图提供多种商业空间模式以适应多变的市场要求。图为商业街局部效果图

street landscape, providing the visitors coming to the shopping mall with a cushion space and an outside relaxing space. On the south of the office building is green plot, forming a relatively independent green space. At the central area close to the 3rd Ring Road, the site of the Fortune Plaza is planned to create concentrated green space by being combined with the green belt. The green space provides the complex with green landscape and is open to the urban space, leading the citizens to the complex of Fortune Plaza; it acts as the green relaxing space for the citizens and the green lungs for the large-scaled building complex. The design of the ouside green space closes the distance between the huge volumn of the super skyscrape and the dimension of human body.

In the nightfall, the tower of Fortune Plaza stands under the huge canopy of the sky with its glass wall reflecting the light of the setting sun, forming a fantastic and maganificent sightview.

傍晚中，财富中心大厦矗立在巨大的天幕下，玻璃幕墙反射夕阳光照，形成了一幅奇异瑰丽的景色

It feels new and fresh to be at the gold and glittering elevator lobby. 置身泛着金色光芒的电梯间，感觉别样不同

The verdent indoor bonsais enhance the elevator lobby of the appartment, enabling the users to enjoy the natural beauty of the plants in the plain environment created by the mordern techonology. 室内苍翠葱郁的盆景为公寓电梯空间增色不少，在现代科技构筑的平板环境中，还能一赏自然植物的动人姿容

Every detail, suach as rattan-weaved basket and paper floriculture, sends the lifestyle full of delicate and taste to everyone. 藤编小筐，纸扎花艺，点点滴滴，将人们对生活的那份精致和情趣，带往各处

The tablewares placed in order interact with the impresionistic rose paint, creating a tranquil and pleasant dining atmosphere. 摆放整齐的餐具在红玫瑰的写意画幅映衬下，恬静舒适的餐饮氛围正浓

The hall of the Fortune Plaza

财富中心大堂

The display of the bedrooms is in the simple manner without useless redundancy, providing the users with a free and leisure sleeping atmosphere.

卧室摆设干脆简单，少去了枝枝蔓蔓和拖沓累赘，给住户一个轻松自在的睡梦环境

The golden wall and the charater logo echo with each other, indicating the connotation of "fortune" both in appearance and in the subject matter.

金色墙壁与字体标识相互辉映，从外观和内容上突现了“财富”的内涵

The red sofa and the window screening gauze generate a light smell of wine. When wandering in the hall, looking at the blowing window screening gauze, you would be dawn in certain dreaming mood, which tastes sweet and lingers on.

红色的沙发和窗纱氤氲出一股股淡淡的酒红，徜徉客厅，看窗纱飘拂，仿佛沉醉在某种萦绕不断的情结上，甜丝满口，回味无穷

Shanghai Huamin Empire International Plaza

上海华敏・翰尊国际广场

Client: Shanghai Huayou Real Estate Developing Company
Architect: Shanghai Shengnuo Consultation Company
Location: No.726, West Yanan road, Changning District, Shanghai, China
Building area: 135,000m²

开 发 商：上海华友房地产开发有限公司
建筑设计：上海圣诺咨询有限公司
项目地址：上海市长宁区延安西路726号
总建筑面积：135,000m²

What office building needs is efficiency, taking no tolerance for dreaming. Standing for the city's economy system and wealth, the location for the office buildings are the coordinates for the city's economy system. Huamin Empire Plaza is an intelligentized combination of age-crossing informationization theme, and the office building is 5A level with international standard, leading the most updated development of office building in Shanghai and providing successful CEOs with a perfect business intelligentization. The office building has floor automatic controlling system, public broadcasting system, security system, garage management system, building information management system, central fire control and alarm system, communication atomized system, international conference center and four languages simultaneous interpretation system.

Huamin Empire Plaza located in the flourishing downtown of Shanghai on the north of west Yanan road and the east of Jiangsu road, surrounded by lots of office buildings and high end uptown, meanwhile the Yanan road trestle is going through the site. After its completion, this building will become a very important

写字楼要的是效率，不宽容梦想。作为城市经济脉络和财富的体现，写字楼地理位置就是城市的财富坐标系。华敏・翰尊国际是结合了超时代信息化主题的智能化5A国际级写字楼，拥有楼层自动控制系统、公共广播系统、安全防范系统、停车库管理系统、大楼信息管理系统、消防报警系统、通讯自动化系统、国际会议中心四国同步翻译系统，引导上海办公新时尚，让成功的CEO享受完美的商务智能化。

华敏・翰尊国际位于上海市中心繁华地段，延安西路北侧，江苏路东部，周边是已成规模的办公区和高档住宅区，延安路高架桥在此经过。本项目建成后将成为该地区乃至上海重要的城市景观。其主体为两栋28层纯办公性质的姊妹楼，由延安路红线后退20m后，形成绿化分隔带，既丰富了都市景观，又可将高架桥上的噪声对办公楼造成的景观及环保上的冲击降至最低点。办公楼之间的中心庭院及裙房上的屋顶花园，不仅使高架桥的第五立面看上去与众不同，也为城市提供了一个开放的绿色庭园。办公楼中间以挑高二层的大厅与玻璃顶的中心庭院作为两栋立体建筑之间的联

urban landscape not only in this district but also in Shanghai. The main buildings are 28-storey twin towers only for office use. Backward 20m from red line of Yanan road, green belt is designed to enrich the urban landscape and reduce the noise and environmental pollution impact from the trestle on the building. The central yard between the two office buildings and the roof garden on the annex can not only provide the city with a public open garden but also make the fifth façade of the trestle appear to be more unique. The two-floor high hall and the central yard with glass roof in the middle of the office buildings are the connections of the twin buildings, below which are the two-floor underground parking space which can contain almost 500 automobiles.

The Huamin Empire Plaza will be first class 5A office building, which means it will be qualifie by BA (building automation system), OA (office automation system), CA (communication automation system), SA (security automation system) and FA (fire automation system). This building is elegant, classical and stylish, which adopts the typical three-segment drawing, consisting of building base (1st-3rd floor), main body (4th-28th floor) and equipment room (28th-top floor). All these make the building stand out in this district with a unique character and rise up a new and vivid Postmodernism culture. The exterior wall of the building base is mainly covered by the dark brown Tan Brown and the brown Baltic Brown stones and decorated by horizontal strip, combined with the colonades made of white limestone, which is rigid and elegant. The exterior wall of the main building is covered by the warm colored cherry blossom red stones, has windows on each sides, and in the middle there is a gorgerous glass wall of 19-floor high. The two materials match perfectly and enhance each other, cool versus warm and heavey versus light. Coordinating with the structure of the equipment room, the roof is mainly designed to be vivid and even cornice and covered by aluminum-plastic plates, adding some vigorous elements to the elegance and steedy charater of the building. In some detailed parts, the designers properly decorate the dark brown column aluminum plats with some golden aluminum plates, representing the noble air of the office building. Between the office buildings is the central yard with glass roof, stretching from outside to inside, where the huge palm trees, high and low, create a strong tropical taste. The stuffs under heavy work load will acquire a romantic spiritual enjoyment when they are in the central yard with a cup of fresh tea in hands.

系，其地下二层均为停车库，可停放机动车近500辆。建成的华敏·翰尊国际为5A级的甲级办公楼，即达到BA（BUILDING AUTOMATION SYSTEM建筑设备自动化系统），OA（OFFICE AUTOMATION SYSTEM办公自动化系统），CA（COMMUNICATION AUTOMATION SYSTEM通信网络系统自动化），SA（SECURITY AUTOMATION SYSTEM安保系统自动化），FA（FIRE AUTOMATION SYSTEM火灾报警系统自动化）。

整体造型古典大气、张弛有度，采用了典型的三段式构图，由大楼基座（1~3层）、主体（4~28层）及顶部机房（28至顶层）组成，使其在周围地带具有独立明确的建筑个性，倡导一种崭新而又生动的后现代建筑文化。基座外墙以深褐色的英国棕和棕色的菊花岗石材贴面加水平状饰带为主，辅以坚挺有力、落落大方的烧毛泽山白花岗岩柱廊。主体建筑外墙亦采用暖色的樱花红石材贴面，四周开窗，正中更气度非凡地设计了19层高的玻璃帷幕墙。两种材质一冷一暖、一重一轻，配合得丝丝入扣、相得益彰。屋顶配合机房的造型，以灵动、错落有致的飞檐为主，外包铝塑板，为整体典雅稳重的建筑增添了一些跳动的因素。在某些局部，设计师更恰如其分地在深棕色柱状铝板上点缀了一些金色铝板造型，臻显了办公楼高贵的气质。办公楼之间为玻璃顶的中心庭院，由室外延伸至室内，高低错落的巨大棕榈树营造出浓郁的热带风情，置身其间，品一香茗，对紧张工作着的白领人士不失为一种具有浪漫情致的精神享受。

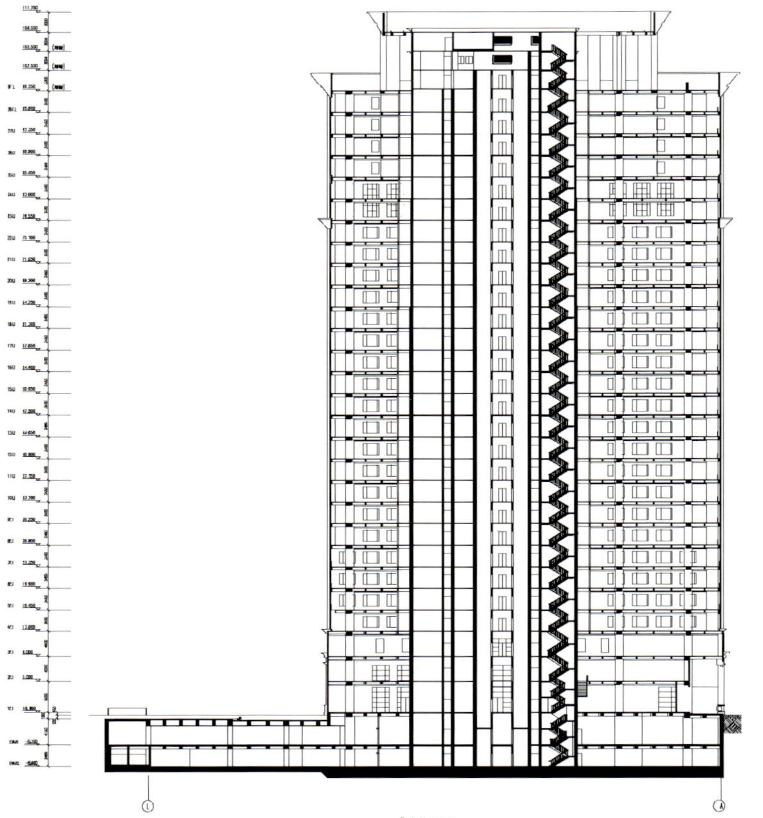

剖面图 1:200

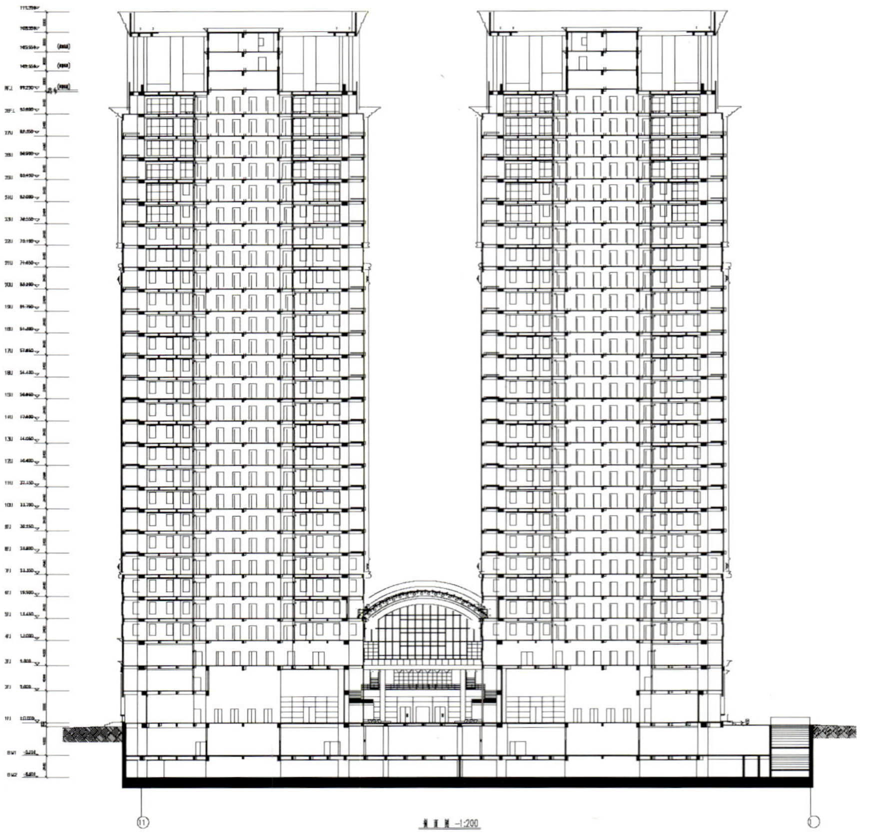

剖面图 1:200

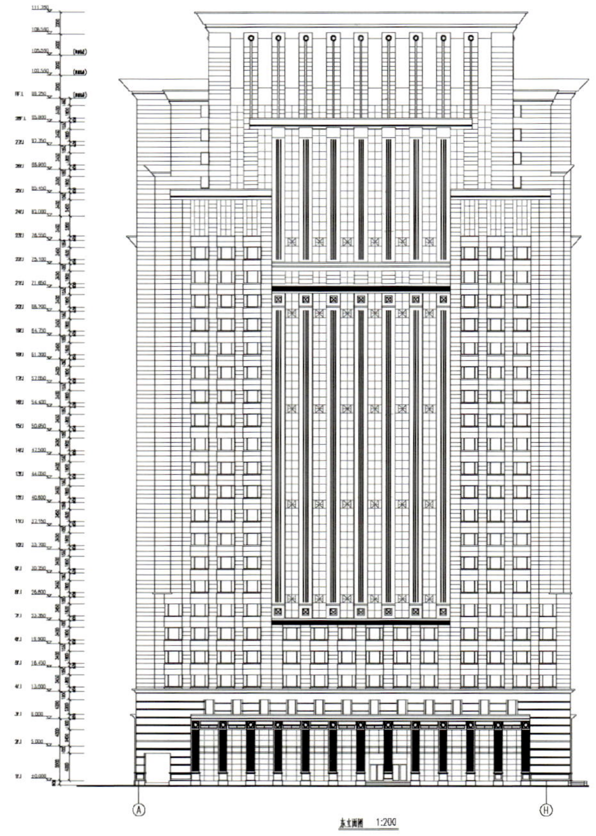

东立面图 1:200

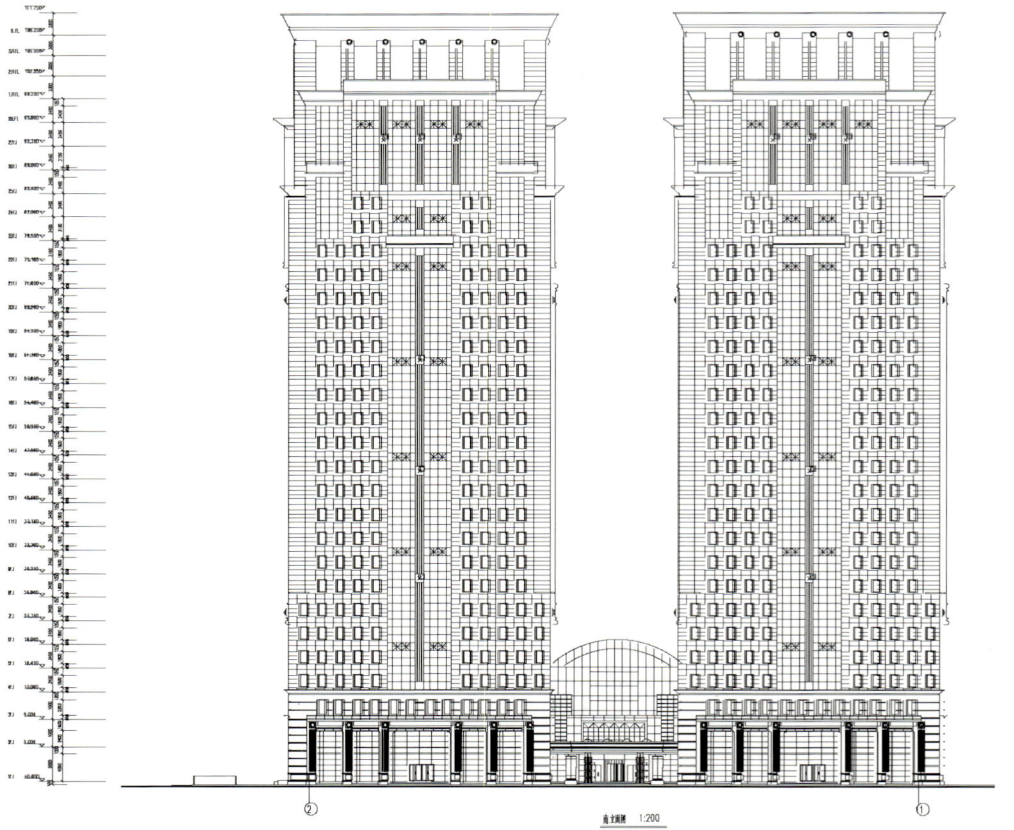

南立面图 1:200

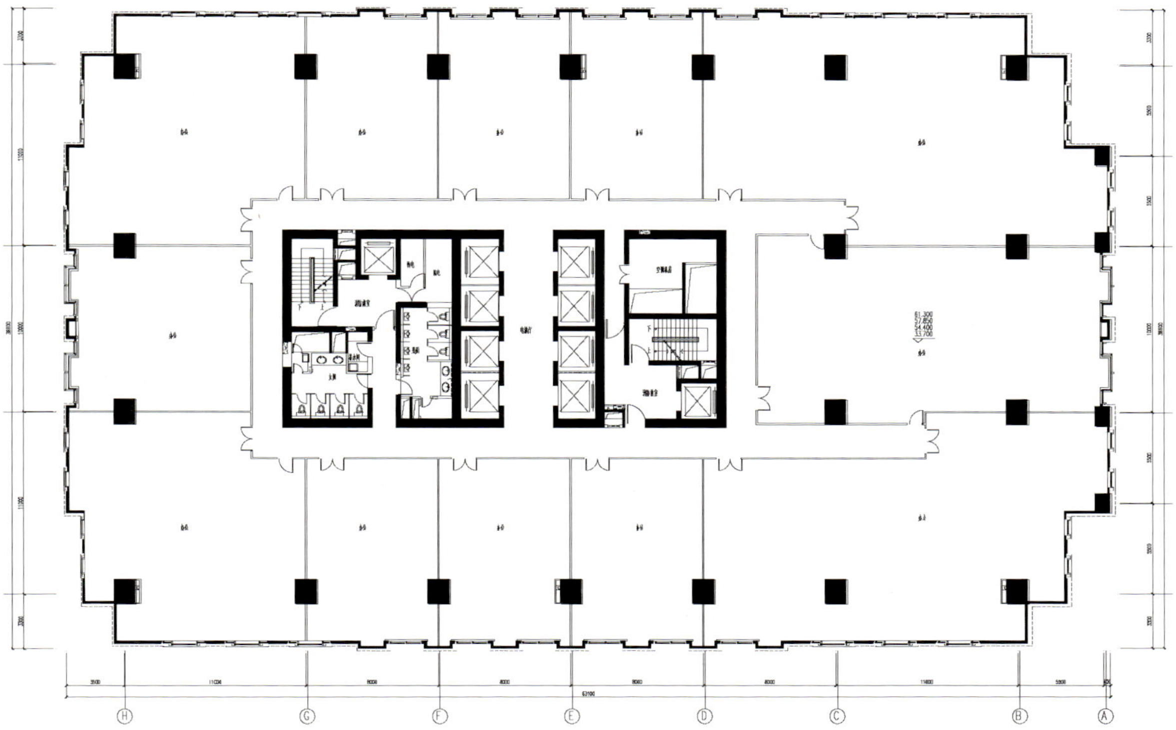

标准层平面图 1:100

728

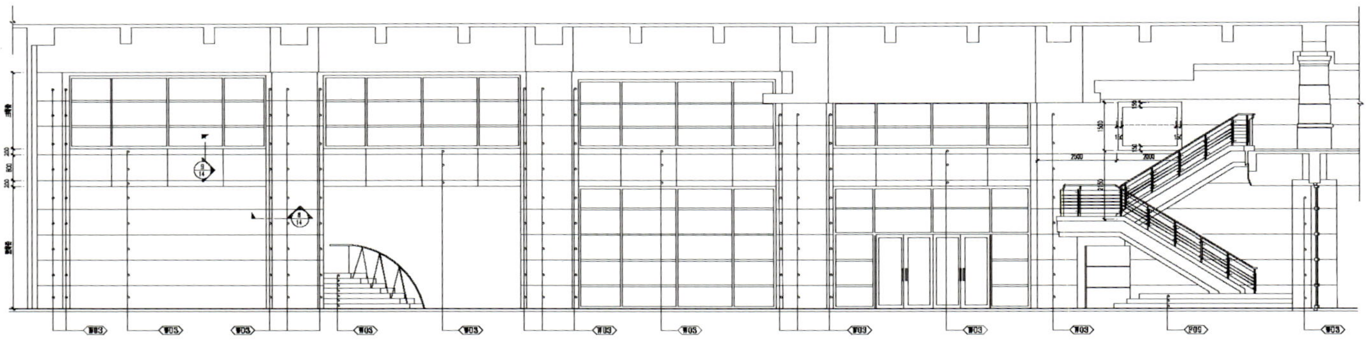

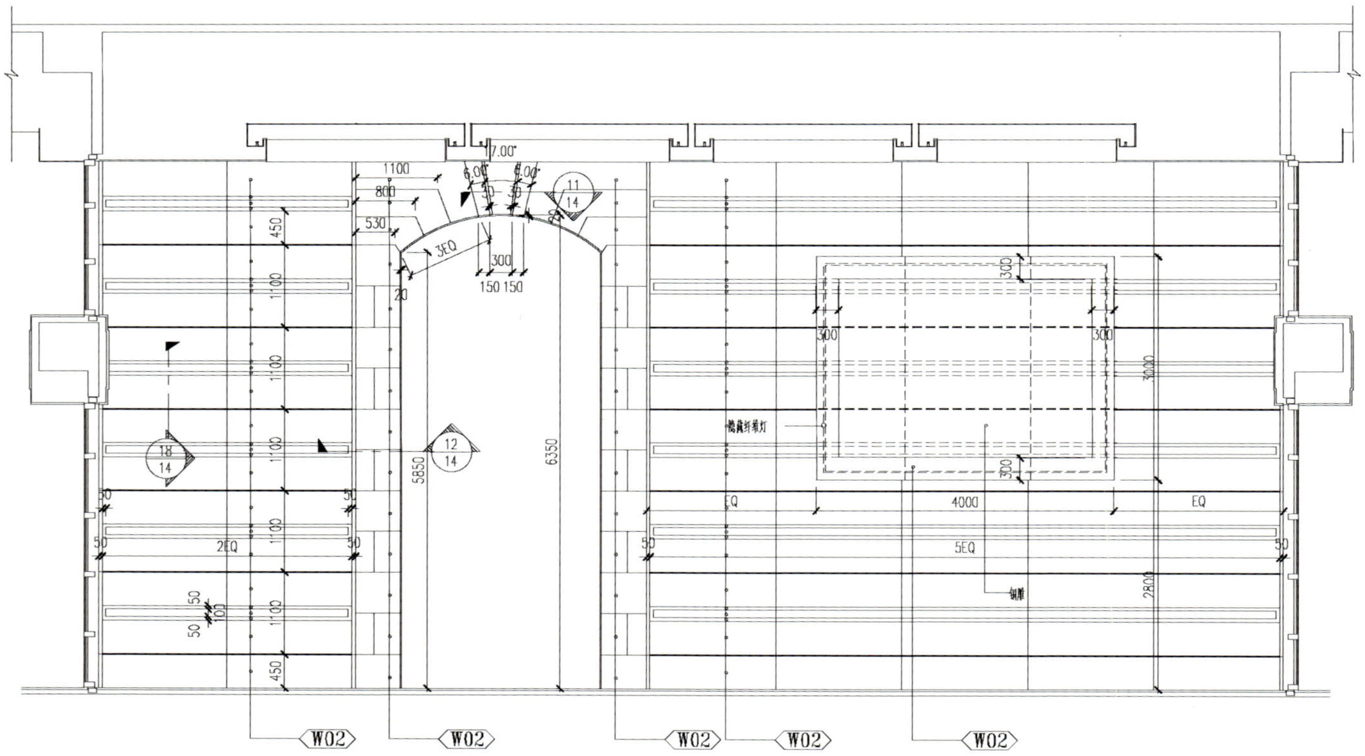

3EQ
2EQ
10EQ
W02
W02

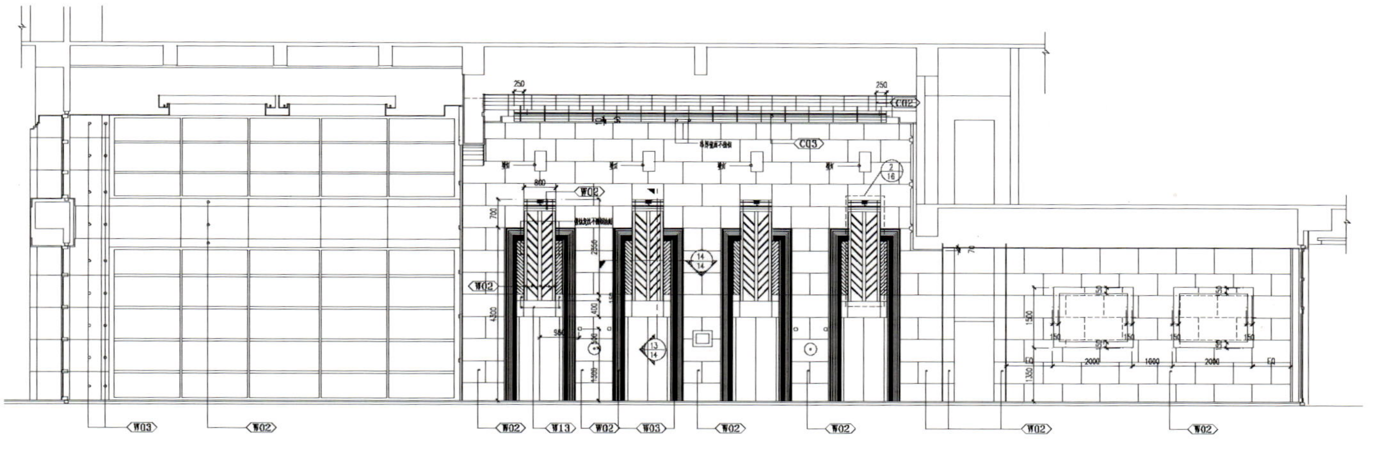

West Ring Plaza

西环广场

Real Estate Developer: Financial Street Construction Development Co., Ltd.
Architecture Design: China Architecture Design & Research Group
Co-operation Organization: France AREP
Project Site: North-west Corner of Xizhimen Overpass of Beijing City
Gross Site Area: 59,900m²
Gross Building Area: 294,800m²

开发商：金融街建设开发有限责任公司
设计单位：中国建筑设计研究院
合作单位：法国AREP
项目地址：北京西直门立交桥西北角
总占地面积：59，900m²
总建筑面积：294，800m²

This project is located in the north-west corner of Xizhimen Overpass of Beijing City, whose structure indicates the consistency of functions and formations from all aspects, the functionality of each conpoment is separated clearly. South and north façades of the tower, and the shape of roof follow the tradtional architectural structure of Beijing, as well as add something creative, which provides a series of fresh urban contours for Beijing City.

The design of public spaces make sure perfect connection bewteen the building and urban transportation network; the west leads to the stone-material front lobby of the tower, and the south leads to the lobby of the commercial center. The designation core of interior spaces is to obtain daylight as much as possible, the huge courtyards inside the podiums allow daylight shines on each space of the 6-floor commercial center directly. The design of façades, especially that of tower is handled distinctively, which manages to combine the complicated functionalities within a relatively simple building for convenient application. The façades are made of stone, glass and metal materials, Low-e hollow glass

本工程位于北京西直门立交桥西北角，建筑体形综合体现了功能与形式的统一，各个部分的用途划分清晰。塔楼建筑的南北立面、屋顶形式继承了北京传统的建筑屋顶形式，又有所创新，为北京提供了一道崭新的城市轮廓线。

公共空间设计确保建筑与城市交通网有着很好的连接：西侧通向塔楼的石材前厅，南侧通向商业中心的大厅。室内空间设计的的核心是尽可能争取大量的自然光，在裙房内设大型中庭，使日光可以直接照射到商业中心六层各空间。外立面设计，尤其是塔楼立面的设计作了特别的处理，力图将复杂的使用功能统一在一个相对简单的建筑体型内，方便人们使用。立面选用石材、玻璃和金属等材料，Low-e中空玻璃作为主要立面材料，玻璃幕墙用在裙房中庭部分和整个西立面，黑色密封干挂花岗岩板用于不透明墙面。在建筑外立面的施工技术选用上做到既耐久又快速方便，采用现代风格设计传统建材，实现传统现代自然融合。三座塔楼的夜晚灯光设计有如三座魔幻般的灯塔映在北京的夜空中，裙房中庭和换乘大厅中闪烁的灯光体皆体现出建筑的无限活力。

The novel and peculiar shape and designation of the 3 podiums in West Ring Plaza makes it become the landmark of Beijing as well as achieves its unique sytle.

西环广场的三座塔楼以新颖奇特的造型和设计为其城市地标形象奠定了独一无二的特质

West Ring Plaza is located in the north-west corner of Xizhimen Overpass of Beijing City, whose structure indicates the consistency of functions and formations from all aspects, the functionality of each component is separated clearly.

西环广场位于北京西直门立交桥西北角，建筑体形综合体现了功能与形式的统一，各个部分的用途划分清晰

The night light of the 3 podiums looks like 3 magic light towers in the sky of Beijing City at night.

三座塔楼的夜晚灯光设计有如三座魔幻般的灯塔映在北京的夜空中

is mainly used for façades, glass curtain walls are used for the courtyard of podium as well as the entire west facade, and black sealed dry-hanging granite is used for non-transparent walls. The constructive techniques selected for façades are durable and convenient, i.e. traditional constructive materials are applied to modern design style, which realizes integrations between traditions and modernism. The designation of the night light of the 3 podiums looks like 3 magic light towers in the sky of Beijing City at night, the blinking light in the courtyards and transition lobbies also indicates infinite dynamism of the building.

In addition, barrier-free designation is taken into consideration for the entire architecture, barrier-free elevators and ramps are set on each floor of podium of the tower, the commercial center, the office tower, the lobby of each component, and internal & external transport transition system. Escalators are set in the core of courtyards to lead to each floor of the podium.

The commercial podiums are divides into B, C and D zone from north to south, the main entrance is set in the south, and 2 sub entrances are set in the west. 3 floors underground the podiums, whose bottom floor is connected to the subway station, where is set defense construction, equipment rooms and a parking lot. 2 ring structured vehicle ramps located in the south and north, are set at the entrances of the parking lot, and the top floor is connected to the internal roads. The second floor is also connected to the urban subway stations, where is set a double-layers mechanic duplex parking lot (whose height meets the usage demands), and equipment rooms, B2 of the greening square is used as a common single-layer parking lot. The south of B1 is used as staff restaurant, a commercial street is set underground the east of greening square, and a courtyard is set in the center for daylight. A huge super-market is set under the commercial development zone and the subway station, and an open multiple-functional transition lobby is set on F1, and the super-market, the staff restaurant, the commercial street and the subway stations in horizontal direction, and leads to the greening square, bus stations, subway stations, the department store and national subway stations can be reached through convenient escalators and stairs lead to the ground.

There are 6 floors above the ground of the podiums. The entrances to the main lobbies of T1, T2, T3 and internal staff are set on F1, the elevators lead to the department store and snack bars located on F1 to F5, and the elevators of T1 lead to high-grade restaurants on F6. Besides the department store and boutiques, a snack bar surrounding the 2 courtyards on F2 and F3, leads to urban subway stations. A shopping mall and boutiques are on F4 and F5. A beautiful roof garden, kitchens and high-grade restaurants are on F6.

Office towers T4, T3, T2 and T1 are layout from north to south, correspond to commercial podiums A, B, C and D zones respectively. The standard floors of T1, T2 and T3 are designed as huge and open office spaces, correspond to D, C and B zones, which can be divided in several units with about 100m², the corridors are centered, offices are layout on the south and north sides, and vertical transportation system is set on the east and west sides. F22 is used for equipments and sight viewing, and F23 is used for huge and multiple-functional meeting rooms, where scenery of Beijing City can be overviewed.

T4 is located in the farthest north of the site, faces to Gaiban River and

此外，整幢建筑充分考虑了无障碍设计，在塔楼a裙房的各层、商业中心、办公塔楼、各部分门厅处、室内外交通换乘系统中均设方便残疾人到达的电梯或残疾人坡道。在中庭核心体部分设自动扶梯到达裙房各层。

商业裙房由北向南分为B、C、D三区，南侧设主入口，西侧设两个次入口。裙房地下共三层，最底层与城铁车站部分相连，设有人防工程、设备用房及停车库，停车库有南北两个环行汽车坡道出口，并在首层出地面与内部道路连通；第二层亦连接到城铁车站部分，设有双层机械复式停车库（建筑高度满足使用要求）和设备用房，绿化广场地下二层处车库则为单层普通停车库；地下一层南侧设员工餐厅，东侧绿化广场下方设商业街，中间设采光中庭。商业开发部分和城铁车站地下处设大型超市，同时，一层亦设有一个开放式的多功能换乘大厅，通过这个换乘大厅往水平方向，可直达超市、员工餐厅、绿化广场的商业街和地铁，往上可乘通向地面的自动扶梯和楼梯，抵达绿化广场、公交车站、城铁车站、商场以及国铁车站等处，通行十分顺畅便捷。

裙房地面共六层。一层为T1、T2、T3塔楼主大厅及内部人员入口，电梯可到达一到五层商场及快餐厅，T1塔楼的电梯厅可直达六层高级餐厅。二、三层除布置有商场和精品房外，围绕两个中庭还设有可从城铁车站直接到达的快餐厅。四、五层为大型商场和精品商店。六层设美丽的屋顶花园、厨房及高级餐厅。

办公塔楼由北向南依次为T4、T3、T2、T1塔楼，分别对应商业裙房的A、B、C、D四区。T1、T2、T3塔楼标准层设计为大型开放式办公空间，对应D、C、B三区，可划分成面积约为100m²的办公单元，平面采用中间走廊，南北两侧布置办公空间的形式。垂直交通体布置在东西两侧。22层为设备层及景观走廊，23层为大空间多功能会议室,可俯瞰北京城市景观。

T4塔楼布置在地段最北侧，面向盖板河，对应A区。设计为普通办公楼，具备独立入口，设有普通办公室和会议室，同时考虑安排两个驻站派出所（具体位置及功能安排待定）。建筑设计高度为60m，其塔楼东侧与城铁指挥中心相连。地下两层为设备用房和办公辅助用房，地上1H—2H层设主入口大堂，3H—18H层均为办公层，地下二层设通道与B、C、D区地下二层车库连通。

商业中心及三座塔楼的基座部分称作“裙房”，其4个立面装饰为体量厚重的石材。覆盖在金属钢结构外的黑色花岗石与混凝土楼板连接牢固。所有玻璃窗立樘位于厚重墙体的内侧，以便造成较强烈的光影变化。玻璃窗选用Low-e中空玻璃和深黑色抗氧化镀膜铝合金窗框。裙房里设有阳光中庭，空间开敞，其中有两组垂直交通体。中庭上方为玻璃顶棚，顶棚钢梁由两个齿口钢梁焊接而成，相互之间另由一组金属小梁连接，“I”字型金属桁梁作为辅助支撑。屋顶玻璃要求连接准确，以确保雨水顺利排放。

T1、T2、T3三座塔楼的东、西、南、北立面皆为玻璃幕墙，其中西立面玻璃幕墙一直落地，并沿高梁桥路设置了入口大厅。幕墙采用金属钢结构外覆双层玻璃的作法。“T”型垂直金属构件设在玻璃外侧，垂直方向组件设在两层楼板边缘，每楼层间设三组玻璃窗，中间小窗扇可开启，便于自然通风。塔楼南侧立面从屋顶至裙房屋顶(六层)高度间设遮阳装置，水平遮阳板构成了建筑的第二层外墙装饰。这种遮阳装置外观既现代，又能使

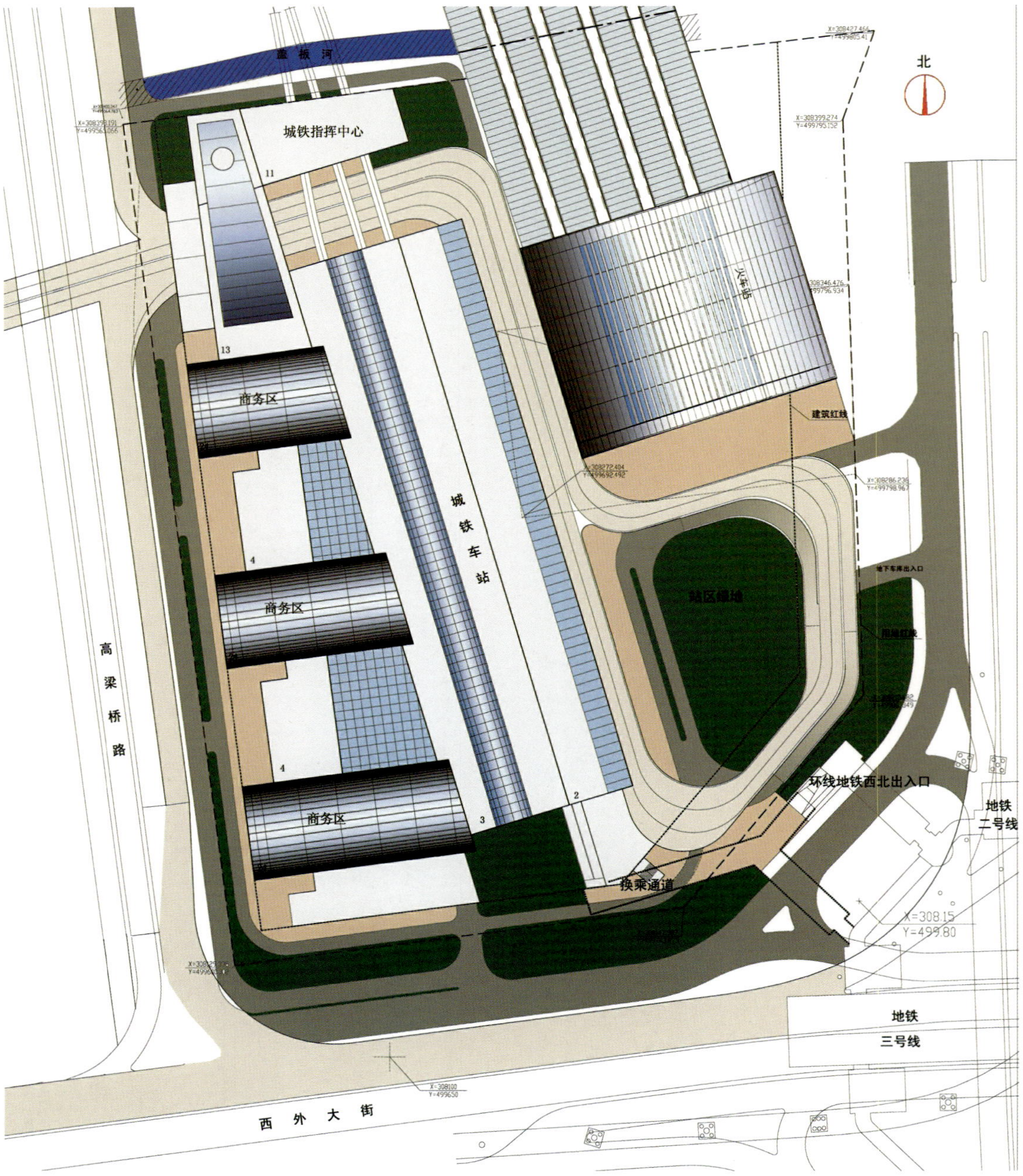

General layout plan

总平面图

The constructive techniques selected for façades are durable and convenient, i.e. traditional constructive materials are applied to modern design style, which realizes integrations between traditions and modernism.

立面选用石材、玻璃和金属等材料，外立面的施工技术选用上则做到既耐久又快速方便，采用现代风格设计传统建材，实现传统现代自然融合。

Characteristic shape of the façade 颇具特色的外立面造型

Vault-shaped roof 形如穹隆的楼顶造型

corresponds to A zone. It is designed as a common office building, which provides independent entrances, common offices and meeting rooms, and 2 police stations (whose locations and functions are still pending). The architecture height is 60m, whose east side is connected to the direction center of urban subway. The 2 floors underground are used for equipment rooms and office auxiliary rooms, 1H - 2H above the ground provides a main entering lobby, 3H-18H are all used for offices, and the passage on B2 is connected to the parking lot underground B, C and D zones.

The commercial center and the bases of 3 towers are called "podiums", the decoration of its 4 facades adopts thick stone materials. The black granite covered on the metal steel structures and secured the concrete plates. All the vertical frames of glass windows are inside the thick walls, in order to create impressive shade and shadow alter-nativities. Low-e hollow glass and dark black anti-oxidant filmed aluminum alloy frames are used for glass windows. A sunshine courtyard is set inside the podiums, whose space is quite open, with 2 sets of vertical transportation systems. The glass ceiling is above the courtyard, whose steel beams are soldered with 2 grooved ones, and connected with each other through a series of tiny metal beams, and are hold with "I-shaped" metal trusses. The glass pieces are connected accurately, in order to drain the rain smoothly.

4 facades of T1, T2 and T3 are all made of glass curtain walls, the ones on west facades fall to the ground directly, and an entrance lobby is set along the Gaoliangqiao Road. The curtain walls adopt the technique of double-layers glass covering metal steel structures. The "T-shaped" vertical metal components are set outside the glass, along the edges of double-layers plates. 3 series of glass windows are set between every two floors, and the tiny windows in the center can be open for ventilation. A sun-shade device is set between the roofs (6-floor high) of south facades of the towers and podiums, the horizontal sun-shade plates form the second external decoration, whose appearance is modern as well as recalls the memory for traditional architecture roofs of Beijing City. Over viewing the entire facades, distinctive visible contrast is made through the slim and exquisite glass curtain walls on the top and the thick and heavy podium stone walls.

3 distinguished light effects are designed for 3 podiums, including working days, weekends and holidays. The light makes the buildings semi-transparent and blinking at night, which forms excellent visible effects. In order to maintain and clean the windows, 4 sets of horizontal suspended cradles, which rise and descend along the rails, are set on both north and south facades, a horizontal hinge cradle, which scrolls along the rails is set on east and west facades, and a service slide is set on the podium roof for maintaining the top floor.

Stones are used for the walls on podium of T4 and other towers, while Low-e hollow glass is used for the walls of towers. The entire appearance looks like a

人产生对传统北京建筑屋顶的回忆。通观整个立面，上部轻巧细致的玻璃幕墙与下部厚实凝重的裙房石材墙面在视觉上形成了鲜明的对比。

三座塔楼的灯光设计分为3种不同效果：普通时间、周末和节假日，夜间的灯光使建筑物外表面呈半透明状，闪闪发光，产生良好的视觉效果。在维修擦窗系统方面，三座塔楼的南、北立面皆设有四组可沿轨道自由上下的水平支架吊篮，东、西立面各设有一个可在滑轨上自由滑动的水平铰轮吊篮，塔楼屋顶部分设有一个服务滑梯用来维修最后一层。

T4塔楼的裙房与其它塔楼裙房一样采用石材墙面，但在其塔楼部分，则采用了Low-e中空玻璃，其整体外观效果即为一个玻璃盒子坐落在统一的黑色裙房基座之上，其上盖一个金属格栅顶棚以隐藏屋顶设备机房。

本工程的交通设计要旨是：减少各种交通方式之间的换乘时间；减少换乘厅内的视觉障碍；最大限度地提供自然光线，保证旅客的安全感和舒适感；合理区分到达与出发人流；合理组织和设置垂直交通，使旅客方便迅速地找到其来去方向。在工程场地与室内设置了多种交通方式，利用大型高效的空间保障了大量人流在不同交通工具之间的换乘。其中，各种交通方式如国铁、城铁、公共汽车、出租车、汽车、自行车、行人等的换乘集中于本工程地段东部的多功能换乘大厅，设置于绿化疏散广场以西的三层内，即地下一层、广场层和二层。换乘大厅内部的交通组织围绕水平和垂直两条轴线，垂直交通工具(楼梯和自动扶梯)设置于大型中庭空间内,并可通向绿化疏散广场及与自行车换乘流线连接。由于地面层设有多条机动车道，因此国铁和地铁相关的换乘应主要通过地下通道实现，不会在地面层穿越机动车道。

Staff in the office buildings can view the overlapped and well-ordered architectural jungle through clean and bright glass, and is inspired and creates values inside the concrete architecture.

干净明亮的玻璃能让工作中的人们一眼望穿重重叠叠、此起彼伏的城市建筑森林，在高低不一的钢筋水泥构筑物中激发灵思，书写才智

Details of the architecture 建筑细部

glass box located on the consistent bases of black podiums, and the metal grid ceiling covered on the top hides the equipment rooms.

The transportation designation concept of this project is to decreases the time to transition bewteen each transportation methods, to eliminates the visible barriers inside the transition lobbies, to maximizes the daylights, to ensures the security and comfort of passengers, to distinguishes the arrival flows from the departure ones reasonably, to organizes and sets vertical transportations reasonably and to shows the appropriate directs for passengers easily and quickly. Multiple transportation methods are set in the project site and inside the buildings, huge and efficient spaces guarantees the transitions of large flows between transportation tools. Kinds of transportation methods, such as national subway, urban subway, buses, taxies, vehicles, bi-cycles and pedestrians are located in the multiple-functional transition hall, east of the project site, and are set in the 3 floors, i.e. B1, square floor and F2, west of the greening & evacuating square. The transportation network inside the transition hall surrounds the horizontal and vertical axes; the vertical transportation tools (i.e. stairs and escalators) are set inside the huge courtyard, lead to the greening & evacuating square and connect to the bi-cycle transition streamline. Multiple vehicle roads are set above the ground, so the transition between national subway and urban subway realizes through underground passage, which would not pass through the vehicle roads.

Partial construction indicates the technology and complication of modern architectural techniques.
局部构件显示了现代建筑手法的科学和繁复

The curtain walls adopt the technique of double-layers glass covering metal steel structures.
幕墙采用金属钢结构外覆双层玻璃的作法

Transparent glass and metal components form a shade and shadow alternative geometrical patterns, which are vivid and interesting.
透明的玻璃与金属构件形成了光影交错的几何图案，生动有趣

Glass ceilings of the bases　基座的玻璃顶棚

The simple color and fresh style laid on the corridor, makes people feel pure, thick and reliable.

走道面砖铺贴色泽质朴、清新自然，给人以敦纯厚实、安全可靠的心理感受

Distinctive visible contrast is made through the slim and exquisite glass curtain walls on the top and the thick and heavy podium stone walls.

裙楼立面装饰为体量厚重的石材，与轻巧细致的玻璃幕墙在视觉上形成了鲜明的对比

8837
3183

Hangzhou Jiahua International Business Center

杭州嘉华国际中心

Location: Hangzhou ZheJiang
Site area: 11,000m²
Building Area: 56,000m²

项目地址：浙江 杭州
用地面积：11000m²
建筑面积：56000m²

The site of the project is located on the west of the Hangda Road and the south of the Xixi Road in Hangzhou, well equiped with urban infrastructure. The site area is almost 11,000m², and the building area is 56,000m²; the project is a complex architecture, housing the shopping mall and offices, 80m from east to west, 137m from north to south.

The city of Hangzhou gives the site a unique value. With the sightview of mountain and temple, the site is overlooking the skyline of the city and neighbouring the bustling Hangda Road, located at the golden area - the cross of the Xixi Road under planning. In order to create a high end environment of shopping and working for the users, the design takes priority to consider the relationship between architecture and city as well as taking full use of every advantage of the site.

Due to different building times and various styles, the architecture alongside the Hangda Road makes the urban landscape appearing massy and disorder. As a part of the new urban landscape of the Hangda Road, Jiahua International Business

本项目基地位于杭大路以西，西溪路以南，周围城市设施完善。地块总面积约11000m²，总建筑面积为56000m²，东西长约80m，南北长约137m，是集商场、办公于一体的综合楼。

杭州城给了这个基地独特的价值，它拥有山景塔影，饱览参差错落的城市风景线，毗邻繁华的杭大路，并处在与规划中的西溪路交叉口的黄金地点。方案除了充分利用基地各方面优势，为使用者创造一处高品质的购物和办公环境外，更首先考虑建筑与城市的关系。

沿杭大路的建筑景观，因不同建筑时期甚至同一时期建筑风格不甚协调，城市界面也因此显得零乱，作为将成为杭大路新的城市界面一部分的嘉华国际中心，设计中充分考虑城市界面的形成和延续。另外，作为杭大路与西溪路交界处的重要城市关节点，它的建筑形象应具有视觉上和空间上的统辖地位。因此，设计以简洁大气的处理手法强化建筑形象的标志性和独特性。

Center is designed with the consideratioin of forming and the continuation of the urban lanscape. Besides, as the important urban node at the crossing of Hangda Road and Xixi Road, the architectural image of the project should plays a leading role at the aspect of visual effect and space layout. Therefore, the iconic feature and uniqueness of the architecture are created in a simple and elegent manner.

Located at the north of West Lake in Hangzhou, the site of the project is facing the verdant Baochu Mountain and the magnificent Baochu Tower in the south, and overlooking the elegant West Lake. The project has an impulse to talk to the mountain and lake, and interacts with them by the waving front façade, the roof garden facing to the Baochu Tower and the waterfront in front of the square.

Marco Polo described Hangzhou as "the City of Heaven, the most beautiful and magnificent in the world". Hangzhou has the facinating air of peace and gentle; as men of letters depict its beauty, the designers give it a wordless poem with the architectural language, gorgeous colors and affluent and flexible lines.

Surrounded by the casual atmosphere of culture and landscape, the mordern city of Hangzhou has become a bustling metropolin, demanding for a growing quality of life and working environment. On the front public square, delicately cut plants cast the dark shadow; glass ground lights vertical to the architecture are leading people to the interior with funs and appeals; the urban facilities for the visitors create a living room for the city.

The entry square and the roof garden of the office building enhance the quality of the mordern office space, providing the users with a semi-public outside space, where it is cool in summer and warm in winter.

In the view of the master plan, the project featurs strip layout; its east façade is like a golden wave, streches from south to north and forms an architectural intention to Xixi Road at the northeast, firmly combined with the terrain and endowing the architecture with vivid appearance, broad view field, mordern air and elegant quality; the folding-line west façade forms a fascinating contrast with the curved-line front façade, transferring the adverse direction to the advantageous one. Without taking up the commercial building area, the ground

基地位于杭州市西湖北岸，南望苍葱的保俶山、优雅的保俶塔，更会遥想远处空灵淡雅的西子湖，方案不禁有与山水对话的欲望：起伏波动的正立面、视野朝向保俶塔的空中花园、前广场的一条条水面，无不是对风姿绰越的杭州山水的思慕。

马可波罗曾盛赞杭州"美丽华贵之城"，她拥有的恬静、柔美的气质令人迷醉，像无数文人墨客倾尽心思描摹她的美丽一样，设计师用建筑的语言，华贵的色彩、流畅柔韧的线条，为她献上一首无声的诗……

现代的杭州，在悠闲的文化和景观氛围中是越来越紧张的大都市生活，越来越多的对生活和工作环境品质的要求。被尽量突出的前广场上修剪得方正趣致的树木投下浓荫；一条条垂直于建筑的地面玻璃灯带引导人们进入室内，并带来生动和趣致；为游人和顾客布置的城市家具带来城市客厅的氛围。

办公空间的入口内广场及空中花园，使现代办公空间的品质更上一层楼。为使用者提供一处冬暖夏凉的半公共的外部空间。

从总平面来看，设计以条形体量布局为特征，东立面如一段金色的波浪，自南向北掠过，在东北角形成一个转向西溪路的建筑意向，不仅与地形结合紧凑，并且使得建筑造型活泼、视野开阔，更富于时代感和高雅的品质；西侧折线形立面，与曲线型正立面形成有趣味的反差，并将不利朝向完全转化为有利朝向。一、二层商业裙楼，在保证商业建筑面积的前提下，建筑后退形成二层高的骑楼空间，为商业人流和城市提供更多的室外空间。建筑以其完整简洁的雕塑感，形成一段亮丽的城市风景面。

临杭大路结合暗河及绿化带设置商场入口广场，既能解决大量人流的疏散问题，也为杭大路提供一处轻松、活泼的城市空间。上部会所、办公的主要入口广场设在由北侧自动扶梯和大楼梯引至的二层屋顶平台，与商业广场分设在不同的标高上，避免人流的交叉，并为商业和办公提供相对独立的广场空间。另外，在基地西侧设两处独立的公寓式入口，直接由地面标高进入。

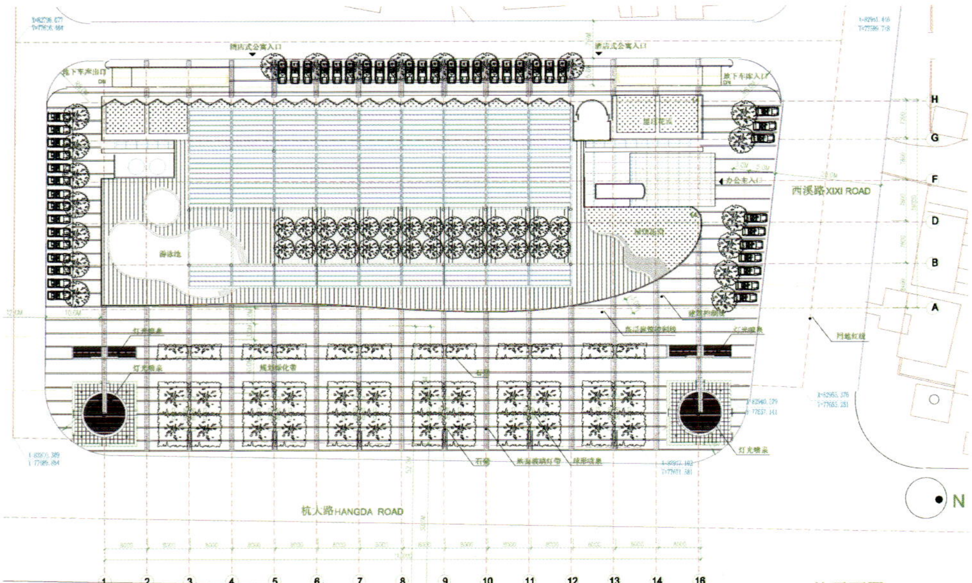

总平面图 1:500

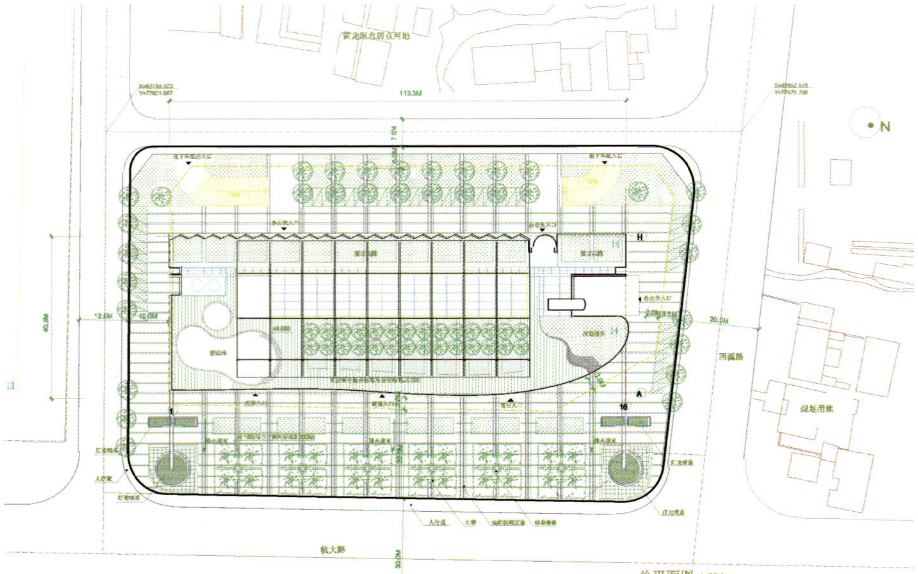

总平面图 1:500

and first floor of the commercial annex move backward and become an arcade architecture space of two-storey high, providing the citizens and the city with more outside space. With its clear structure of sculpture, the architecture forms a magnificent urban sight.

Neighbouring Hangda Road and combined with underground river and green belt, The entry square of the shopping mall solves the problem of evacuation of the mass and provides Hangda Road with a relaxing and casual public space. The main entry square of the club and office building is on the roof deck, led by the automatic staircases on the north and the main staircase. It is not on the same level with the commercial square, so as to avoid the streams of people to interlace and to provide either the commercial or the office space with a relatively independent square space. Besides, the two independent appartment entrances in the west of the site are directly paralled with the ground standard level.

The commercial square and the office space face Hangda Road and have the advantage of southeast direction and sightview of urban landscape. By the folding line of the façade, the small offices on the west side transfer the west direction to advantageous one and open the view field.

Jiahua International Business Center mainly funcions as the complex architecture for business, office and its supporting facilities. On the first and second floor is the shopping mall, and on the third floor is the entry square of office area and its club (including business center, business cafeteria, entertainment, services and

health club). Office area is from 4th to 14th floor, and the roof is equipted with swimming pool and roof garden. On the first and second underground floor are the garage and equipment room with the design of civil air defense.

The entrance and exit of the underground garage is placed on the north and south side of the road on the west of the site, avoiding to intervaine the business stream and the main urban crossing roads.

The façade to architecture is what clothes to people. The front façade of this building features affluent and flexible horizontal lines, champagne curved glass and light grey aluminium plates, presenting the style of magnificence, nobleness and uniqueness; it also defines the urban features of Hangzhou and stretches the urban landscape. The north façade combines with the change of direction and contrast with the front façade. Featuring the folding line shape, it functions as the main entrance to the office area and hotel appartment, forming a 10m notch for the building. The transparent glass wall uncovers the view of atrium, generating an appealing entrance of the building. The north façade forms roof gardens partly on the third, forth and fifth floors, providing the atrium space with natural ventilation most time of the year, which lowers the energy consumption.

Moreover, considering the control of light pollution of buildings by Hangzhou government, the project chooses the low reflective glass (lower than 15%) as the building material; at the same time, the diffuse reflection of the curved front façade reduces the light reflection; and the back façade uses as little glass as possible so as to increase the area of hard wall.

商业广场及办公空间朝向杭大路，具有优越的东南朝向与景观面。布置在西侧的小开间办公通过转折的立面设计，化不利的西向为有利朝向，并打开景观面。

嘉华国际中心主要功能为商业、办公及为办公配套的会所。设计将商场布置在一、二层，三层为办公入口广场及会所（包括商务中心、商务餐吧、娱乐、服务、健身等）。四至十四层是办公，屋顶设置泳池及屋顶花园。地下一、二层为车库及设备用房，并结合人防设计。

地下车库由基地西侧规划道路的南北两端出入，避免对商业人流的干扰，并减少与城市主要交叉口的冲突。

建筑立面如同人的外衣，建筑正立面以流畅、柔美的水平向线条为特征，并采用香槟色曲面玻璃和浅灰色铝板为建筑材料，着意体现大气、高贵、独特的风格，并对杭州城市特征加以诠释，对城市界面加以延续。背立面结合对朝向的改变和正立面的对比。以折线型为特征，北立面作为办公及酒店式公寓的主要入口，形成宽10m的建筑凹口。简洁、透明的玻璃幕墙将中庭的景观展现出来，形成一个有吸引力的建筑入口意向。南立面在三、四、五层局部形成空中花园，从而给中庭空间在一年的大部分季节带来自然通风，降低建筑能耗。

另外，设计充分考虑杭州市对建筑光污染的控制,首先选择小于15%反射率的低反射玻璃作为建筑材料,同时在正立面通过曲面的漫射作用，将光反射减弱,在背立面尽量减小玻璃面积,增加实墙面积。

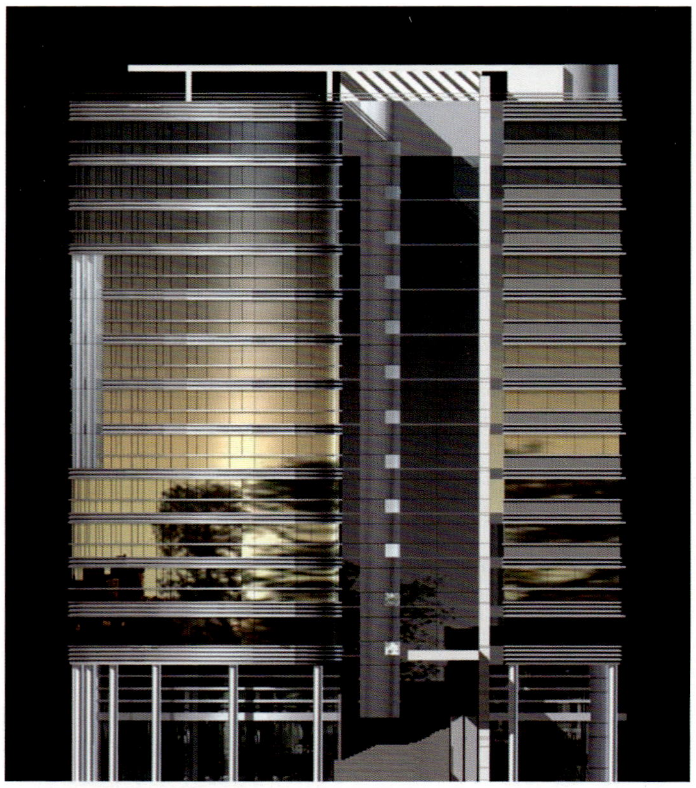

A-H 轴立面图

H-A 轴立面图

1-16 轴立面图

16-1 轴立面图

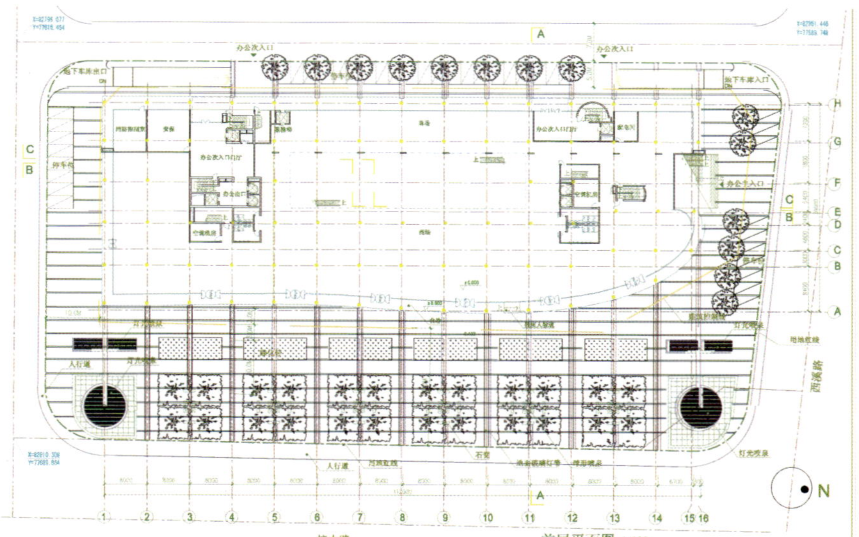

首层平面图 1:400

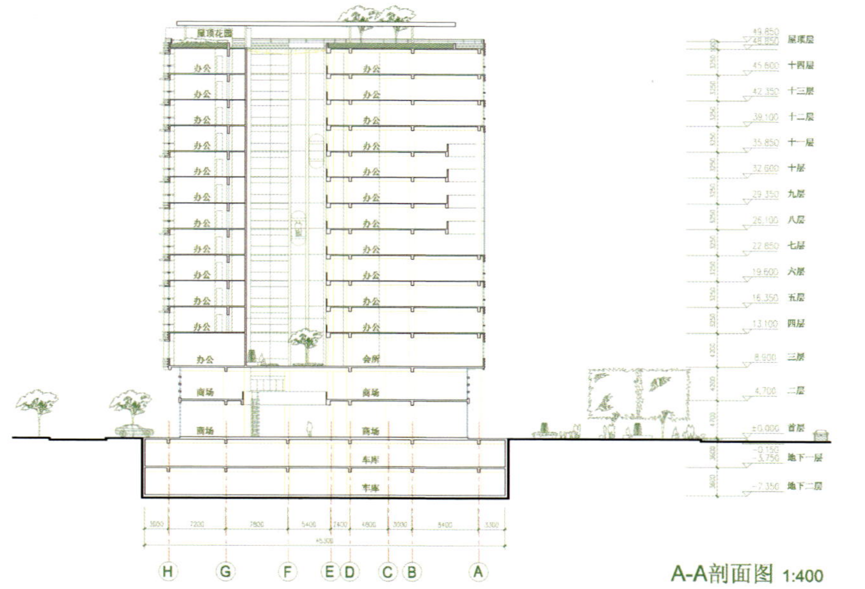

A-A剖面图 1:400

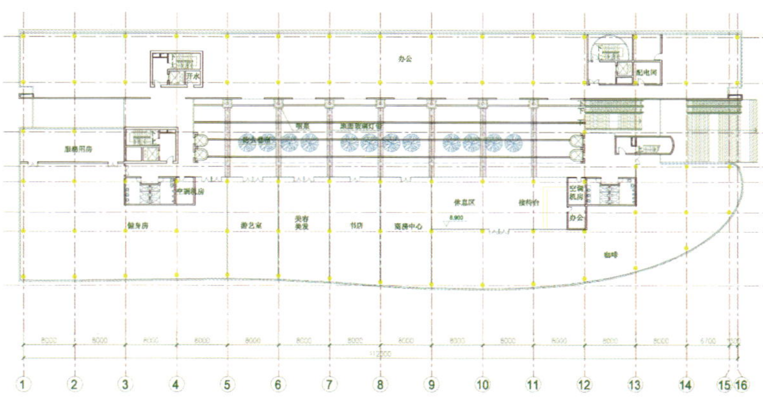

三层平面图 1:400

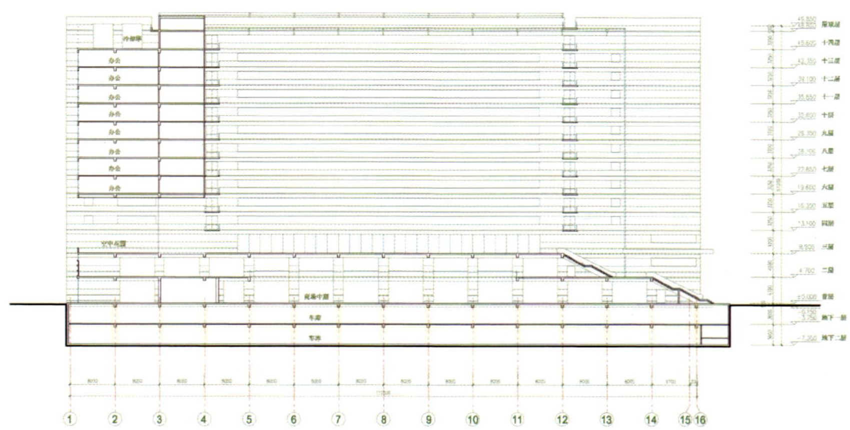

B-B剖面图 1:400

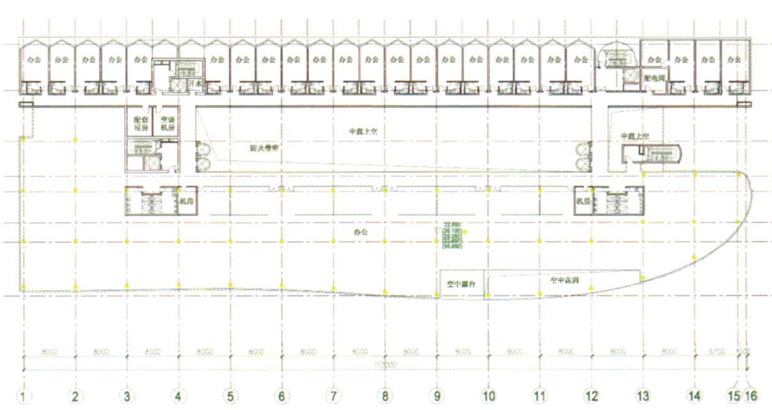

七至十一层平面图 1:400

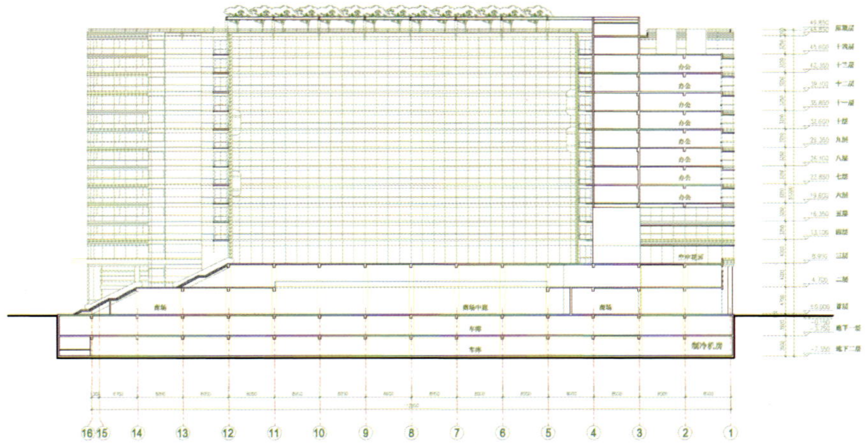

C-C剖面图 1:400

杭大路沿街立面图 1:600

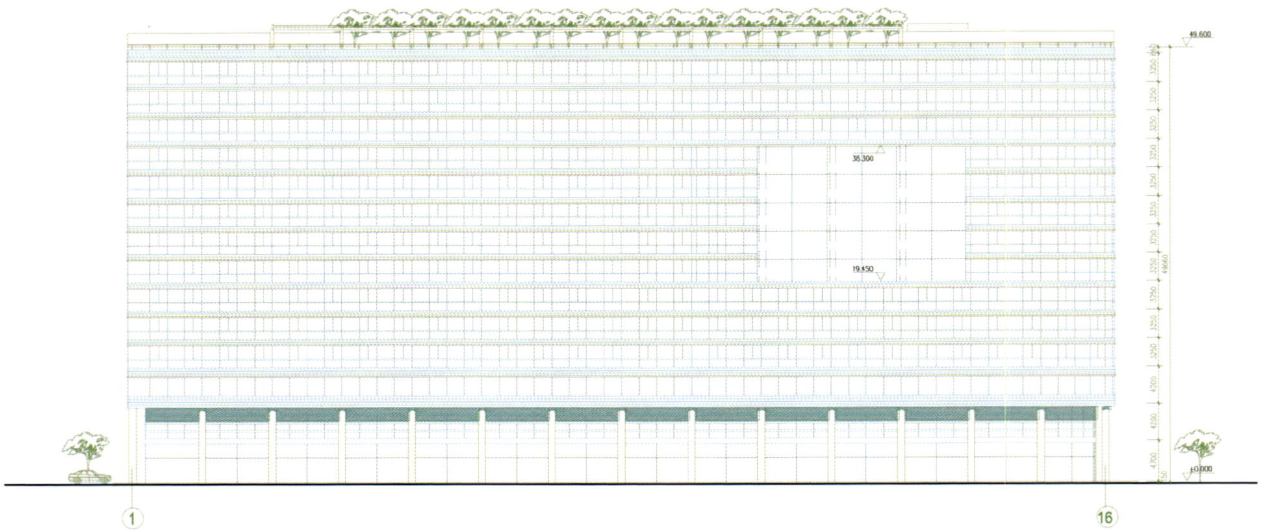

①-16轴立面图 1:400

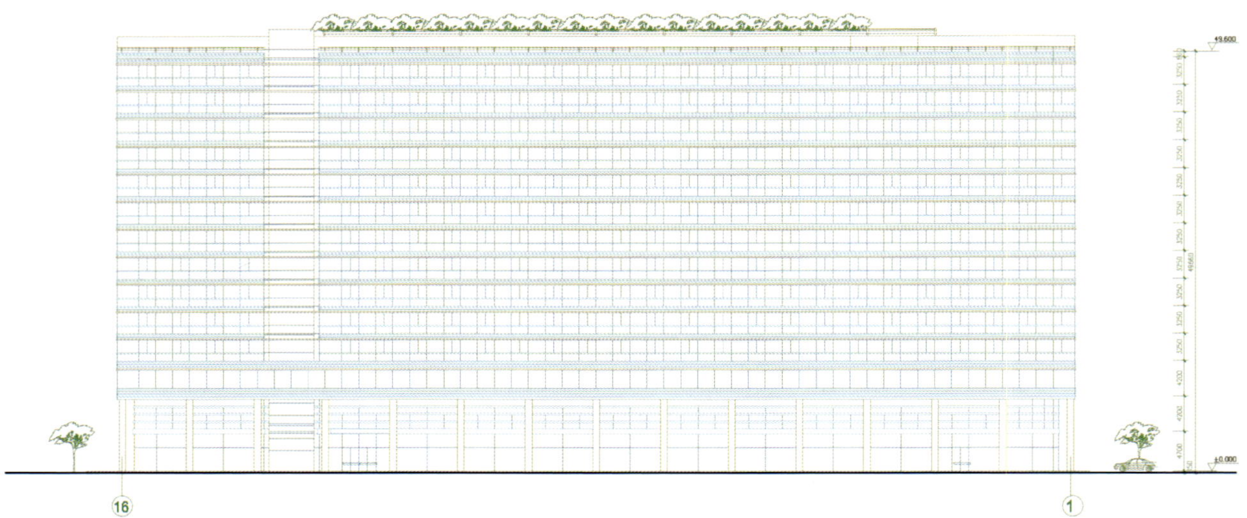

16-①轴立面图 1:400

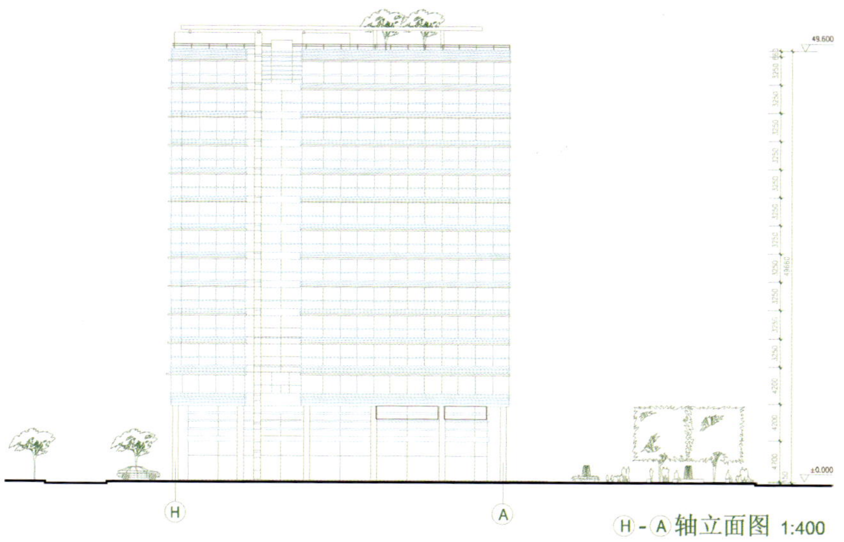

Ⓗ-Ⓐ 轴立面图 1:400

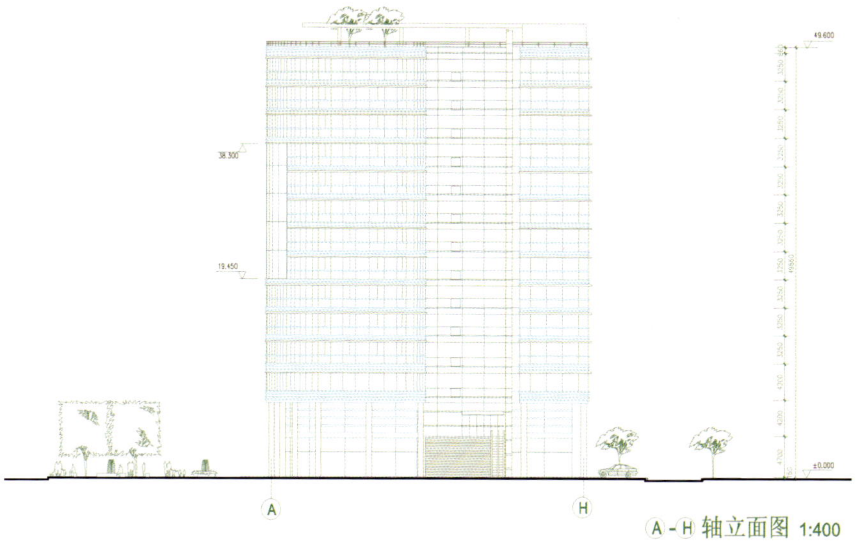

Ⓐ-Ⓗ 轴立面图 1:400

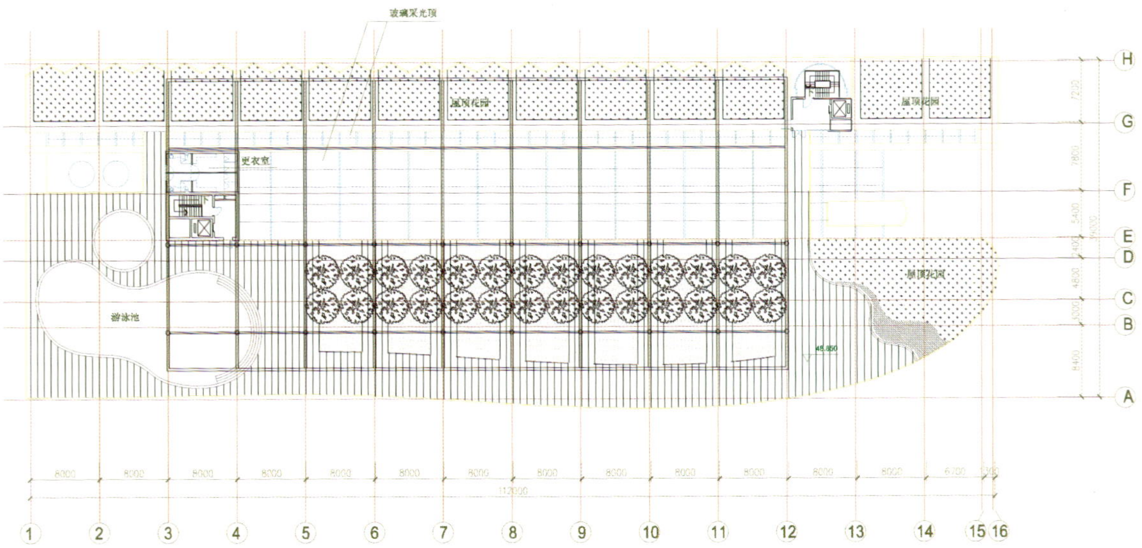

屋顶层平面图 1:400

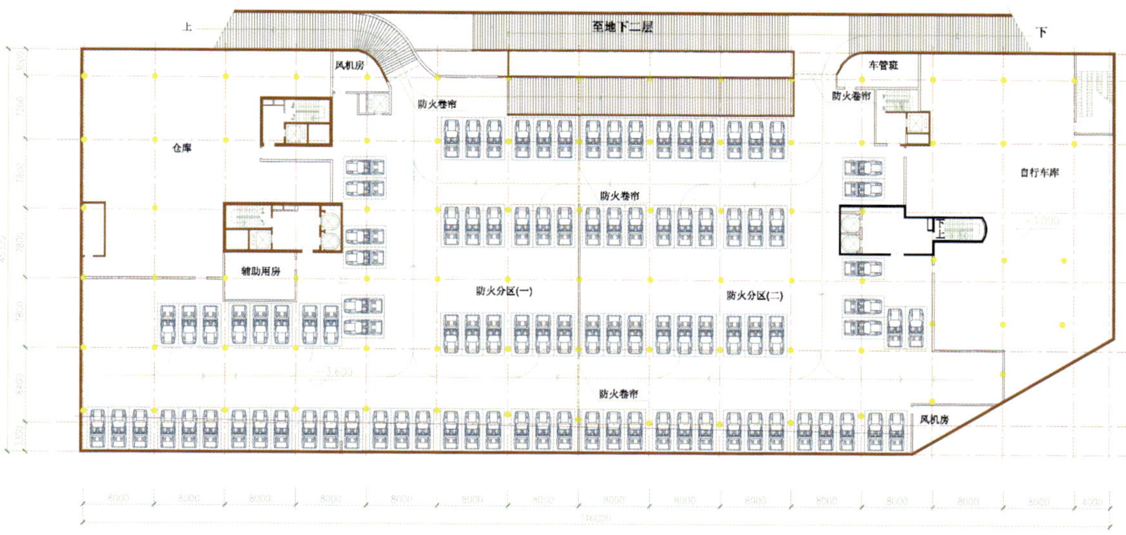

注：地下一层停车99辆，总停车位190个。

地下一层平面图 1:400

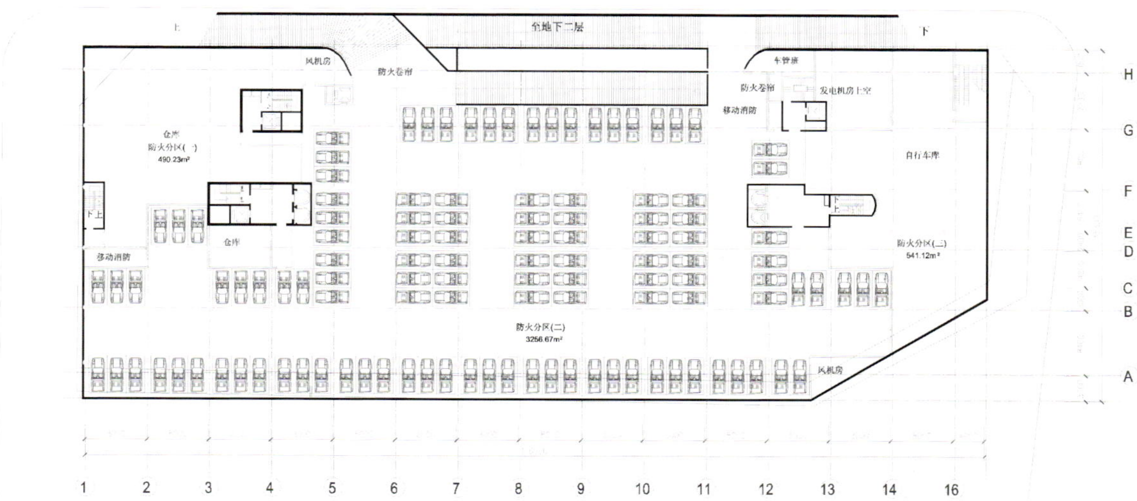

注：地下一层停车117辆，总停车位231个。

地下一层平面图 1:400

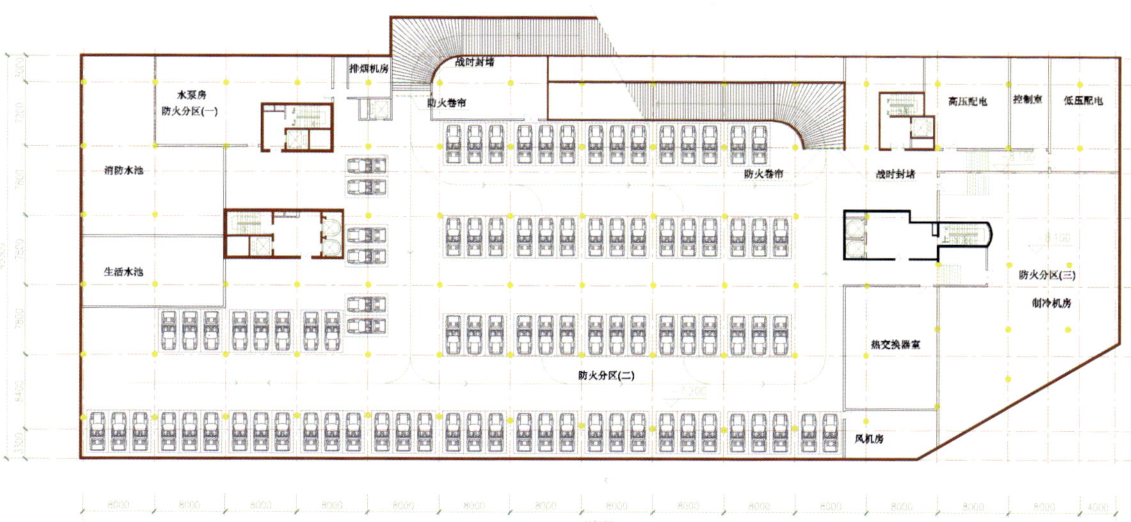

注：地下二层停车91辆

地下二层平面图 1:400

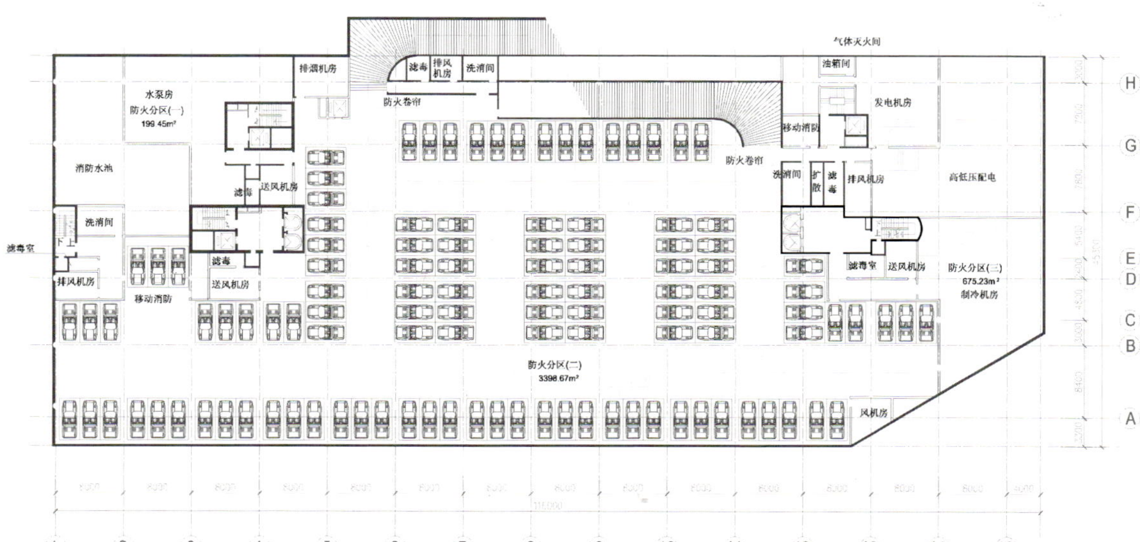

注：地下二层停车114辆

地下二层平面图 1:400

Shenzhen Phoenix Mansion

深圳凤凰大厦

Real Estate Developer: Shenzhen Phoenix Real Estate Co., Ltd.
Architecture Design: Sino-Construction International (Shenzhen) Design & Consultant Co., Ltd.
Project Site: Center District of Shenzhen City, nearby Shenzhen Citizen Hall
Gross Site Area: 11,038m^2
Gross Building Area: 106,436.2m^2
Floor Area Ratio: 9.64

开发商：深圳凤凰置业有限公司
建筑设计：中建国际（深圳）设计顾问有限公司
项目地址：深圳市中心区，紧邻深圳市政厅
总用地面积：11,038m^2
总建筑面积：106,436.2m^2
容积率：9.64

China Phoenix Mansion is located on the front line of the shinning Shennan Avenue of Shenzhen City, which is invested by Phoenix Satellite TV, an authority on international information. The main body of the mansion is a high-grade business office building, it is also composed of Phoenix Administration Mansion, Phoenix International Block, Phoenix Board, Phoenix Telecast Hall and the courtyard shared with the citizens, which provides a complete, efficient international business platform the enterprises. The impressive culture and style of the Phoenix Satellite TV shows everywhere from planning design to architectural details. Its modern and international façade indicates the unique image of a phoenix, which provides an excellent signature and high-level of the enterprise. The eye-catching logo on the building is the first remarkable stamp to start with information age for office buildings in Shenzhen. The residence of Phoenix Satellite TV Base, which attracts the attention of 1.5 billion Chinese, will lead information economy of Shenzhen to the globalized rail of high-speed development.

中国凤凰大厦，耸立于星光璀璨的深圳深南大道第一线，由国际资讯权威凤凰卫视荣誉打造。大厦主体为高档商务写字楼，另有凤凰行政公馆、凤凰国际街区、凤凰会、凤凰演播大厅、城市共享中庭等组成部分，为企业提供完整高效的国际商务平台。从规划设计到建筑细节，无不体现凤凰卫视极强的人文关怀与情感寄托。其现代化、国际化的外立面，张扬着凤凰独特的形象，具有绝佳的昭示性与企业级别感。楼体上醒目的凤凰标记，则为深圳中心区写字楼进入资讯时代刻下了第一个深刻的烙印。聚焦15亿华人眼球的凤凰卫视深圳基地的进驻，将引领深圳资讯经济走向高速发展的全球化轨道。

中国凤凰大厦位于深圳市中心区[26-3-4]地块，紧邻深圳市政厅，南临深南大道，东接海田路，西临鹏程五路，北隔中心区规划内部道路与规划中心区绿地相望。项目地理位置优越、地势平坦、交通便捷，各项开发条件成熟。

项目开发内容集传媒制播、商业、办公、公寓、会所于一身。设计要求功能齐全、技术先进、布局合理、环境宜人、典雅大方，又能体现凤凰卫视文化特色的标志性。希望通过本项目的建设，贯彻和深化城市设计，激发地区活力，带动和促进周边写字楼和商业的繁荣发展。

The Phoenix Mansion is located in the 26-3-4 block, Center District of Shenzhen City, nearby Citizen Hall, south of the Shennan Avenue, east of Haitian Road, west of Pengcheng Fifth Road, north of the planning internal road of Center District, and faces to the greenbelt of planning center area. The project is endowed with excellent position, stable terrain, convenient transportation and mature development conditions.

The project developing items combine media production, commerce, offices, apartments, and the club, which require complete functionalities, advanced technologies, reasonable layouts, comfortable environment, and the characteristic signature representing culture of Phoenix Satellite TV. This construction of this project carries out and furthers urban designation, inspires local dynamism, promotes the prosperous development of surrounding office buildings and commercial activities.

According to the requirements of Phoenix Mansion design documents, 3 main functions should be provided by the component above the ground, i.e. the apartment building, the office building and the podium; Phoenix production located in the podium.

The office building integrates 2 functionalities, i.e. Phoenix Satellite TV and affiliated administration & business office building, whose building area above the ground is 32,040m², 99.55m in height, 25 floors in total and is designed as a slab-type building. The administration building is located in the south of the site, faces to the greenbelt of Shennan Avenue, property set-back redline distance is 6m, which meets the requirements of urban designation, as well as arranges the transportation evacuation reasonably.

The apartment building integrates 3 functionalities, i.e. commercial, self-service and business apartments, whose building area above the ground is 33,190m², 99.55m in height, and is designed as a slab-type building. The apartment building is located in the east of the site, faces to urban sub-main road, Haitian Road, property set-back redline distance is 6m, which makes the evacuation of staff and commercial flows goes to Haitian Road, and does not impact the transportation

依照凤凰大厦设计任务书要求，其地上部分综合为三大功能，即公寓楼、办公楼及裙楼。其中，裙楼部分含凤凰制播。

办公楼含凤凰卫视及附属行政办公和商用写字楼两大功能，地上建筑面积32,040m²，楼高25层99.55m，板楼。办公板楼布置在用地南侧，面对深南大道绿化带，后退用地红线15.5m，既能满足城市设计要求，又能合理地安排本楼的交通疏散。

公寓楼含商业、自用服务式公寓和商务公寓三大功能，地上建筑面积33,190m²，楼高29层99.55m，板楼。公寓楼布置在用地东侧，面对城市次干道海田路，建筑物后退用地红线6m，其人员疏散及商业人流疏散均在海田路上，不会对深南大道的交通造成直接影响。从城市空间设计的角度来看，整个中心区26号地块的建筑均沿地块外环布置，用地内部空间虽宏大却略显呆板。公寓楼呈南北向布置，正对26号地块街心绿地，将26号地块内部空间一分为二，但隔而不断。既为街心绿地提供了建筑对景，又丰富了26号地块内部空间的趣味性，赋予其新的活力。

裙楼含凤凰卫视演播制作、传媒商业及会所三大功能，地上建筑面积14,730m²，楼高7层38.55m，为一方形裙楼。业务裙楼布置在用地西北角北侧，后退用地红线6m，西侧后退用地红线9.4m，为机动车交通流线留出空间。裙楼中凤凰制播部分自三层至六层，货梯直通地下室，以满足其复杂的交通要求。商业设置于首层及二层，会所布置于裙楼顶层北侧，与骑楼共同形成商业街，并延续至"城市中庭"内部。空中会所则通过架空连廊方便与办公及公寓楼产生联系。

"城市中庭"为一融会室内外空间，贯通地上地下空间的灰色空间地带，兼具休闲、娱乐、参观、购物等众多功能。在办公、公寓、业务裙楼三者结合部设计"城市中庭"，是出于以下三个方面考虑。一是从丰富城市空间设计考虑，此空间既融会了26号地块的内外空间，使街心花园与深南大道绿化带产生新的景观联系，又有效地打破了现有内部空间的机械与呆板，使整个空间富有流动感和生命活力，二是为整个综合体提供了人性化的休息空间，在此处设计水面、竹林和乔木绿化，波光鳞鳞、竹影婆娑，有效赋予了综合体鲜活的生

of Shannan Avenue directly. From the aspect of urban space designation, the buildings on the No. 26 block of Center District are all layout along the external ring of the block, the internal spaces is huge but inflexible. The apartment is layout along south-north direct, faces to the greenbelt in the center of No. 26 block, and divides its internal space into 2 pieces, which are separated and connected to each other. It provides opposite scenery for the centered greenbelt, and adds more fun to internal space of No. 26 block, as well as endows it with fresh dynamism.

The podium integrates 3 functionalities, i.e. Phoenix Satellite TV telecast production, media business and a club, the building area above the ground is 14,730m², 38.55m in height, 7 floors in total, and is designed as a square-shaped podium. The business podium is layout in the north-west corner of the site, property set-back reline distance is 6m, and that of the west is 9.4m, which reserves spaces for vehicles transportation streamline. F3 - F6 are taken by Phoenix telecast production, the cargo elevators lead to the underground directly, in order to meet the complicated requirements. F1 and F2 are taken by commercial operations, the club is layout in the north of the podium top floor, forms a commercial zone with the arcade, and extends to the internal of "Urban Courtyard". The sky club is connected to the office and apartment building through the overpass corridor.

"Urban Courtyard" is gray space merge internal & external spaces, as well as spaces above and under the ground, which integrates multiple functions, such as leisure, entertainment, sight viewing and shopping. It is designed to locate at the intersection of the office building, the apartment building and the business podium, which takes the followings into consideration. First, to enrich designation of urban spaces, this space merges the internal & external spaces of No. 26 block, forms a novel landscape connection bewteen the garden at the intersection and Shennan Avenue greenbelt, breaks the inflexibility of existing internal space, and endows the entire space with sense of movement and vitality. Secondly, to provide a human leisure space for the comprehensive body, water surface, bamboo forest and arbors greenbelt here endow it with vitality and cultural taste. Finally, to provide the designation of the comprehensive body with open and friendly characteristic, it provides funny leisure & shopping spaces for the citizens, and takes care of both social and economic benefits.

According to the requirements of urban planning, a transparent arcade is set in the north of the site, along the podium, "urban Courtyard" and north of the apartment building, which is 6m in width and 9m in height. It meets the requirements of urban planning, enhances the integrity of the comprehensive body, adds fun of the spaces, set "gray" space for connection between internal & external spaces, and makes them both separated and connected to each other.

The designation of greening system is composed of peripheral common greening, public greening for "Urban Courtyard", special greening for the roof of podium and public greening sharing with external city; the greening ratio is up to 27%.

Dotted greening system is designed along the building periphery according to the functional layout and organization of transportation streamline, the comprehensive designation for paving, greening, landscapes and entrance & exit evacuation are organized reasonably, manage to be well-ordered, tiny and exquisite, provide fine, elegant and generous external gardening interface, and create perfect environment.

The public greening of "City Living Room" is the core of ground greening designation for this project, represents the open, generous and positive characters, adopts the design concept of mergence of internal & external spaces, co-existence of dynamics & statics, purity and simplicity. It blurs the boundary between outdoor and interior greening, makes use of simple geometrical water-features, bamboo forest, floor paving and sharing with the external greening, associates with the commercial vertical transportation system, creates pure, natural, squish and separate gardening environment, shares landscapes with outdoor greening, and is harmonious with the interior greening pavement.

Special greening for the roof of podium is designed to meet the requirements of the club and internal customers in particular, associates with the swimming pool, flower shelves, gardening landscapes, water features, pavement, arbor planting

态活力和更高的文化品味；三是使整个综合体设计具有了更为开放与亲民的特质，为市民提供了富有趣味的休闲购物空间，兼顾了社会与经济效益。

按城市规划要求，在用地北侧，沿裙楼、“城市中庭”和公寓楼北侧设计了通长的通透式骑楼，宽6m、高9m，既满足了城市设计要求，加强了综合体的整体性；又增加了空间的趣味性，为内外空间的联系设置了隔而不断、通而不透的“灰色”界面。

绿化系统的设计分为周边普通绿化、“城市中庭”公共绿化、裙楼屋面专用绿化和外部城市公共绿化借用四个体系，绿化率达27%。

结合功能布局、交通流线组织沿建筑物周边设计了大量见缝插针式的绿化，合理地组织铺地、绿化、景观和出入口疏散综合设计，力争做到疏密有致、小而精致，为综合体提供精致、淡雅、大方的外部园林界面，创造良好的环境氛围。

“城市客厅”部分的公共绿化是本设计地面绿化设计的重点，也是本项目开放、亲民特质的积极体现，采取内外融汇、动静并存、至纯至简的设计方针。模糊室内外绿化的分界，利用简单的几何水景、竹林、地面铺装和外部绿化的借用及结合商业竖向交通组织，营造纯净自然、动静有致、小中见大的园林氛围，并同外部绿化互为借景、互相辉映，与室内绿化铺装浑为一体。

裙楼屋面专用绿化专为会所和本项目内部客户而设，结合会所泳池、花架、园林小品、水景、铺装、乔木种植及草坪铺设，营造一个自然天成、趣味盎然、郁郁葱葱的空中花园，为客户提供一个静谧自然，都市村庄式的园林休闲空间，再配以简洁现代的空中连廊，赋予建筑轻盈飘逸、纯净自然的生态氛围。

由于本项目用地较紧张，绿化条件有限，明智的做法就是将自身的绿化系统同城市公共绿化紧密融合。本设计通过周边普通绿化设计和“城市中庭”景观廊的运用，巧妙地将道路绿化和街心花园绿化借用过来，再配以空中花园，以达到立体、系统、小中见大、至简而繁的绿化效果。

为简化平面设计，提高平面使用率，最大限度地照应影视制作、办公、商业、停车不同功能对柱网的要求和为施工快捷创造条件，本设计所有平面布置均有8m×8m的规则柱网内组织，并优先考虑采用新型组合钢结构体系。

办公楼主要使用功能为凤凰卫视办公和商务办公两部分，竖向分布为1~9层为凤凰卫视用房，10层及以上为商务办公面积。为此本设计结合板楼平面处理手法在楼内设置了两个交通核心筒，西侧核心筒电梯只通到10层，为凤凰卫视专用；东侧核心筒电梯为1~25层商务办公专用。两个核心筒内的消防电梯及楼梯均贯通全楼，满足消防需求，平时依靠物业管理区分。办公层层高3.5m，楼高25层99.95m，为组合钢框架结构。办公层内部除功能有特殊要求外均采用大空间开敞式设计，赋予空间最大的灵活性和创造性。办公隔间100~300m²，可灵活划分。在办公楼东端更设计了8套250~300m²的跃层办公单元，满足不同的办公个性空间需求。

公寓楼首层为大堂及商业，二、三层为商业，四层以上为公寓用房，其中4~10层为服务式公寓（其中9、10层为凤凰专用服务式公寓），11层及以上为商务公寓。公寓层高均为3.15m。公寓标准层设两个交通核心筒分别为两个公寓单元服务，7层服务式公寓设有专用交通体系。结构形式为钢筋混凝土框支结构。商务公寓套型面积、服务式公寓套建筑面积60~120平方米不等，户型为二至三居，可适应不同商务需求，并可互相组合。

裙楼部分主要含凤凰卫视演播制作用房和商业及会所用房两大类，功能上互为独立。裙楼楼高7层38.55m，结构形式为组合钢架排架结构。3~6层分别为各类大中小型演播厅。会所设于7层，商业部分设计在裙楼1、2层，与业务部分截然分开。商业及会所部分各自有独立的交通体系。凤凰业务用房按观众、内部工作人员及后勤、美工布景道具、外来参观等不同功能各自设计了不同的楼电梯交通体系。并且于三层及以上有与凤凰办公楼通过连廊相连，方便凤凰内部联系。会所则通过屋顶花园和连廊与办公用公寓相连，共同形成功能相对独立又互为补充、交通联系快捷又方便的综合体。

地下室均采用8m×8m柱网来组织设计。地下一层布置有凤凰地下配套用房及库房和凤凰转播车库。地下二、三层则为地下车库及设备用房。

采用至简至美的极少主义设计原则，通过简洁的造型、几何体和线条组

and lawns, creates a natural, funny, and luxuriant sky garden, provides quiet, urban village-hood gardening leisure space for the customers, and gives the building a graceful, pure and ecological style with the simple and modern sky corridor.

Due to the limitation of site area and greening condition, it is wise to integrate the private greening system with the urban public greening system. The designation borrows the road planting and garden at intersection technically through peripheral common greening and application of "Urban Courtyard", and archives the 3D, systematic, exquisite and simple greening effect associating with the sky garden.

To simplify graphic design, increase the rate of plane usage, optimize the functions of media production, offices, commercial operations and parking lots, meet the requirements against column network and create convenience for construction, the layouts of all the planes in this designation are equipped with 8m×8m regular column network system, and prioritize the application of new combined steel structural systems.

The main functionalities are Phoenix Satellite TV office building, which takes F1 - F9, and business offices, which takes the floors above F10. 2 transportation core tubes are set inside the building associating with the techniques to handle a slab-type building; the one in the west can get to F1- F10, which is for Phoenix Satellite TV only, and the one in the east is used for business offices from F1 to F25. The fire elevators and stairs inside the 2 core tubes passing through the entire building meet the requirements of fire-extinguishing and are managed by property separately. Each floor is 3.5m in height, and the building is 99.5m, 25 floors in total, and adopts combined steel structures. The interior of offices adopts open designation with large spaces except for some with special requirements, and endows the spaces with optimized flexibility and creativity. The areas for isolated office rooms are 100~300m^2, which can be divided as necessary. 8 sets of duplex-type office units whose area are 250~300m^2 are set at the east end of the office building, in order to meet personal space requirements for offices.

F1 of the apartment building is for a lobby and commercial operations, F2 and F3 are all for commercial operations, F4 and the above are for apartments. F4 - F10 are service apartments (F9 and F10 are special for Phoenix Satellite TV), F11 and above are business apartments. Each floor of the apartment building is 3.15 in height. Standard floors are equipped with 2 transportation core tubes serving the 2 apartment units, a special transportation system is set for F7 service apartments. It adopts steel and concrete frame structures. The building areas for business apartment suites and service apartment suites are 60~120 m^2, each suite provides 2 or 3 rooms to meets various business requirements and can be combined with each other.

The podium part integrates 2 functionalities, i.e. Phoenix Satellite TV telecast production and commercial rooms & the club, which are independent to each other. The podium is 38.33m in height, 7 floors in total, adopts combined steel frame & bent frame structures. Various types of cast halls are set on F3 - F6. The club is set on F7; the part commercial operations are set on F1 and F2, which is separated from the business part. Separate transportation systems are provided for commercial operations and the club. Different stairs and elevators transportation systems are set according to different functions of Phoenix business rooms, such as audiences, staff, logistics, art and stage setting, props management, and visitors. In addition, the transportation systems are connected to the corridor of Phoenix office building above F3, for convenient internal connection. The club is connected with the business apartment building through the sky garden and the corridors, whose functions are relative independent, supplemented to each other, and form a comprehensive body with convenient and fast transportation systems.

8m×8m column network are adopted for basement designation. B1 is for Phoenix underground equipments rooms, a parking lot and the Phoenix broadcast garage. B2 and B3 are for underground parking lot and equipments rooms.

Design principle of simplicity and beauty is adopted for this project, combination of brief shape, geometry and lines, collation of elegant and gentle colors, exquisite and detailed designation achieve perfect unity and interpretation of

the various complicated functions. Design technique of simplicity and beauty is adopted to represent the globalized operation concept of modern media, which is efficient, vast, justified and inclusive.

3 entering and exiting streamlines for pedestrians are set in the office building, the streamlines for Phoenix Satellite TV office and business are the main ones, which are set in the south of Shennan Avenue, and are independent to each other. The set-back relined distance for the office building is increased to 15.5m from 10m according to the planning requirements, and meets the requirements to evacuate staff in the office building. A VIP entrance is set for the office building, where provides special parking corridor, lobby and elevator. It is located in the west of the office building adjacent to Pengcheng Fifth Road, through which the VIPs take elevator to get to each studio, and are isolated from the other staff.

The pedestrian transportation of the apartment building is composed of the main entrances and exits for business apartment, service apartment and those for commercial operations part. The entrance and exit to apartment is set in the middle of the apartment building, faces to Haitain Road, and 3 special lobbies as well as 3 elevator halls are set there. The main entrances to commercial operations are set on south and north side as well as the middle of the building, which has independent vertical transportation system. The backup entrances and exits to both the apartment and commercial operations part are set in the west of the apartment building, facing to and for convenient connection with "Urban Courtyard".

There are plenty of entrances and exits for pedestrian in the podium, such as those for commercial operations part, the club, Phoenix telecast production, Phoenix VIPs and visitors. Audience entrance and exit is set in the west of the podium, faces to Pengcheng Fifth Road, and the stairs leading to audience waiting room and telecast production visiting corridor are set there for convenient evacuation of flows. 2 sets of vertical transportation systems are set inside the courtyard, used for staff and visitors, which are well-organized and independent, both 2 streamlines can get to sky garden directly. The entrances and exits for commercial

合，典雅柔和的色彩搭配，精致简练的细部设计，达到对各种繁杂功能综合的完美统一和诠释，用至纯至简的设计手法表达现代传媒高效、博大、公正、包容的全球性运营理念。

办公楼共设置3个人行出入流线，其中凤凰卫视办公流线及商务办公流线两个主要流线均设在沿深南大道一侧，并互相独立。办公楼后退红线距离由规划要求的10m扩大为15.5m，较好地满足了办公部分的人员疏散要求。办公楼另设有贵宾入口，并在入口设有专用的停车门廊、专用门厅及专用电梯，位置设在办公楼西侧临鹏程五路处，贵宾经此门厅乘梯下达各个摄制棚，与其他人员流线完全隔开。

公寓楼的人行交通分为商务公寓、服务式公寓主出入口及商业用出入口两大类。公寓出入口设在公寓楼中部，面临海田路，设有3个专用大堂及3个电梯厅。商业主入口则设在南北两端及中间，具备独立的竖向交通体系。公寓楼西侧面临"城市中庭"处均设有公寓及商业备用出入口，与"城市中庭"联系方便。

裙楼部分的人行出入口众多，分为商业、会所、凤凰制播、凤凰贵宾及参观等。观众出入口设于裙楼西侧，临鹏程五路处，方便大量人流的出入及疏散，设有楼梯直达观众等候厅及制播参观廊。中庭内分设两组竖向交通体系，分别用于工作人员和嘉宾，组织有序、互不干扰，两个流线均可直达屋顶花园。商业出入口设在裙楼四周，设有专用竖向楼电梯交通体系，与公寓楼商业及地下商业共同形成完美的休闲购物氛围。会所出入口设在裙楼东北角，设有专用竖向交通体系，既方便空中会所与地面联系，又增加了"城市中庭"的休闲和文化品味。裙楼内部设有道具布景垂直运输流线，设有一部货梯、一部楼梯，可在不受其它流线干扰的情况下极为方便地将道具布景自地下一层制作部运输到每一个摄制棚。

"城市中庭"人行流线主要考虑为自深南大道到用地北侧支路，贯穿整个用地。此部分流线与商业流线既有交织又相互独立，与办公、公寓及凤凰业务流线则截然分开。

operations part are set around the podium, the special vertical stairs, elevators and commercial operations part of the apartment and underground form perfect and leisure shopping environment. The entrance and exit to the club is set in the north-east corner of the podium, a special vertical transportation system is set there for convenient connection with the sky club, and adds leisure and cultural taste for "Urban Courtyard". Special vertical transportation streamlines for properties and staging sets only is set inside the podium, a cargo elevator and a set of stairs transport them from production department on B1 to each studio easily without disruption of other streamlines.

The pedestrian streamlines in "Urban Courtyard" are mainly used for flows from Shennan Avenue to the sub-road in the north of the site, which pass through the entire site. These streamlines interact with those for commercial operations part, but they are independent to each other and completely isolated from the streamlines for office, apartment and Phoenix business.

Shenzhen Unicom Mansion

深圳联通大厦

Real Estate: Shenzhen Branch of China Unicom Co., Ltd.

Architecture Design: Sino-Construction International (Shenzhen) Design & Consultant Co., Ltd.

Project Site: Center District CBD of Shenzhen City

Site Area: 4674.3m^2

Gross Building Area: 39,143.91m^2

Floor Area Ratio: 6.13

开发商：中国联通有限公司深圳分公司

建筑设计：中建国际(深圳)设计顾问有限公司

项目地址：深圳市中心区商务区内

用地面积：4674.3m^2

总建筑面积：39,143.91m^2

容积率：6.13

The planning of this CBD is quite detailed, emphasizes the concept of Main Street, requires an arcade which is 3m in depth, set-back platforms are set at 40m and 80% height, the limited height of the mansion is 100m, and the façade helps to restrict the rate of opening windows.

Unicom Mansion is located in an important position, and the specified geography promotes the importance of it in Center District. It impacts the space landscape and city interface of the entire district. On one hand, the style should be harmonious with the entire Business District, indicate atmosphere of Business District and High-tech (Citizen Hall, Crystal Island and High-tech Exhibition Hall), and function as transition center outside this district; on the other hand, China Unicom acts as the leader of domestic IT industry, its headquarter should have the appropriate appearance and culture, which represents not only the most advanced science & technology, but also the most advanced concept, i.e. co-existence bewteen human science & technology and the nature.

中心商务区做了很详细的规划，强调主街的概念，要求本地块南面设置3m进深的骑楼，40m及80%高度上有退台，大厦限高100m，立面更有开窗率的限制。

联通大厦位于一个十分重要的位置，特定的地理位置无形中提升了联通大厦在中心区的重要性：它对于全区城市空间景观及城市界面有十分重要的影响。因此，在设计者看来，一方面，它既要与整个商务区的风格相协调，又要发挥商务区与高科技风格（市政厅、水晶岛、高科技展馆）极其明显的区外大环境的转换枢纽作用；另一方面，中国联通作为国内IT业的顶尖企业，其总部大楼也需要具备与之气度相应的外观与内涵，不但应代表着最先进的科技，也应包含着最先进的观念，即人类科技与自然共生的概念。

综合以上考虑，结合当代高层建筑的发展趋势，三个方面的意念油然而生，成为本案的特点。

Taking the above into consideration, and associating with the development trend of modern high-rise buildings, concept from 3 aspects is proposed, and becomes the unique character of this case.

The site faces to the east of Greening Park in Center District across the Yitian Road, is adjacent to Citizen Center in the northeast, the opening park in Business District in the south, and becomes the background building located in the north of Business District Park. Within the atmosphere of emphasizing environment conscience in CBD and ecological environment in architectural field, the mansion cannot interact with the surroundings, and west-exposure and bad orientation problems occurs, if the traditional construction method is followed. Therefore, the designers take the interaction between Unicom Mansion and the surroundings into consideration, break the restriction of traditional structure and change the internal spaces completely.

First of all, the core tube of Unicom Mansion is moved to the west, which optimizes visible effects and interaction with the landscapes surrounded, solves the problem of west-exposure and bad orientation, and shows the character of human design. The extension and movement towards outdoor of transportation also highlight several advantages of designation, the plane spaces of the high-rise is more variable, the layout of shaft is more flexible and suitable for the intelligent design of mansions in information age; vertical transportation spaces are above the façade, break the image and limitation of high-rise buildings.

Moreover, the position change of the core tube makes possible for the existence of high-rise courtyard spaces. The floors above F14 are taken by China Unicom, and the courtyard establishes both horizontal and vertical relationship with the mansion. According to the research of sustainable development, courtyard spaces help a lot in ecological designation, i.e. courtyard and semi-courtyard have become the essential methods to adjust internal micro-environment without artificial intervene, such as light, wind and water. The courtyard of this case forms perfect greening spaces, becomes the lung of the building. The semi-open designation of it establishes visible connection spaces between interior and outdoor, and makes people feel comfortable. Therefore, Unicom Mansion establishes close relationship between human and the nature through the courtyard.

Simple and power shape as well as the exquisite and elegant texture shows the extraordinary style, up straight vertical lines form the main texture of the mansion. The distance between the vertical components outside the curtain walls and the glass is 150mm, forms impressive shadow effects with the sun-shield plates, light blue-gray low-e glass is quite energy-saving, matches the concept of sustainable development and energy-saving. Exquisite and vivid vertical texture, blinking glass windows and lamps hidden behind, and the courtyard full of daylight indicate the characters of virtual world with intensive future style. It

此地块与中心区绿化公园东隔益田路相对，东北部临近市民中心，南面更紧挨商务区内的开放性公园，为商务区东公园北侧底景建筑。在当今强调环境意识的中心商务区和建筑界日益重视生态环境的大氛围里，如果按传统方式建造，无法使大厦与周围环境形成良好互动，并会出现如西晒严重、朝向差等问题。在此，设计师充分考虑到联通大厦与周围景观的互动关系，设计使联通大厦摆脱了传统结构的束缚，内部空间发生了根本的变化。

首先，联通大厦的核心筒移至西侧，使大厦在视野效果和与周边景观的互动上达到最优，并解决了西晒和朝向等问题，体现了人性化设计的特点。交通核进行扩散和外移，也突出了许多设计上的优越性：核外移使高层的平面空间更加灵活多变，空井的布置更加自由，更加适合信息时代大厦的智能化设计；垂直交通空间出现于立面上，从而令联通大厦突破了传统的高层形象和局限。

其次，联通大厦的中庭空间由于核心筒的改变，使高层中庭空间成为可能。大厦的14层以上作为中国联通自用部分，其高层中庭与大厦建立了平面及垂直的空间关系。而且，在可持续性发展的课题里，中庭空间十分有益于生态设计—中庭及半中庭已成为在不依赖人工干涉的情况下调节包括光、风、水等室内微环境的不可或缺的手段。本案的中庭形成了很好的绿化空间，成为了建筑的肺；而中庭通过半开敞的设计还直接建立起室内外的视觉沟通空间，给人愉悦的感受。在此，联通大厦通过中庭建立了人与自然的亲密关系。

简洁有力的形体及细致优雅的肌理是体现其不凡气度的方式。挺拔向上的竖向线条构成大厦的主要肌理。幕墙外竖向构件距玻璃150mm，与遮阳板共同形成较强烈的阴影关系，浅蓝灰色low-e玻璃有较好的节能效果，充分契合了可持续发展和节约能源的时代思维。细腻而生动的竖向肌理、闪烁隐约其后的玻璃窗和灯光，以及充满阳光的中庭都通过建筑语言共同构成了未来气息浓重的虚拟世界特征。这与SOM城市设计中关于建筑外表限高100m等要求相吻合。同时，建筑体量的变化也遵循了SOM的要求。

建筑塔楼部分通高的中庭空间螺旋向上，有效地将室外景观与光线渗透到建筑的各个部分，并与顶部高耸明亮的玻璃大厅融为一体，成为一首高亢的尾曲。从室外看，它使大厦充满时代感和国际感；从室内看去，透过共享大厅可以俯瞰市中心，并可以远眺耸立在市中心的簇簇摩天大楼，充分体现了联通大厦卓尔不群的气度，从而使人、建筑和环境有机结合，形成完美体验。

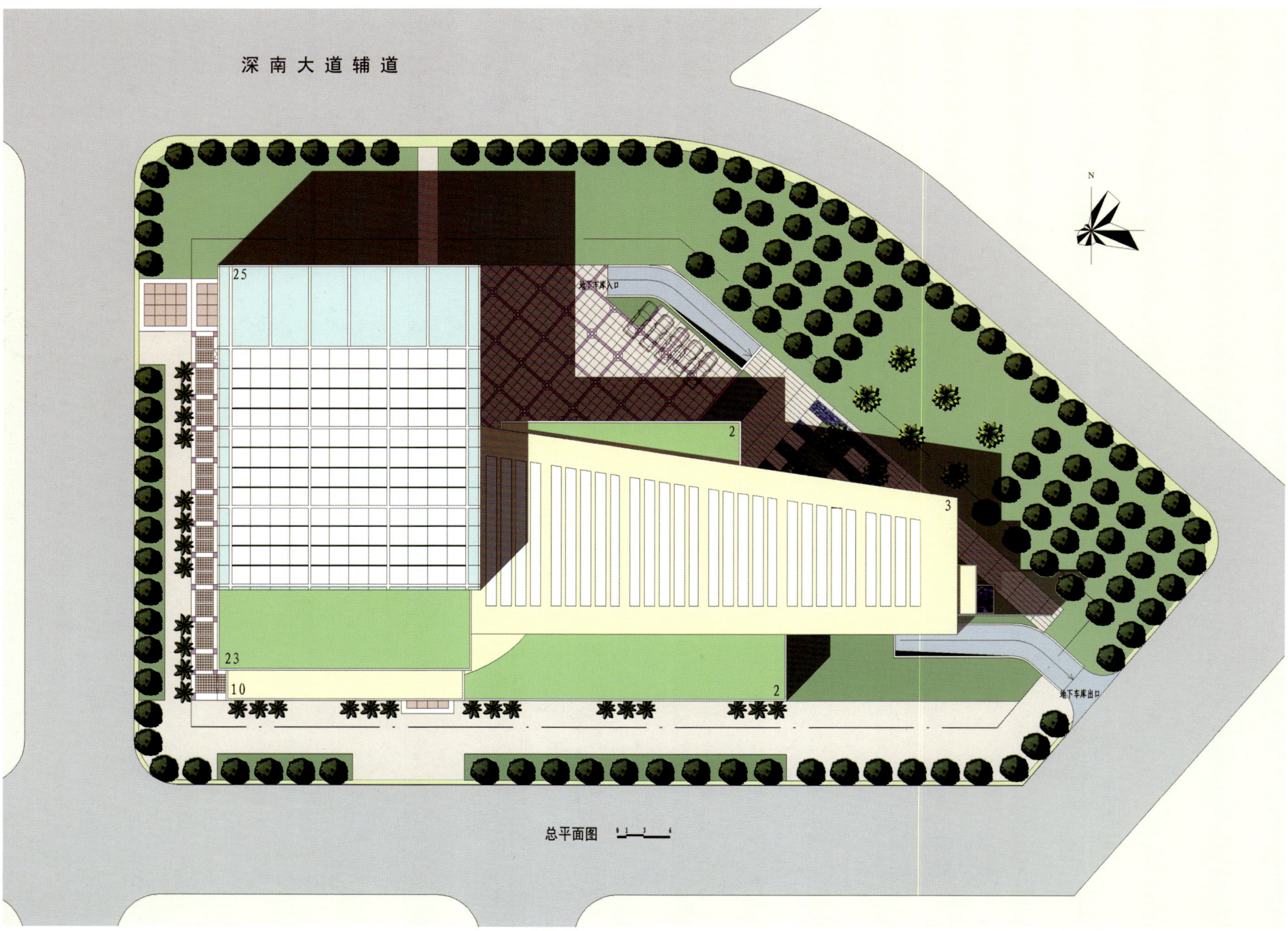
深南大道辅道
N
25
地下车库入口
2
3
23
10
2
地下车库出口
总平面图

meets the requirements that limited height of building surface is 100m according to SOM urban design. Meanwhile, the change of building volume follows the requirements written in SOM.

The spiral rising courtyard space inside the podium permeates outdoor landscape and daylight into each component of the building, integrates the straight and bright glass hall on the top, and makes the entire project perfect and completed. Viewing from outdoor, it makes the mansion full of modern and international sense. Viewing through the shared hall inside, the urban center and various skyscrapers located there are all visible, it represents the noble style of Unicom, connects human, the building and environment organically, and becomes perfect experiences.

Shenzhen Truth Mansion

深圳求是大厦

Real Estate Developer: Shenzhen Huaminghui Real Estate Co., Ltd.
Architecture Design: Shenzhen Yuepeng Architecture Design Co., Ltd.
Project Site: Bamboo Forest nearby Shennan Road of Shenzhen City
Gross Site Area: 5,309m^2 (with FAR)
Gross Building Area: 49,000m^2
Floor Area Ratio: 8

开发商：深圳市华明辉置业有限公司
建筑设计：深圳市粤鹏建筑设计有限公司
项目地址：深圳市深南路旁竹子林
总用地面积：5309m^2（计容积率）
总建筑面积：49,000m^2
容积率：8

Truth Mansion is located in the north of Shennan Avenue, a international avenue of Shenzhen City, east of the powerful business platform of Center District, west of high-grade community of Chinese Town, south of excellent seascape in Shenzhen Bay, and north of Antuo Mountain and Tanglang Mountain. Truth Mansion is one of the "Business Apartment" approved by the government, as well as the rare business apartment with 70 years ownership in Shennan Avenue. The project is characteristic, and endowed with the façade of A-grade office building, a 3D club, a swimming pool with seascape, multiple-functional business meeting rooms, a sky garden on each three floors, and a business service center, which meets various business needs. Truth Mansion attracts the attention of international financial groups with its charms and quality. Hong Kong the Zha Financial Group was attracted when launching construction, and bought the East Block of the mansion. As the business platform operated with international investment, Truth Mansion is endowed with much more international culture, and shows the style of globalization progress of Shenzhan thanks to the unique

求是大厦位于深圳国际化大道——深南大道北侧，东依中心西区强大商务平台，西靠华侨城高尚社区，南望深圳湾绝色海景，北眺安托山、塘朗山优美山景。据悉，求是大厦是政府批出的首批以“商务公寓”正式分类的项目之一，也是深南大道上罕有的70年产权商务公寓。项目个性品质突出，甲级写字楼的外立面，三重立体会所，海景泳池，多功能商务会议室，每三层一个的空中花园，商务服务中心等等满足了商务的不同需求。求是大厦以自身的品质魅力引起了国际财团的关注，早在求是大厦动工之初，项目就被香港查氏集团一眼相中，一举买下大厦东座。作为国际资本运作的商务平台，求是大厦因查氏家族的独具慧眼和求是基金的高科技化、人文化而注入了更多的国际化内涵，彰显着深圳国际化进程的气质。这种国际化背景加上国际资本的注入与运作，将为更多企业提供高品质的商务平台。

项目位于深圳市福田区竹子林中区深南大道北侧，交通枢纽大厦以东，用地性质为高层综合楼。南为深南大道，可远眺深圳湾海景和红树林

自助银行

judgment of the Zha Family and high-tech & humanity of Truth Foundation. The globalization background and international capital inflow & operation provide a high-grade business platform for more enterprises.

The project is located Bamboo Forest Central Area, Futian District of Shenzhen City, north of Shennan Avenue, east of Transportation Transition Mansion, and is supposed to be a high-rise comprehensive building. The sea landscape of Shenzhen Bay and Red Forest Natural Protection Zone are visible from Shennan Avenue, the south of this site. Chinese Tourism Scenery Zone is in the west, Futian Central District is visible from the east, and Antuo Mountain is visible from the north. The environment is beautiful and the location is excellent.

This project is a high-grade, small dwelling-size, high-rise apartment comprehensive building targeting white collars and investors. From the aspect of landscape design, the program emphasizes harmony with the surroundings. On the premise of urban design, modern technologies and materials are applied during designation, and create an architectural image with high-tech characters. Ecological, energy-saving and environment friendly concepts are taken into consideration during designation, associating with climate of Shenzhen, techniques such as space sharing, gray space and misconnection of volumes are applied to create a full-ventilation, full-daylight, environment friendly comprehensive apartment and office building.

The architectural layout is based on existing terrain, follows the frame of the city and takes the relationship with urban roads into consideration. The main body is composed of 2 "E-shaped" buildings, which misconnect to each other and form an "S-shape", in order to optimize the orientation and ventilation. In addition, the "misconnected layout" enhances the horizontal continuity and integrity along the interface of Shennan Avenue. The advantage of misconnection is to open the notch of the building, optimize natural ventilation, daylight and realize the concept of energy-saving and ecological building. Meanwhile, it helps to form an appropriate "flowing space" at the south-east intersection of the site, and reliefs the pressure of this building against the crossroad. The tower in the west of the site is aligned to Transportation Transition Mansion nearby, and achieves the continual space effect; set-back layout of the west tower is aligned to the public buildings pending to construct. The south-east angle between 2 towers is formed through connection of 4-floor podiums. In summary, "misconnection layout" enhances the integrity of building group, indicates and directs the community paths in the east of the site, makes the façade alternative, profound and stereoscopic, and achieves the ideal unity of functions and forms.

The entire building group is composed of a 4-floor commercial podium, a single-floor overhead sky garden and a 26-floor tower. The entrances and exits for commercial operations part are layout along the east, south and north-east streets of the site, and those for the apartments are set in the north-west of the site, which do not interfere with each other.

PS: Truth Mansion was named "Zhucui Bay Comprehensive Building" originally. And Hong Kong "Truth Foundation" was attracted by its excellent design concept during the construction, then bought half of it to use as the office building for Truth Foundation headquarter, therefore renamed it "Truth Mansion".

自然保护区，西为华侨城旅游风景区，东可远眺福田中心区，北可远望安托山，环境优美、地理位置优越。

本工程主要是针对白领阶层和投资人士的高品质小户型高层公寓综合楼。方案设计从景观设计的角度出发，注重与周边环境的相互协调。以城市设计为前提，设计中充分运用现代技术、材料，创造出高科技特色的建筑形象。充分考虑生态、节能、环保等设计理念，结合深圳气候条件，运用共享空间、灰空间、体量错接等设计手法，创造全通风、全采光的环保综合公寓、办公楼。

建筑布局结合现有地形，顺应城市架构，照顾与城市道路的关系，主楼呈两栋"匚"形楼错接布局而成，形成水平"S"形，以争取最大面宽的朝向及通风。此外"错接布局"加强了沿深南大道界面横向的连续性和整体性。错接的最大好处是打开建筑凹口，最大限度利用自然通风、自然采光，创造节能、生态的建筑概念。同时"错位布局"有助于对基地东南侧的路口构成适当的"流动空间"，减少建筑对十字路口的压迫。基地西侧塔楼与相邻的交通枢纽大厦基本对齐，起到空间延续作用，东侧塔楼退后布局，与待建公建对齐。两塔楼所形成的东南夹角由4层裙房相连而成。综上所述，"错接布局"既有利于建筑群的整体感，又有利于对基地东侧小区道路的明示与引导，使立面自然有变化、有深度和立体感、功能和形式高度统一的理想境界。

整组建筑由4层商业裙房、1层架空屋顶花园及26层塔楼构成。商业出入口沿基地东、南、东北侧临街布置，公寓出入口位于基地西北侧，两者互不干扰。

备注：求是大厦原名"竹翠湾综合楼"，由于良好的设计理念，在建设中被香港"求是基金会"看中并购买一半作为求是基金总部办公楼，改名为"求是大厦"。

小心地滑

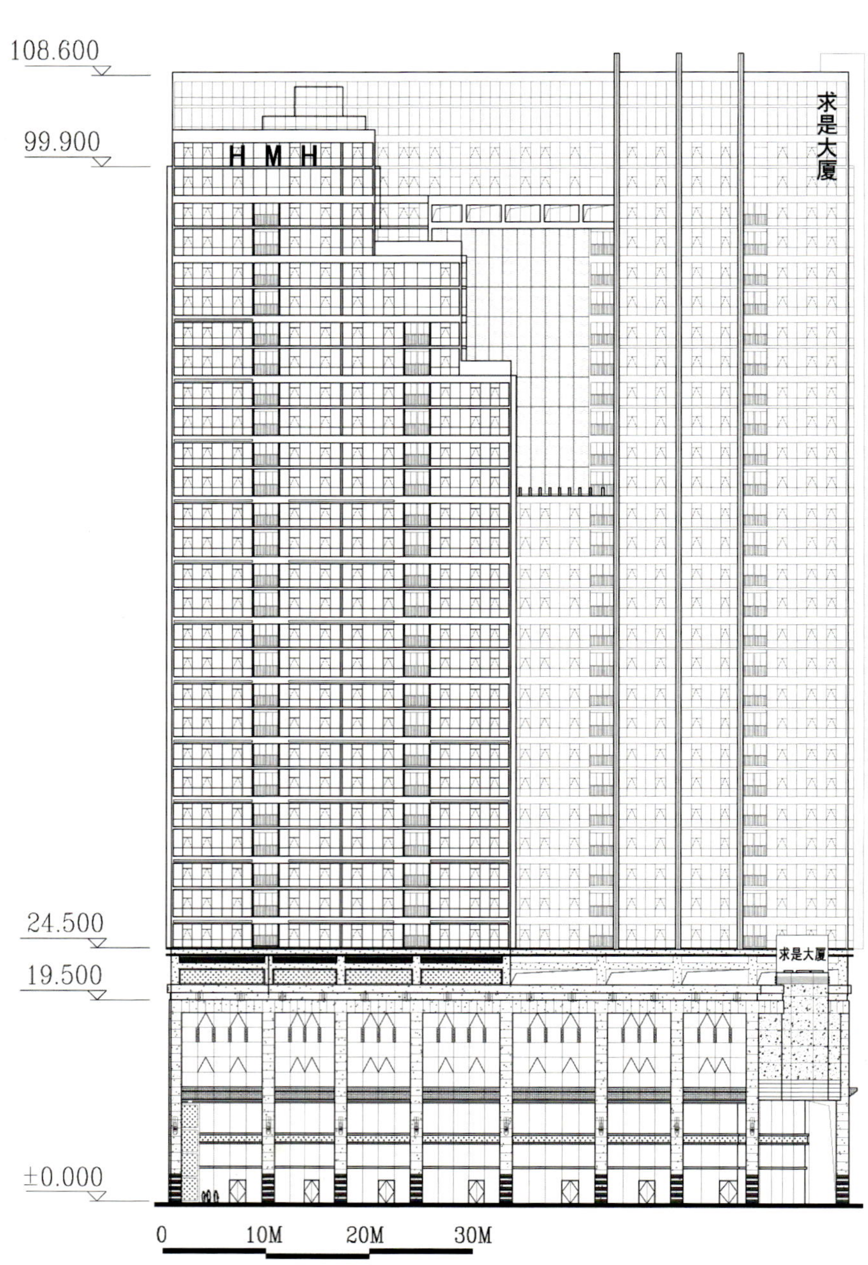

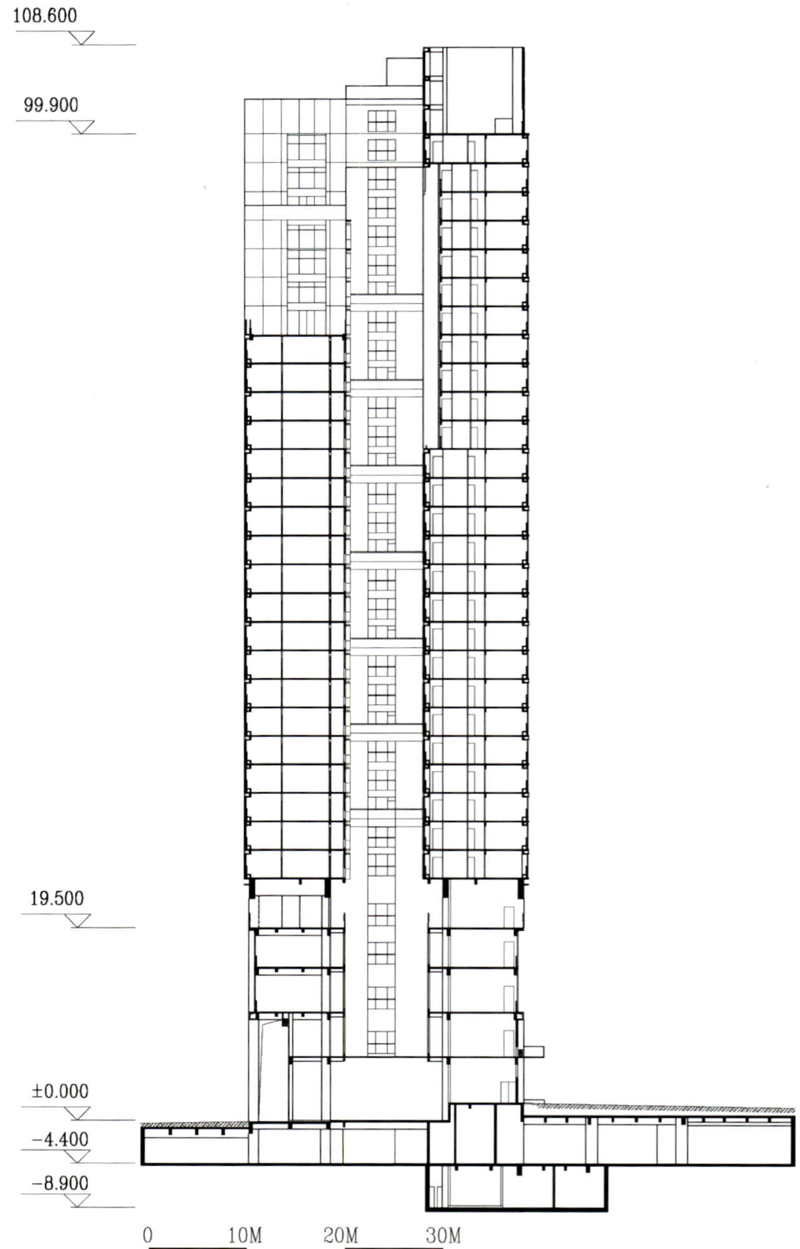

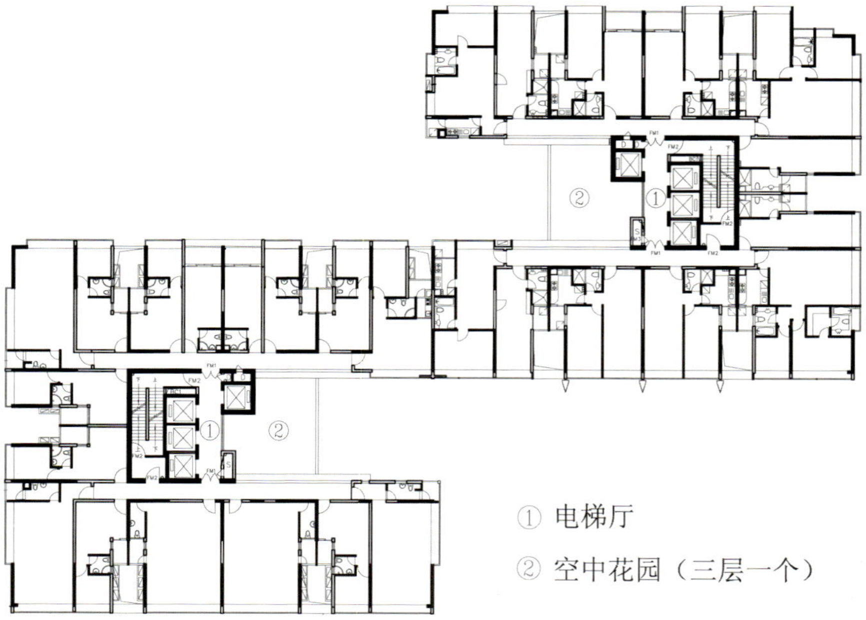

① 电梯厅

② 空中花园（三层一个）

LOTUS YOGA
Merry Christmas
City Professional Hub

求是大厦·西座
6/F -- 11/F
12/F -- 17/F
18/F -- 24/F
25/F -- 31/F
求是大厦
现楼出售
求是大厦西座信报箱

Shenzhen Excellence Era Plaza

深圳卓越·时代广场

Real Estate Developer: Shenzhen Excellence Group
Architecture Design: Shenzhen Xiepeng Construction & Project Design Co., Ltd.
Project Site: Shenzhen City
Site Area: 7,295m^2
Building Area: 118,927m^2

开发商：深圳市卓越集团
建筑设计：深圳市协鹏建筑与工程设计有限公司
项目地址：深圳市
占地面积：7295m^2
建筑面积：118，927m^2

Shenzhen CBD is endowed with hard and honorable mission by globalization trend and remarkable development of regional economics. Located on the south axis of Central District in Shenzhen City, Excellence Era Plaza stands profoundly at the core area of CBD with 218m height; it senses that urban business is going to face new challenges. The innovative business aesthetics matches the concept of urban business and strategic judgment of CEOs properly. Business aesthetics promoted by Excellence Era Plaza associate business values of CBD, architectural style of office building with development of urban economics and cultures closely. Business aesthetics of Excellence Era Plaza manifest business efficiency, broadness, and radiation repeatedly. It is the time that CBD leads development of business strength and the expression of Shenzhen Business steps towards globalization, Excellence Era Plaza realizes the followings predictably: Business should focus on science and technology, culture, keep step with global trend, and the business territory of the city should proceed towards more profound concept of globalization.

国际化浪潮、区域经济的历史性发展，赋予了深圳CBD艰巨而光荣的时代使命。位于深圳中心区南轴线，以218m的绝对高度傲立CBD核心区域的卓越·时代广场率先洞悉城市商务即将面临的变革，其独创的商务美学全方位地吻合城市商务意义，并与CEO的战略眼光不谋而合。从大定义来看，卓越·时代广场所提倡的商务美学开创性地将CBD商务价值、写字楼建筑气质与城市经济、人文发展紧密联系。商务效率、商务广泛性与商务辐射力，在卓越·时代广场的商务美学中被反复贯穿。在这个CBD统领深圳商务力量发展、深圳商务表现形式正向国际化深度迈进的时刻，卓越·时代广场预见性地意识到了这一点：商务应注重与科技、文化以及全球化同步，城市的商务版图应向更广阔的全球概念迈进。

卓越·时代广场位于深圳CBD南中轴，市政府和会展中心之间，紧临福田香格里拉酒店和国际商会中心大厦，是商务核心区域内的双地铁物业。目前是深圳CBD南区最高的写字楼，在深圳超高层写字楼中居前十位。

The project is located on the south axis of Shenzhen CDB, between Civic Center and Exhibition Center, nearby Futian Shangri La Hotel and International Chamber of Commerce Tower, owns bi-subway properties within CBD. So far, it is the highest office building in the south area of Shenzhen CBD, and is among the top-ten super high-rise office building in Shenzhen.

The project integrates 7 major municipal supporting resources provided for the north area of CBD, such as Citizen Center, Music Hall, Library and TV Center, and optimizes interaction with the resources from the south area. It is surrounded by several large shopping malls, such as Central Walk Shopping Mall, Crystal Island International Plaza, CoCo Park, and Rainbow Department Store, as well as worldwide top-class 5 stars Hotels, such as Hilton Hotel, Qinian Hotel, Ritz-Carlton Hotel and Shangri La Hotel.

The facades are tall and straight, a 20m-high sunshine courtyard, which is unique in Central District of Shenzhen, 3 lobbies and cross-shaped core tubes inside make the public spaces more broad, improve the business quality and provides convenient communication platforms. 20 more Mitsubishi-branded elevators meet the business demands of high, medium and low zones, and promote the image of enterprises comprehensively through the innovative design concept of cross-shaped core tubes.

项目既整合了CBD北区市民中心、音乐厅、图书馆、电视中心等七大市政配套资源，同时又与南区资源形成最大程度的互动。周边有怡景中心城、晶岛国际、购物公园、天虹商场等大型商业购物中心，更有希尔顿、祈年酒店、丽思卡尔顿、香格里拉四家世界顶尖的五星级酒店环绕。

立面外形俊朗挺拔，内部拥有深圳中心区绝无仅有的20m高阳光中庭、三大堂和十字核心筒设计，使公共空间更加宽敞，不仅提高了商务品质，更提供了多个从容的交流平台。20部高速三菱电梯，充分满足了高、中、低区的商务需求，并结合十字中央核心筒的超前设计理念，全面提升了企业形象的展示面。

建筑设计推崇商务美学办公理念，首创CEO俱乐部，倡导一种全新的商务理念，为全球CEO搭建一个管理、咨询、交流和寻找商机的平台，真正打造CEO们的“尖端商务文化圈”。同时源于对人本精神和细节的尊重，时代广场层高净高2.9m，标准跨度11m，1700m²的标准层将为企业创造最舒适的商务空间。同时每层特设8处总裁盥洗室，使商务空间也充满细节之美。

卓越·时代广场拥有便利的交通，从地下二层的中心区四通八达的运输系统，地下一层的商业街、地铁站，地面城市主干道与西部通道、广深高速相连。楼宇内竖向交通直到楼顶设置的直升机停机坪，从空中到地下共拥有六维交通体系，四通八达，充分体现效率优先的商务理念。

金中环

The architecture of this project promotes business aesthetics concept, innovates CEO clubs, advocates fresh business concept that it provides worldwide CEOs with a management, consultant, communication and investment platform, and creates a "top business cultural circle" for them. Upon humanism and admiration for details, Era Plaza is 2.9m in net height, 11m in standard span and 1700m² for each floor, which creates the most comfortable business spaces for enterprises. 8 CEO-grade rest rooms endow the business spaces with exquisite beauty.

Excellence Era Plaza provides convenient transportation, the vehicle system of central district on B2 extends to all directions, Commercial Street on B1, subway stations and main roads above the ground link to West Channel and Guangshen Highway. Vertical transportation inside the buildings and landing field on the roof, establish 6D transportation systems extending to all directions, manifest the business concept of efficiency-first.

Nanning Diwang International Commerce Mansion

南宁地王国际商会中心

Location: Golden Lake Square of Nanning City in Guangxi Province
Site area: 15,157.1m²
Building Area: 158,732.65m²
FAR: 11.16
Designer: Zhongyan Kan; Yiqun Jin
Developer: Guangxi Longmenju Real Estate Developing Company

建设地点：广西省南宁市金湖广场
总用地面积：15,157.1m²
总建筑面积：158,732.65m²
建筑容积率：11.16
主创设计师：阚忠彦、金逸群
开发商：广西龙门居房地产开发有限责任公司

Located at the new CBD-Golden Lake Square, Diwang International Commerce. Mansion is the tallest building in Nanning with the total height of 256 meters, 53 floors and 267 meters high to the top of the tower spire. The synthesis is consisting of four parts - commerce, government offices, business offices and a five-star hotel. By properly separating every functional area, the four different types of pedestrian traffics are well arranged to avoid mutual interference. The skirt building is the shopping centre; the affiliated building is the Nanjing Municipal Construction Bureau. Above the 40th floor of the main tower is a five-star hotel, below that is the business office area. The 2nd floor of the skirt building is the hotel lobby. The architecture of the building intimately responds to Golden Lake Square in term of urban design. The four corner glass boxes make the main tower powerful and uplifting while their difference in height expresses their functions respectively. The tallest tube turned into a tower at the building top. A sightseeing elevator inside it directed people to the oval shaped roof top observation platform which also serves as landing and take off pad for helicopters. The structural frames under the platform are deliberately exposed to articulate a unique iconic modern feature.

南宁地王国际商会中心位于南宁新中心商务区——金湖广场，系广西南宁最高建筑物，总层数为53层，总高度为256m，塔尖高度267m。综合体由商业、政府办公、商务办公和五星级酒店4个功能部分组成。通过合理的功能分区妥善解决了4种人流的流线，避免了人流交叉；裙房为商业中心，附楼为南宁市建设局大厦，主塔楼的40层以上为五星级酒店，四十层以下为商务办公区，酒店大堂设在裙房二层。建筑造型与金湖广场的城市设计密切呼应，通过四个角部玻璃筒体不同高度的处理，使建筑更具有苍劲、挺拔之感，而不同高度的筒体又反映了建筑的功能特征——最高的筒体在顶部形成高塔，其中设置的观景电梯将人流导向屋顶观光平台，椭圆形的观光平台兼作直升机起落平台，平台下的网架全部暴露，形成了独具风格、现代感极强的地标特征。

地王国际商会中心

GIADA
PIAGET

祥福酒楼

云鼎观光

文德大厦

高区客梯B3~B1 40~54
High Area Elevator

Shenzhen Rongchao Business center

深圳荣超经贸中心

Location: CBD of Shenzhen City
Site area: 7,137.7m²
Building Area: 114,943.0m²
FAR: 12.9
Desiger: Xiaobing Mao; Zhongyan Kan
Developer: Shenzhen Rongchao Real Estate Company

建设地点:深圳市中心商务区（CBD）
总用地面积:7,137.7m²
总建筑面积:114,943.0m²
建筑容积率:12.9
主创设计师：毛晓冰，阚忠彦
开发商：深圳市荣超房地产有限公司

Shenzhen Rongchao Trade Centre located in CBD, 200 meters in height, which is one of the landmark buildings in CBD of Shenzhen nearby the Citizen Center. The space between Rongchao Trade Center and Citizen Center is public urban space becoming a personalized street square. The diversified polyhedron jointed by several glass prisms symbolizes "Crystal Tower". Vertical lines on elevation strengthened to rise. The reinforced concrete structure of frame outside & lift core inside, a crisscross core creates an appropriate zoning.

深圳荣超经贸中心位于深圳中心商务区（CBD）北区，毗邻深圳市民中心，总高度达200m，是深圳中心区标志性建筑物之一。荣超经贸中心在面对市民中心方向留出了开放城市空间以形成具有人性化尺度的街边广场。建筑造型通过几个玻璃棱面体和玻璃面的尖锐交接，形成富于变化的多面体，以引喻“水晶塔”式造型；建筑立面强调竖向线条以加强向上升腾的感觉；结构体系采用内筒外框钢筋混凝土结构；核心筒采用十字形核心筒布局，分区合理、平面使用系数较高。

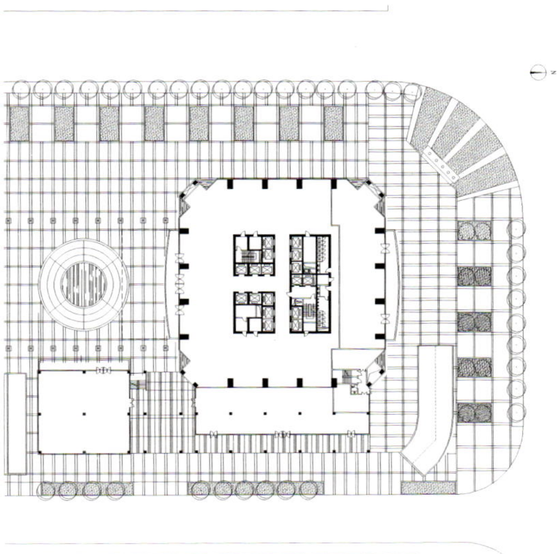
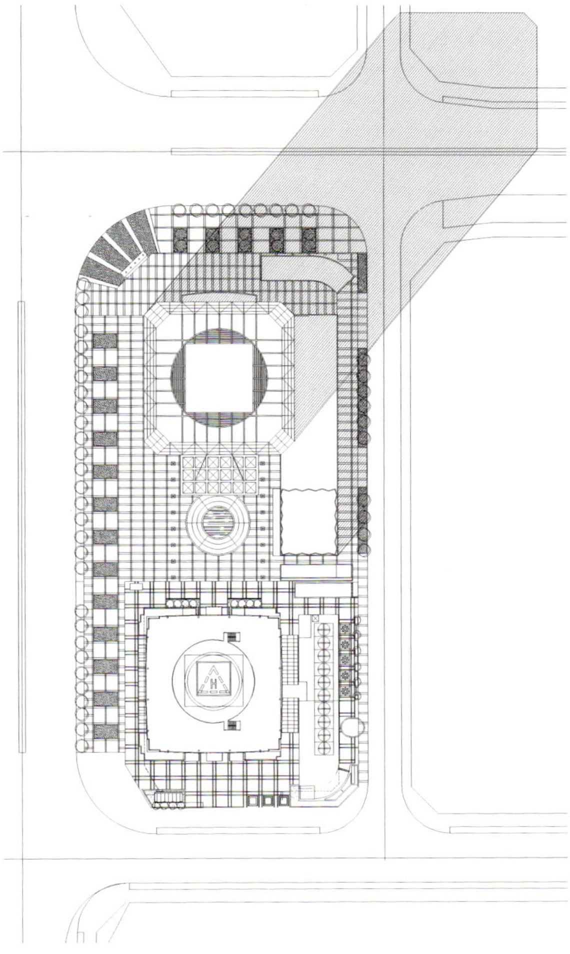
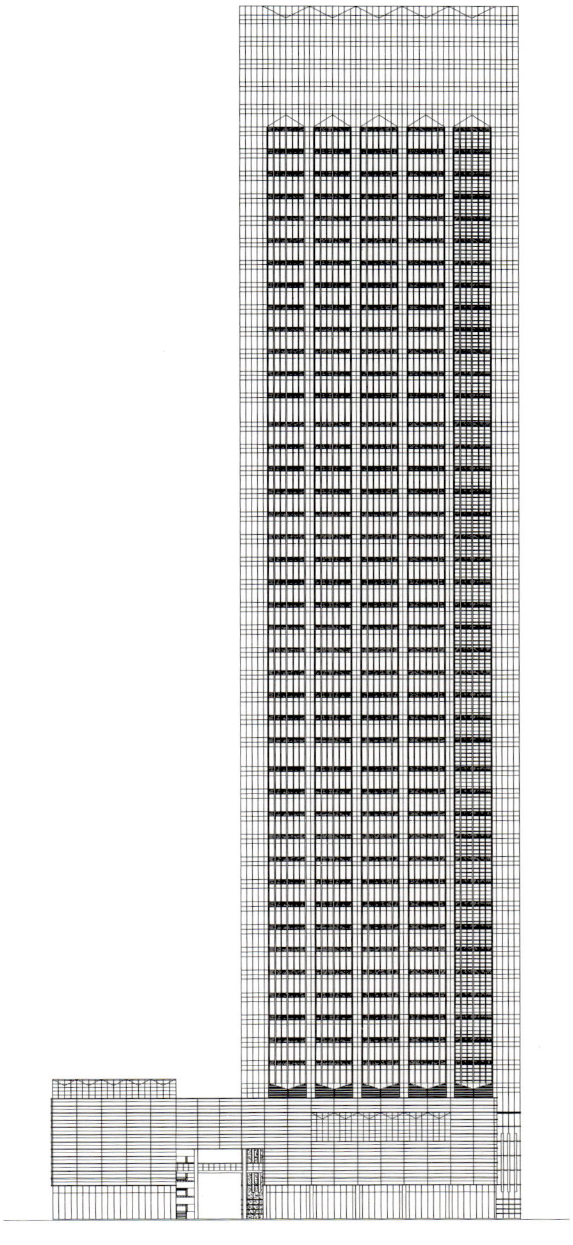

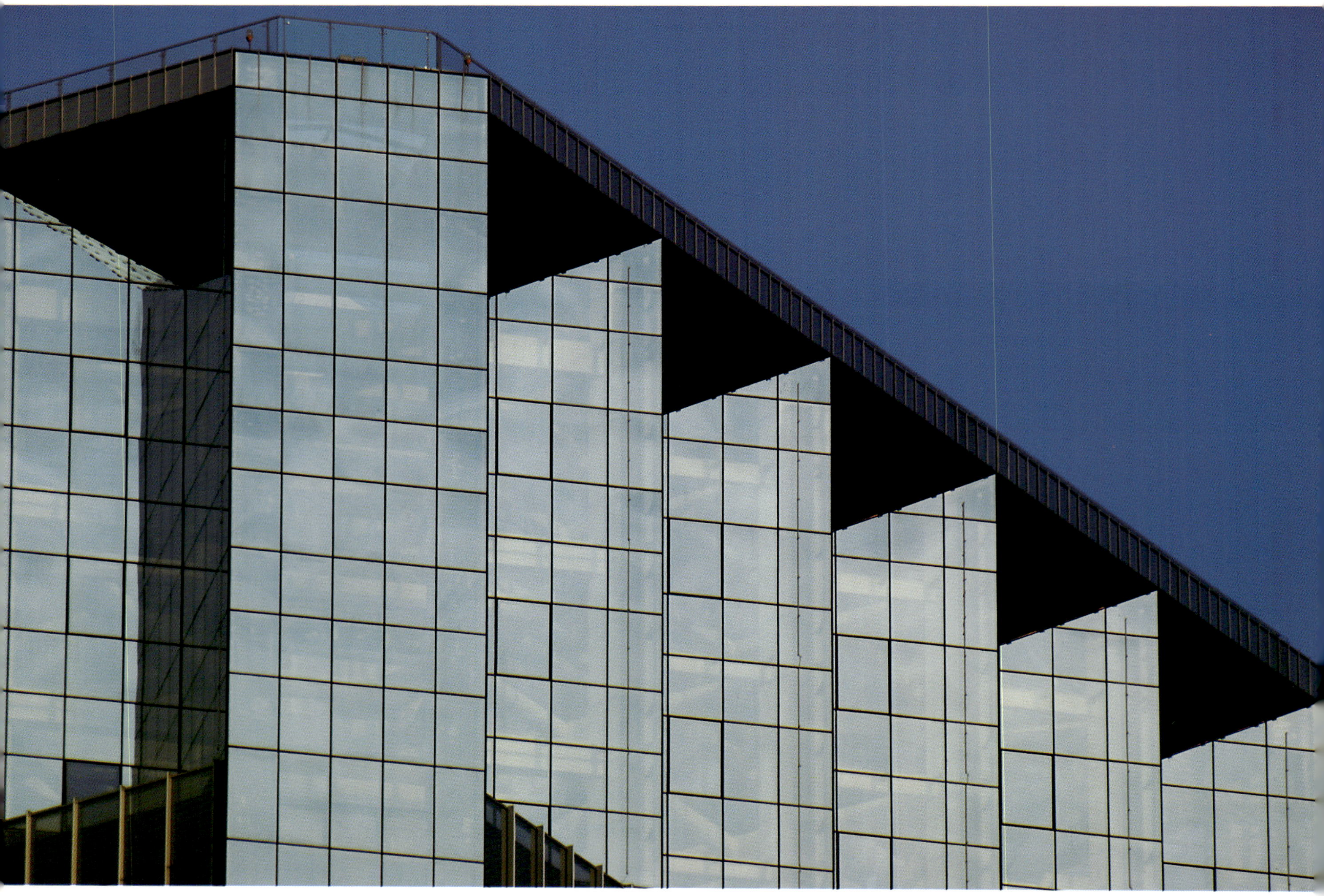

Shenzhen RC Costal Plaza

深圳荣超滨海大厦

Location: Central Baoan District of Shenzhen city
Site Area: 12,676.6m²
Building Area: 64,100m²
FAR: 3.7
Developer: Shenzhen Rongchao Real Estate Company

建设地点：深圳市宝安中心区
总用地面积：12,676.6m²
总建筑面积：64,100m²
建筑容积率：3.7
开发商：深圳市荣超房地产开发有限公司

Located in Central Baoan District of Shenzhen, the project occupies two blocks facing the City Park in the west and enjoys convenient transportation and superior scenery resources. It is fully fitted on the site in north-south direction. The height of the north-west building A is 80m (limit). The differentiating in height and floor area created a contrast of the two building in massing while maintaining the original corner positioning concept. The offset between two buildings ensures the view to the park from building B. The two towers are harmoniously combined and mutually enriched from different view points. The contrast between textures, materials in light and shadow of the wall and glass enhances the expression of the building. The two layered wall frame makes the windows more graceful. Warm and cool grey is adapted to express the secure, peaceful and elegant style of the building. The incorporating of many solid wall, aluminum sheet and jalousie reduces the exterior wall cost. Insulating sheets for outdoor air-conditioner over the posts and beams created a unique aesthetic style by offering abundant variation of shadow while improving the efficiency of the sunshade. The edges are finished with steel channels which marked quite a technical beauty and elegance.

项目位于深圳宝安中心区，中轴线西侧，由两个地块组成，西临城市公园，交通便捷，环境条件好，景观资源优越。是理想的办公用地。本案将两块地南北向布置满，将西北角A幢的高度设计为80m（限高），减小标准层面积，使两幢建筑高度及体形有对比，同时保持原有限定角位的理念，并通过提升高度使城市该节点得以加强。两幢建筑的塔楼前后错开，有利地保证了B幢朝西向公园的视线，减小相互之间的视线干扰。双塔依据相同的构成关系组合而成，相互呼应，而从不同角度观察，又有较为丰富的变化。利用实墙与玻璃强调虚实对比、材料对比、光影对比，使建筑更有力度，双层实墙框架的设计，使开窗比例变得更为修长，暖灰、冷灰色彩运用较为符合建筑内敛、平和的特征，形成典雅精致的风格。由于对造价的控制，本着尽可能少用玻璃幕墙，采用实墙，仿铝板外墙及百叶的组合墙体，同时在柱外及梁边考虑空调室外机的处理，横竖隔板既为造型语言，形成比较丰富的光影变化，又提高了遮阳节能的效率，其外沿贴凹型钢板收口，也反映了技术的美感及精致。

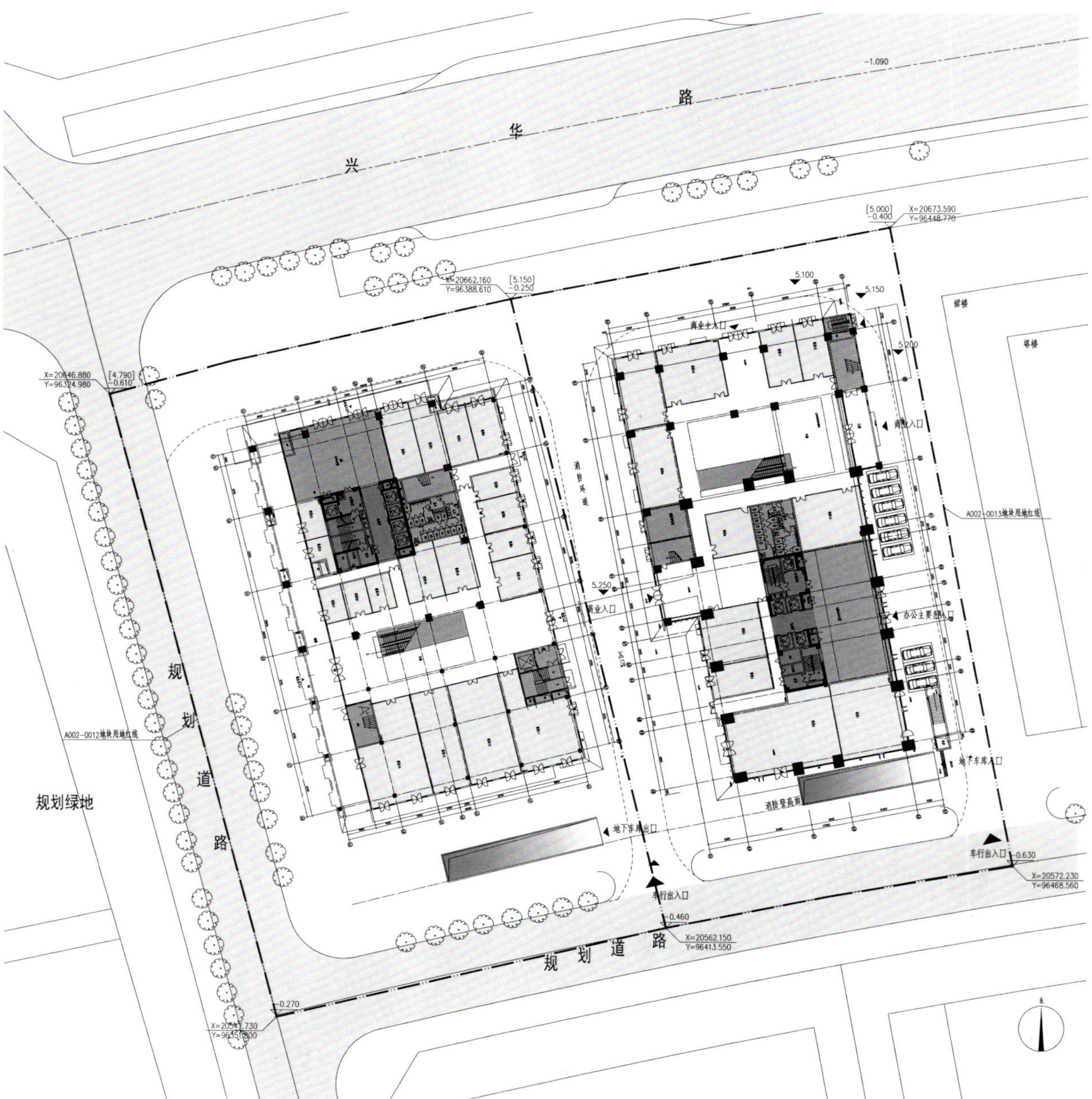

一层总平面图

0 5 10 15 20

Shenzhen longguang century plaza

深圳龙光世纪大厦

Location: Central Baoan District of Shenzhen city
Site area: 17,203.43m^2
Building Area: 79,800m^2
FAR: 3.3
Developer: Shenzhen Longguang Real Estate Company

建设地点：深圳市宝安中心区
总用地面积：17,203.43m^2
总建筑面积：79,800m^2
建筑容积率：3.3
开发商：深圳市龙光房地产开发有限公司

This project locates in the northeast corner of Central Square in Baoan Center, facing the Municipal and Sports Center, and enjoys the central park and ocean scenery towards the southwest. The original general plan calls for two towers connected by a commercial podium. An ocean view corridor is formed between the two control towers and buildings along north and south blocks. The site is divided into six parts; tower buildings are placed at the northwest and southeast corners to define the urban node, reduce mutual visual interference and open up the phase-I tower's view to the city center. The other two site corners opposite to the towers are made into two squares. The northeast one is a high traffic commercial square and the southwest one is a quiet business square with a visual extension of the central landscape area. The two towers are composed with a harmonious discipline while offering rich visual variations viewed from different points. The horizontal and vertical texture of the curtain wall enhanced the contrast of textures and materials in light and shadow and further articulate and slender the tower. The cool grey color is adapted to express a secure, peaceful and elegant building style.

本项目位于宝安中心区中心广场东北角，与宝安区行政中心和体育中心遥遥相对，西南侧为中心公园以及海景景观，所处地理位置十分优越，按照原城市规划原则，在总体布局上，分两栋塔楼设置，中间以商业裙房相连，与该街区南北地块建筑形成以两边为控制塔楼，中间留出海景方向视线走廊的城市空间形态。将地块划分为6个基本块，将西北角与东南角布置塔楼，南北错开的两栋塔楼限定角位，使城市节点得以加强，也保证了一期塔楼朝向城市中心地带的视线方向，减小相互之间的视线干扰。与塔楼相对应的另两个角设置两个对角广场，东北角广场吸引人流的商业广场，西南角广场处理成相对安静的商务广场，是中心绿带的景观延伸。双塔依据相同的构成关系组合而成，相互呼应，而从不同角度观察，又有较为丰富的变化。利用横向幕墙与竖向幕墙强调肌理对比、材料对比、光影对比，使建筑更有力度，塔楼比例变得更为修长，冷灰色彩运用较为符合建筑内敛、平和的身份特征，形成典雅精致的风格。

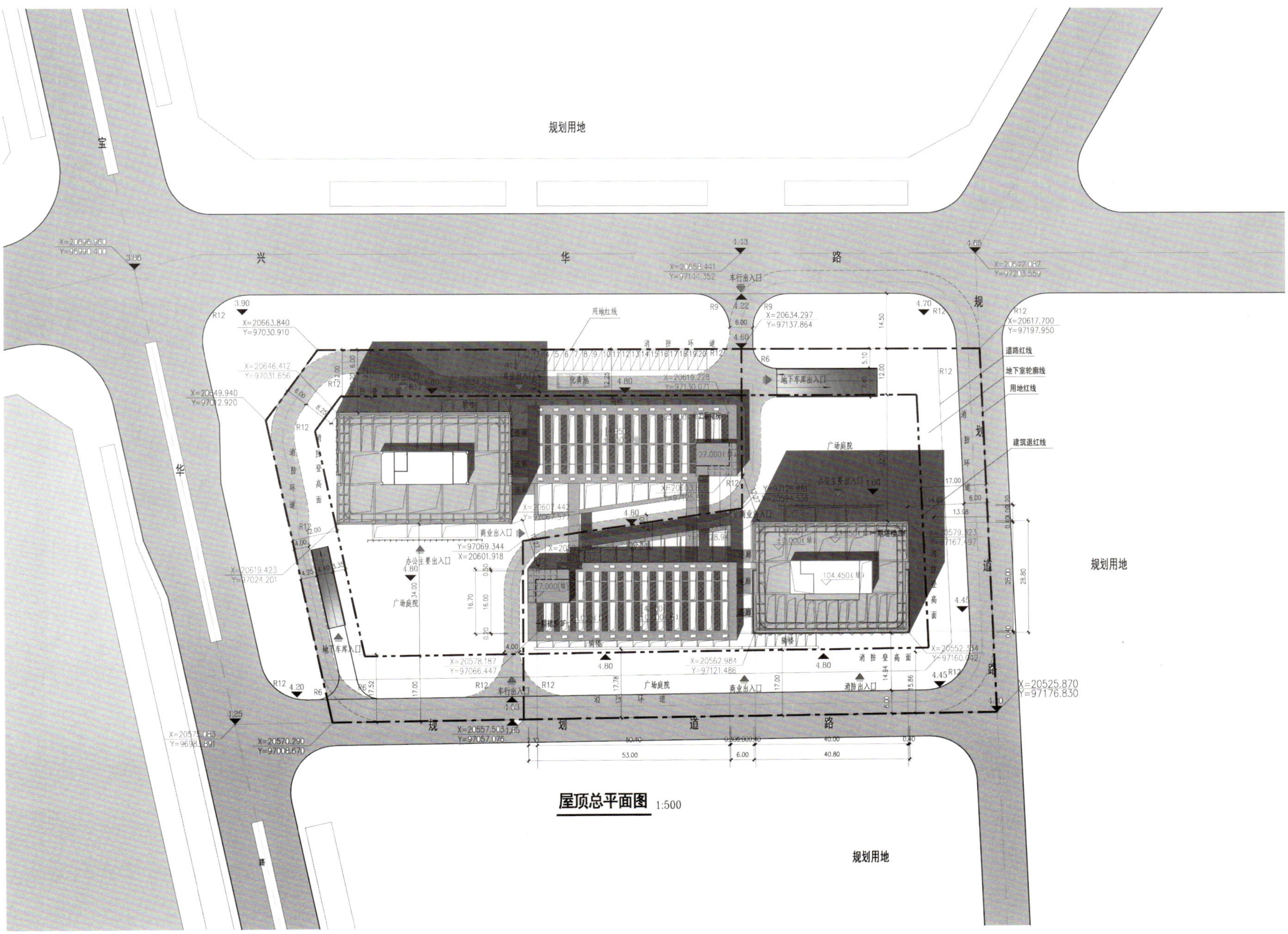
规划用地
宝
兴
华
路
规
划
道
路
华
规划用地
规划用地
用地红线
道路红线
地下室轮廓线
用地红线
建筑退红线
车行出入口
地下车库出入口
广场庭院
办公主要出入口
商业出入口
消防环道
消防登高面

屋顶总平面图 1:500

Shenzhen Max Smart Center

深圳万骏中心

Project Location: Baoan center of Shenzhen
Site area: 5,440.39m^2
Building Area: 32,532.68m^2
Floor Area Ratio: 4.5
Real Estate Developer: Shenzhen Wan Jun Real Estate Co., Ltd.

建设地点：深圳市宝安中心区
总用地面积：5,440.39m^2
总建筑面积：32,532.68m^2
建筑容积率：4.5
开发商：深圳市万骏房地产开发有限公司

Located in Baoan center of Shenzhen, west of the middle line and edge of CBD, the project is endowed with excellent location, commercial environment and scenery resources. With a not quite ideal site that is long on south-north and flanked with buildings on both narrow sides, the project is made narrow in east-west and long in north-south, which results in barely satisfactory orientation. Therefore, the tower is set in the south to avoid the shadow of east tower and to have an overview of the urban central park. The podium, complying with the neighborhood planning, runs along Xinghua Road and forms a continuous Commercial Street with neighboring podium structures, as well as makes it an organic component of certain Commercial Street. In order to makes the building itself more dynamic, the building blocks are interlocked to enrich the quality of the building as well as the urban interface. Whole glass blocks are embedded in northwest corner of the building to emphasize the exposure and vitality of the building, as well as enhance its shape ratio. Steel structures and glass curtain walls are constructed in a simple style embodying building's delicate beauty and exquisiteness. Cool grey color makes the building introvert, peaceful, elegant and generous.

本项目位于深圳宝安中心区，中轴线西侧，其地理位置十分优越，处于核心商务区的边缘地带。有着相当不错的商务环境和景观资源。由于用地南北狭长，且夹在两侧建筑中间，先天环境不是很理想，建筑走向必然为东西窄、南北长，朝向较差，与相临地块建筑的视线干扰较严重，为此，塔楼尽可能的往南布置，错开东侧塔楼的视线遮挡，获得东侧城市中心公园的最大景观视角；裙房结合该片区规划，尽量沿兴华路布置，与相临地块商业裙房形成连续的商业街墙，使其成为该商业街区的有机组成部分。为了更好地获得建筑本身的动态感，造型处理上采用建筑体块的相互穿插，既获得建筑自身品质，又丰富城市界面。作为建筑门脸的西北角，则以通长的玻璃体，强调建筑公共性和高耸的气势，改善建筑的体型比例关系。外立面上采用竖挺的钢构和光洁的玻璃幕墙，力求简洁大方，也反映了技术美感及精致。冷灰色彩运用较为符合建筑内敛、平和的身份特征，形成典雅精致的风格。

CHUN YIP PLAZA

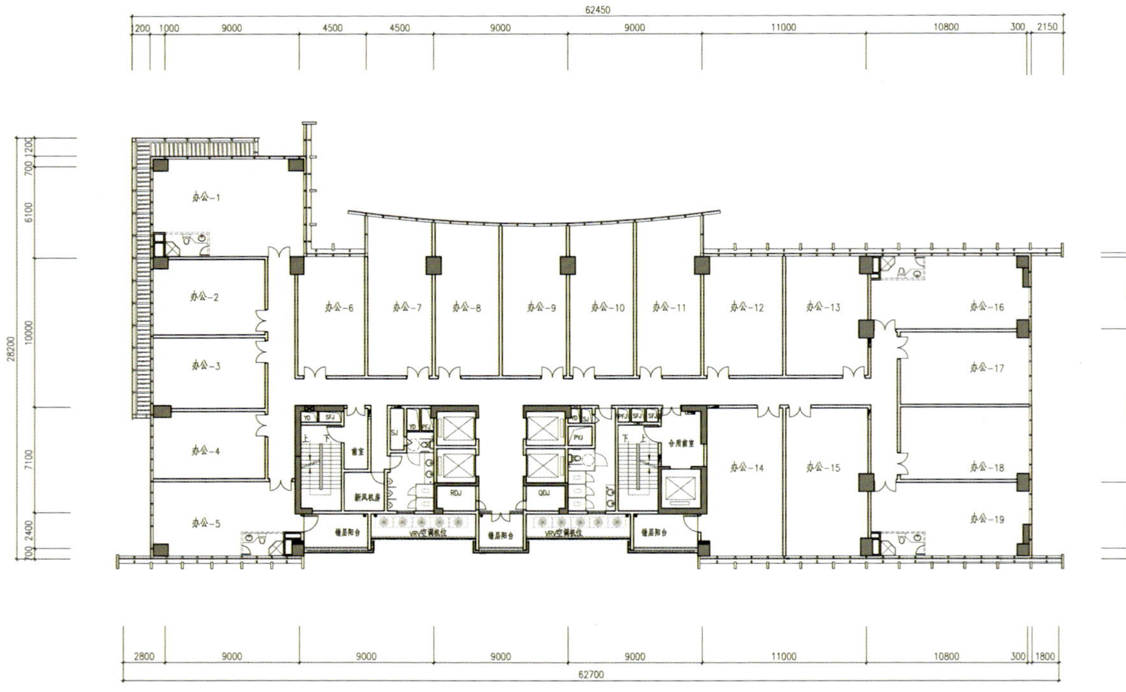

标准层平面图 1:150

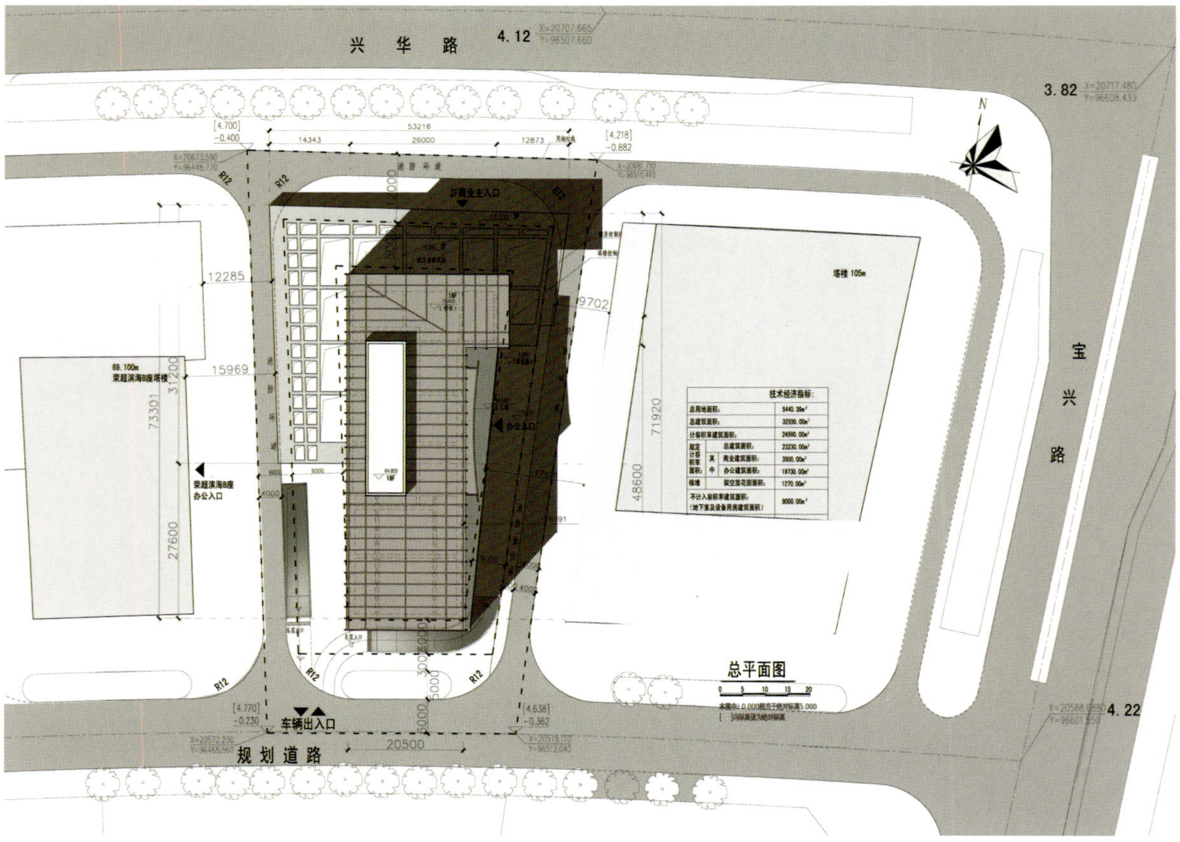

总平面图

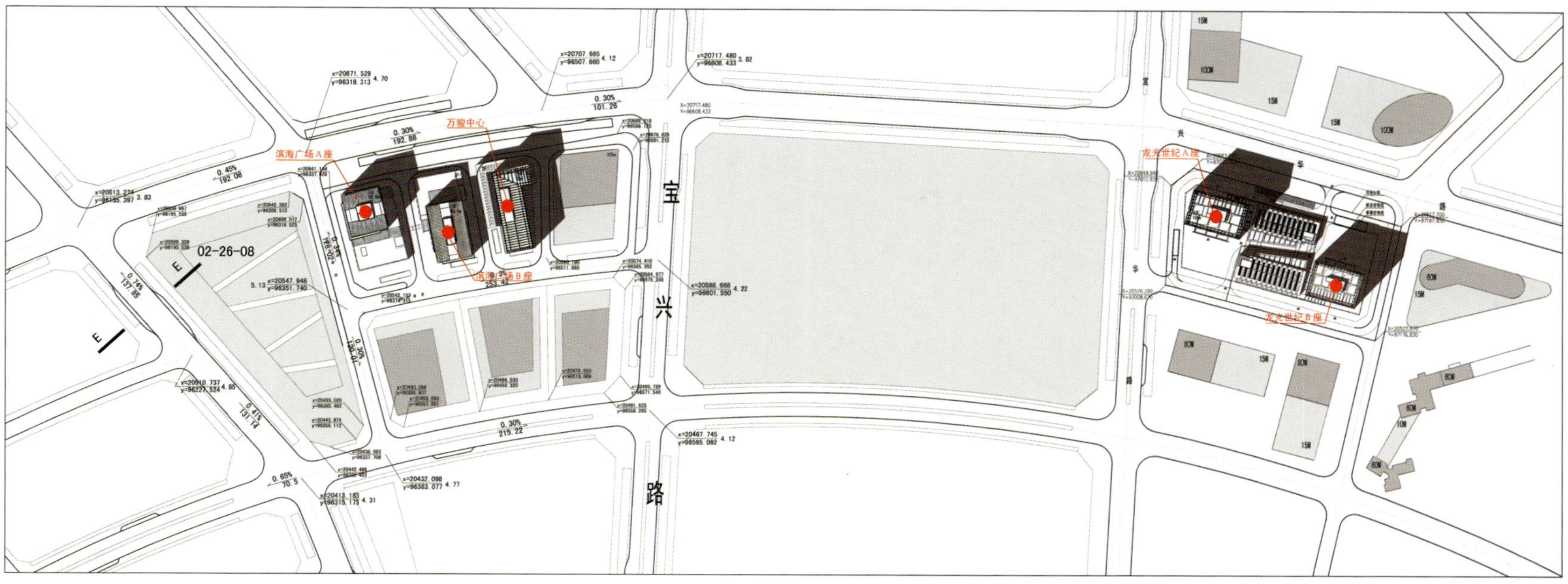
滨海广场A座
万骏中心
滨海广场B座
龙光世纪A座
龙光世纪B座
宝
兴
路
02-26-08

万骏经贸大厦
29786888

万骏经贸大厦
29786888
3333
9888

万骏经贸大厦
29786888

Shenzhen Wanlida Mansion

深圳万利达大厦

Project Location: High-tech Development Zone, Nanshan District of Shenzhen
Site Area: 6,381.4 m^2
Building Area: 77,736.0 m^2
Floor Area Ratio: 9.51
Real Estate Developer: Shenzhen Wanlida Electronic Industry Co., Ltd.
Architecture Design: AECOM Architecture Design

建设地点：深圳市南山区高新科技开发区
总用地面积：6,381.4m^2
总建筑面积：77,736.03m^2
建筑容积率：9.51
开发商：深圳市万利达电子工业有限公司
建筑设计：AECOM建筑设计

Wanlida High-tech Mansion is invested by Shenzhen Wanlida Electronic Industry Co., Ltd., and located in High-tech Development Zone, Nansha District of Shenzhen City, with Shennan Avenue in south. The building faces south and north that makes full use of natural light and city scenery. A series of sky gardens on east and west sides are set to form a vertical ecological green system embodying modern office concept. Its styling is simple but elegant, implicative but outstanding. The tower adopts a large area of glass screen to express a single and pure beauty in its shape and material. The grounding curtain wall forms contrasting horizontal and vertical lines enhancing the variation in light and shadow. The gridded curtain wall symbolizes the character of the electronic industry.

万利达科技大厦工程由深圳万利达电子工业有限公司投资兴建，位于深圳市南山区高新科技开发区中区，南临深南大道。建筑物布置以南、北朝向为主，充分利用自然采光及城市景观。为改善办公环境，东西两侧设计了一系列的空中花园，在高空中引入绿色，提供了办公人员的交流场所，体现现代办公理念，屋顶空中花园形成了一个立体的生态绿化体系。在造型设计上，建筑形体力求简洁精致，含蓄而醒目，塔楼采用大面积玻璃幕墙，展现单一纯净的形体美、材料美。塔楼幕墙直接落地，挺拔而端正，横竖线条的对比增强了幕墙的线条变化和光影效果，网格状的玻璃幕墙隐喻了电子行业的特征。

夜景透视图

鸟瞰图

南向透视图

三工程局
建

科技大厦欢迎您！
P 停车场

P 停车场

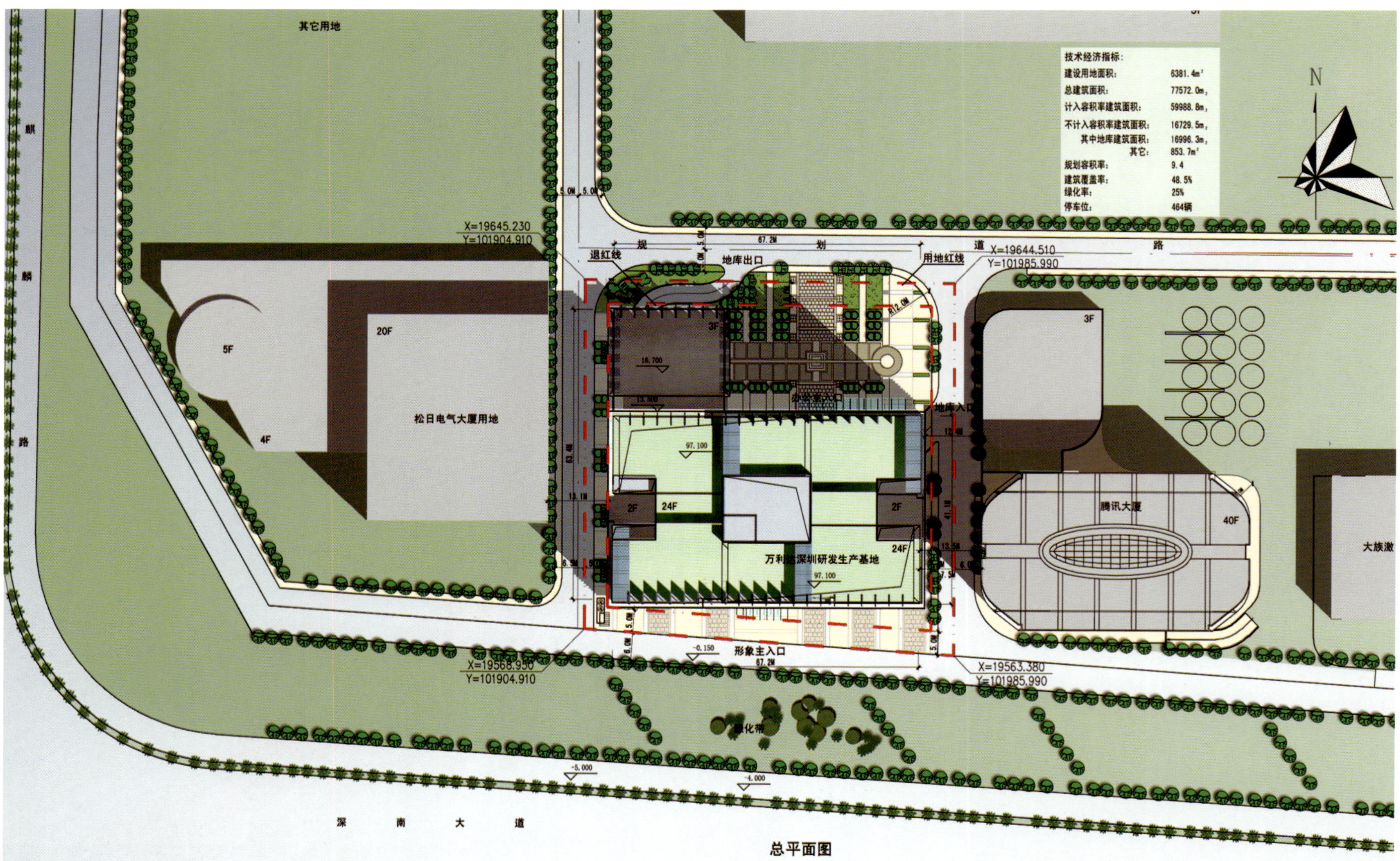
其它用地
技术经济指标：
建设用地面积： 6381.4m²
总建筑面积： 77572.0m，
计入容积率建筑面积： 59988.8m，
不计入容积率建筑面积： 16729.5m，
其中地库建筑面积： 16996.3m，
其它： 853.7m²
规划容积率： 9.4
建筑覆盖率： 48.5%
绿化率： 25%
停车位： 464辆
N
X=19645.230
Y=101904.910
X=19644.510
Y=101985.990
规 划 道 路
退红线
地库出口
用地红线
5F
20F
4F
松日电气大厦用地
3F
地库入口
2F
24F
万利达深圳研发生产基地
腾讯大厦
40F
大族激
形象主入口
X=19568.950
Y=101904.910
X=19563.380
Y=101985.990
深 南 大 道
总平面图

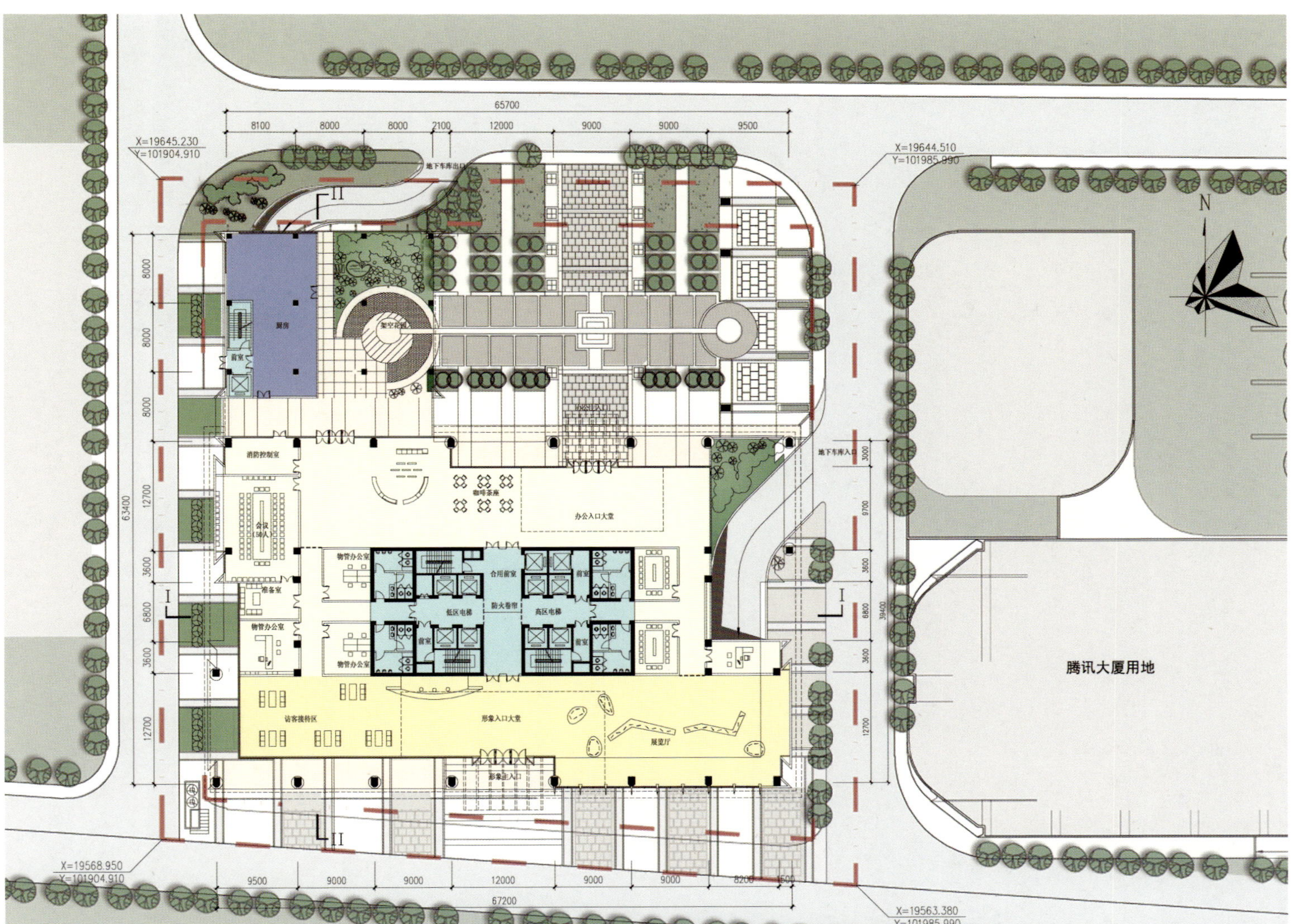

一层平面图　建筑面积：2692.20m²

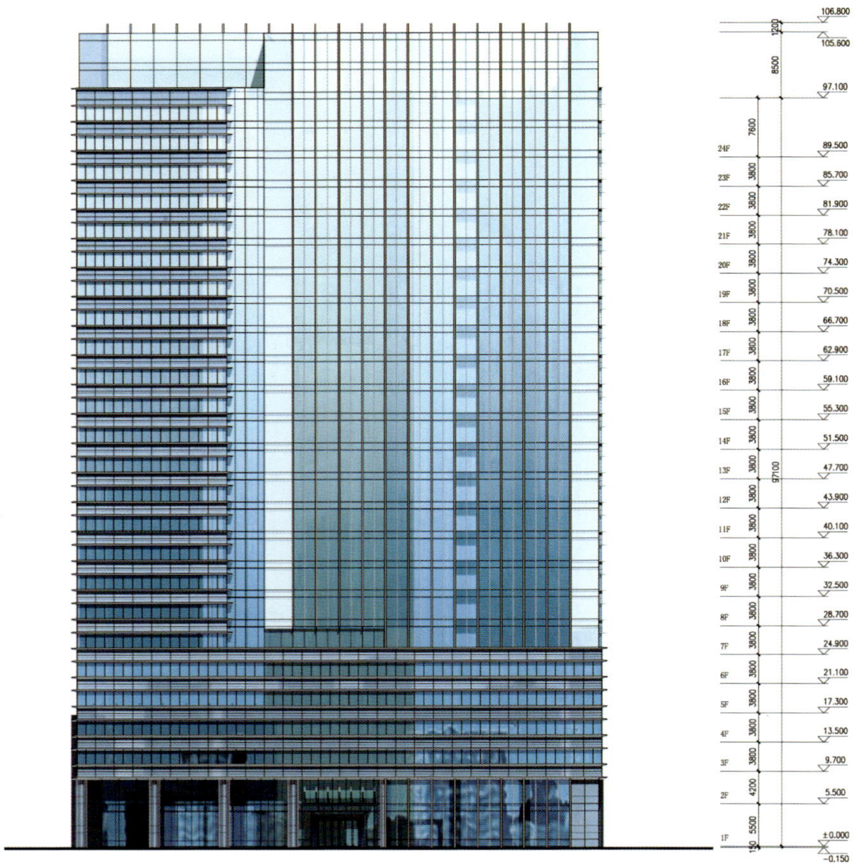

南立面图

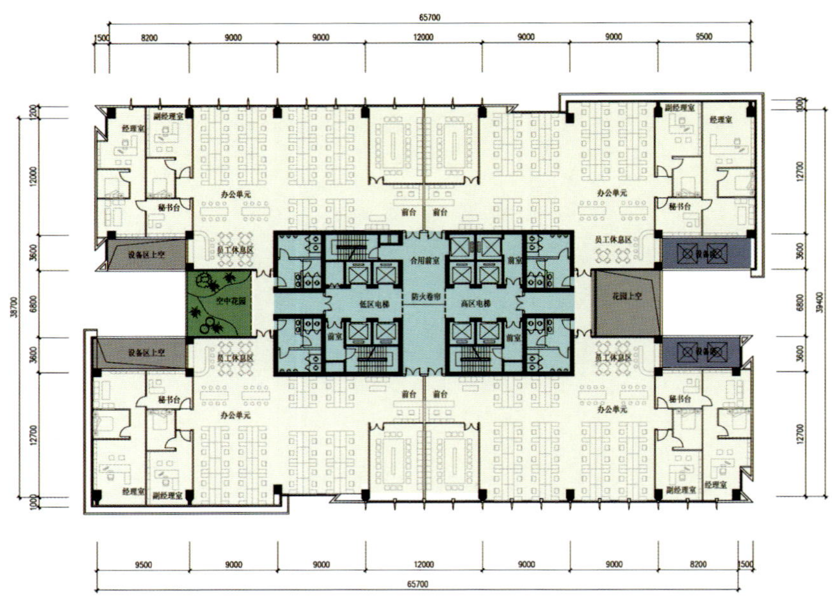

八~二十二偶数层平面图

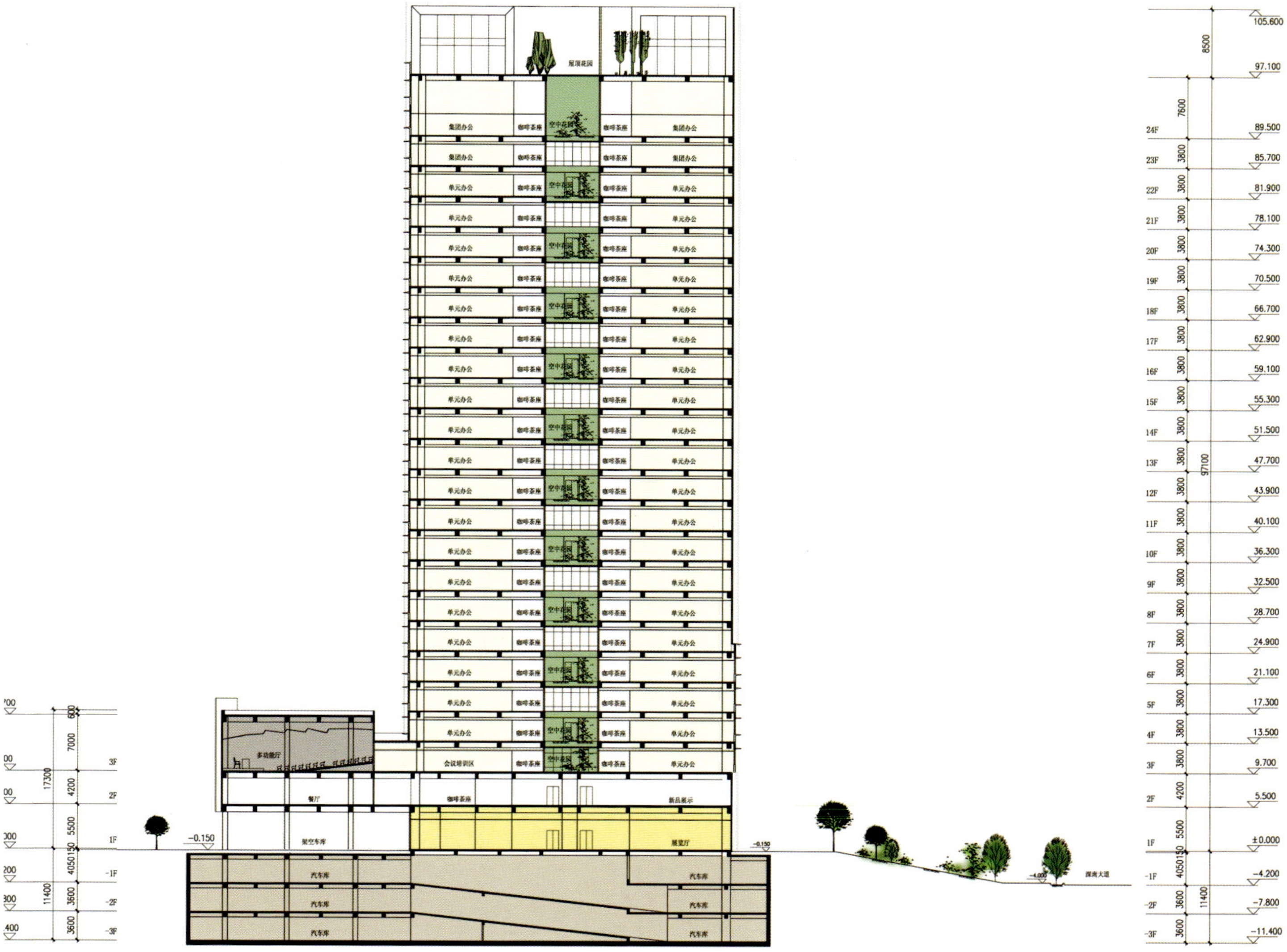

2-2 剖面图

Tencent Mansion

腾讯大厦

Building Area: 88,180.38m²
Office Area: 69,796.76m²

建筑总面积：88,180.38m²
办公面积：69,796.76m²

Tencent Mansion is located in North District of Shenzhen High-tech Park, north of Shennan Avenue. The total height of building is 193m, 39 floors above the ground and 3 floors underground. Functioning as Tencent headquarter base, Shenzhen Tencent Mansion was launched to use yesterday, which means Tencent achieve the business strategic layout successfully in Internet field. With novel shape, complete internal functions, and perfect cultural environment, this ultra high-rise building has become a fresh landmark in Shennan Avenue. Meanwhile, the estimated revenue of Tencent Company, which was established 11 years ago will be more than 10 billion, and makes its leadership of Internet enterprises in China; as well as builds another "landmark" in domestic Internet industry.

腾讯大厦位于深圳市高新科技园北区，深南大道北侧。楼体总高193m，地上39层，地下3层，楼体总高193m的腾讯全国总部基地——深圳腾讯大厦已在深全面启用，标志着腾讯在互联网领域全业务战略布局的顺利完成，这座造型新颖、内部功能齐全、人文环境优良的超高层建筑已成深南大道上的新地标。与此同时，成立仅11年的腾讯公司今年预计收入将突破百亿元，成为中国互联网企业第一家，创国内互联网行业的另一个"地标"。

腾讯

Jiangxi Highway Mansion

江西公路大厦

Jiangxi Highway Mansion is located in Honggutan District of Nanchang City, west of Red Valley Avenue, faces to Red Valley Mansion. This project is invested by Jiangxi Highway Development Headquarter, functions as the head office of highway development headquarter, which combines office, scientific research, business and apartment. It comprises 3 high-rise buildings and an administration meeting center. The interior of the building is simple and elegant, combines glass certain walls and stone materials. Its style is simple and extensive, as well as harmonious with the general plan of the Honggutan central district.

江西公路大厦位于南昌市红古滩区红谷大道西侧，会展路北面，与红谷大厦想对应。项目由江西公路开发总公司投资兴建，作为未来公路开发总公司未来的总部，集办公、科研、商业、公寓为一体。包括三栋高层建筑和一个行政会议中心。

建筑里面简洁而素雅，采用玻璃幕墙与石材相结合的形式，力争做到简洁大气，同时又与红谷滩中心区的总体规划和谐统一。

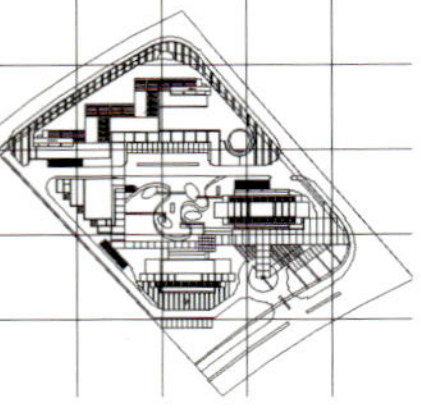

沿規劃路商務辦公樓效果圖

PLANING DESIGN

江西公路開發總公司科研測試基地

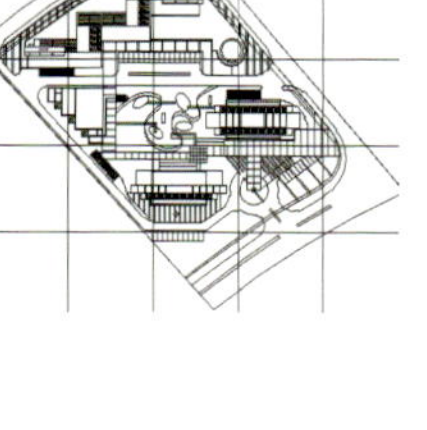

內庭院效果圖

方案設計

PLANING DESIGN

方案設計 PLANING DESIGN

鳥瞰夜景效果圖

江西公路開發總公司科研測試基地

方案設計 PLANING DESIGN

鳥瞰效果圖

江西公路開發總公司科研測試基地

TECO

Shenzhen Lord Center

深圳诺德中心

Real Estate Developer: Shenzhen Sino-Railway Investment Co., Ltd.
Architecture Design: Shenzhen Fangjia Architecture Design Co., Ltd.
Project Location: Futian District of Shenzhen City
Site Area: 4,980m^2
Building Area: 90,149m^2

开发商：深圳市中铁建投资有限公司
设计单位：深圳市方佳建筑设计有限公司
项目地址：深圳市福田区
用地面积：4,980m^2
建筑面积：90,149m^2

Located in municipal CBD planed by Shenzhen City currently, this project is adjacent to Pengcheng Fifth Road in the east, Jintian Road and Citizen Square in the west, Fuzhong third Road in the south, which is very close to the Anlian Mansion under construction, and ultra high-rise building under planning in the north.

The project aims to become a simple, elegant and functional office building, which provides customers with business space featured effective, easily re-decorated and impressive. Thanks to 150mheight Anlian Mansion which is located in the south, ultra high-rise office building under planning which is located in the north, and Citizen Square as well as Lotus Mountain which is located in the west, this building obtain a great view of scenery. In order to take the best advantage of landscape area, the long side of this building is made in east-west direction; therefore, it provides the office area with excellent view of the landscape in Citizen Square and Central Square.

本项目地处深圳市新规划的市级商务中心区，东边为鹏程五路，西临金田路和市民广场，南为福中三路，靠临正在兴建的安联大厦，北面为规划中的超高层建筑。

项目定位为简洁、高雅、实用的高档纯写字楼，为客户提供一个实用率高、易于分隔和高贵气派的办公空间。由于地块南侧一路之隔是150m高的安联大厦，北侧是规划中的超高层写字楼，西侧为市民广场及莲花山，拥有最高景观价值。为了最大限度地利用景观视线区域，建筑的长边沿东西走向设计，使办公区域可充分领略市民中心和中心区广场绿化的景观。

建筑的东西轴线正对市民广场，其主入口、裙房大堂和主体建筑的中庭空间均沿此方向布置，与东面的城市空间和西面的市民广场从视觉上相连，使办公建筑与市民广场互为借景。南北轴线沿建筑核心筒设置，连接了建筑两端的空中花园，令轴线与城市规划的“龙”脊线相吻合，达到城市空间交流和轴线规划的统一。

The east-west axis of this building faces to Citizen Square, and its main entrance, podium lobby as well as courtyard of the main building are all set in the same direction, visually connect to the entire city of the east and the Citizen Square of the west; the landscape of this office building and Citizen Square interact with each other. The south-north axis is set in accord with the core tube of this building, connects the sky gardens on both sides, makes the axis of this building and urban planning match each other, and achieves the harmony of city space communication and axis planning.

The main function of this project is business office, which comprises 41 floors. Floor 1 to 3 functions as administrative lobby and business, offers various convenient facilities such as banks, bonds and postal service; floor 4 41 is used as business office, whose design takes many layouts into consideration, in order to offer various office types for companies with different scales.

The plan layout forms a square, with little curve on the four sides. The core tube is set in the middle of the mansion, forms tube inside and frame outside structural system, makes the most large area become office area. The column grids and glass curtain walls are divided according to modern office style. The first floor is a 12mheight greening lobby, the green plants can be viewed through transparent glass curtain walls. The perfect internal garden and spectacular Citizen Square in the south forms a throughout green visual corridor, as well as enhance the impressive style of entrance and consistency of the entire space.

The core tube equips 14 guest elevators and 2 cargo and fire elevators, which offers vertical transport service. Guest elevators serve high, medium and low floors separately, 5 ones for low and medium floors and 4 ones for high floors.

项目的主要功能是商务办公，楼高41层，其中首层至3层，作为办公大堂和商业用途，提供完善方便的银行、证券、邮电等配套设施，4~41层为商务办公，商务办公层的设计充分考虑多种风格方式，为不同规模形式的公司提供丰富的办公空间类型。

平面布局成正方形，四个面略带弧形，设计上将核心筒布置于大厦的中部，形成中筒外框架的结构体系，让出了最大面积的外窗面作为办公区域。其中，柱网和玻璃幕墙按现代办公模数分格。

首层为层高12m的绿化大堂，配置了绿色植物，通过透明的玻璃幕墙，使优美的室内园林与南面壮观的市民广场形成内外贯通的绿色视觉走廊，增强入口门厅的气势及空间连续性。

核心筒内有14部客梯及2部货梯兼消防电梯提供垂直交通服务。客梯分高、中、低三区，低、中区各五部，高区四部，分别到达塔楼的高、中、低区办公室。此外还增设一部电梯与两部消防电梯直达地下库。另外，在裙房内设置了一组自动扶梯，供一至二层使用，使商业面积区域的竖向交通更为便捷。设计中还充分考虑了残疾人的需要，于各主要入口处设残疾人坡道，使轮椅可进入电梯和残疾人卫生间等。

设计以简约主义作为本项目的建筑语言，以现代简洁作为建筑风格，追求形式反映功能的建筑理念，注重功能和艺术的有机结合，摒弃一切纯粹为造型而设计的多余构件，以精致的细部、合理的比例、适中的体量，创造了一个简洁明快、高雅挺拔、富有强烈时代感的现代建筑。

建筑横向与竖向线条的处理流畅统一，精细而富有细腻的质感。外形稳重大方，平面方正实用，与体形丰富色彩鲜艳的市民中心形成对比，隐

In addition, 1 specific elevator and 2 fire elevators go to underground directly. A set of escalators in podium serve customers who head to floor 2 from floor 1, and make the vertical transport of business area more convenient. The design of this building takes the need of disabilities into consideration, and set ramps at each main entrance to enable wheelchairs enter elevators and toilets.

Simplicity is the architectural language of this project, whose style is modern and simple, pursues the concept that form reflects function, emphasizes organic combination of function and art, removes extra component serves shape only, and creates simple, elegant and modern architecture with exquisite details, reasonable ratio as well as appropriate volume. The horizontal and vertical lines of this building are smooth and consistent, exquisite and textured.

The generous shape and functional plan contrast with the various shaped and colorful Citizen Square, and hide itself among the axis of urban planning. The top of the building follows the simple style, and makes blinded metal blades on the top of the roof garden, which functions as outside observation platform with excellent view, as well as decreases the temperature of the roof surface to save energy.

External wall is made of high quality glass and aluminum curtain, works with the metal sunshade frame; low-reflection glass, aluminum and metal lines form a transparent light high-tech image. The project uses low-radiation filmed hollow glass, take its advantage of illumination and de-conduction, remove the cold feelings of common reflective glass curtain wall, prevent the neighborhood from being polluted by light, as well as decrease the load of air-conditioners and consumption of internal lamps. The hollow glass isolates noises from busy transportation of urban main roads, indicates the environment friendly and energy saving concept.

Design of environment follows the entire Architecture Design concept, which is composed of pure geometrical surfaces. The exquisite decoration is harmonious with the entire building, simple and neat environment is dotted with few sculptures in the pond, which makes the project much more dynamic.

Only one entrance for vehicles is located in the east of Pengcheng fifth Road, which is accord with the principle of splitting people and vehicles. Vehicles take customers at the entrance of the site, and go to underground parking lots through ramp, that makes the extinguishing vehicles reach the four sides of the building directly under emergent condition. Main entrances for people are located in the west and south of Jintian Road and Futianzhong Third Road, which make the people go to office in the most convenient way. Taking the location of subway entrance and exit into consideration that most people enter from the west of the site a square for vehicles is added at the southwest corner in order to meet the needs of splitting people and vehicles.

于城市规划的“龙”脊线之中。建筑顶部处理仍沿袭简约的设计手法，在屋顶花园顶部设计了百叶金属板。屋顶花园既可作为视野开阔的室外观景台，又可降低屋顶表面的温度，利于节能。

建筑外墙为高质量的玻璃与铝板相间的幕墙，结合金属遮阳构架，以低反光度玻璃、铝板和金属线条建立一个通透轻盈的高科技形象。本方案采用“低辐射击镀膜中空玻璃”，取其透光不透热的特殊功效，避免了普通反光玻璃幕墙冰冷的感觉和对周边环境的光污染，同时也有助降低空调负荷和室内灯具耗电量。中空玻璃的设置隔离了城市干道繁忙交通的噪音，反映本项目的环保节能概念。

环境设计配合整体建筑设计理念，以单纯的几何平面构成，精细的质感铺装与建筑浑然一体，简洁整齐的环境结构中点缀少许水池雕塑，使整个项目显得更加生机盎然。

以人车分流互不交叉的原则为指导，根据规划要点，仅在东侧鹏城五路中设一个车行出入口，车流进入地块后可在大堂直接落客，然后通过坡道进入地下车库。大厦周围结合环境设计设置环形消防车道，使消防车在紧急情况下，能不受阻碍地接近建筑物的四个扑救面。主要人流入口设于西面和南面的金田路与福田中三路，以最便捷的方式让人流进入办公楼。考虑到地铁出入口的位置，主人流方向是从地块西面进入，又在地块西南角设车行广场，满足人车分流的需胥求。

高区热销
83116666

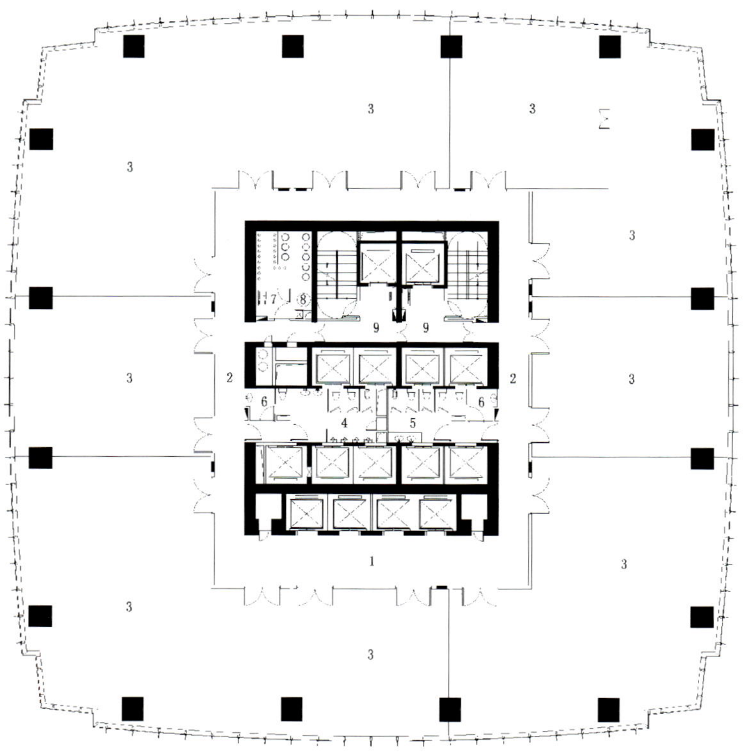

1. 电梯厅
2. 走道
3. 办公
4. 男厕
5. 女厕
6. 行政人员卫生间
7. 开水间
8. 清洁间
9. 合用前室

三至十四层平面图

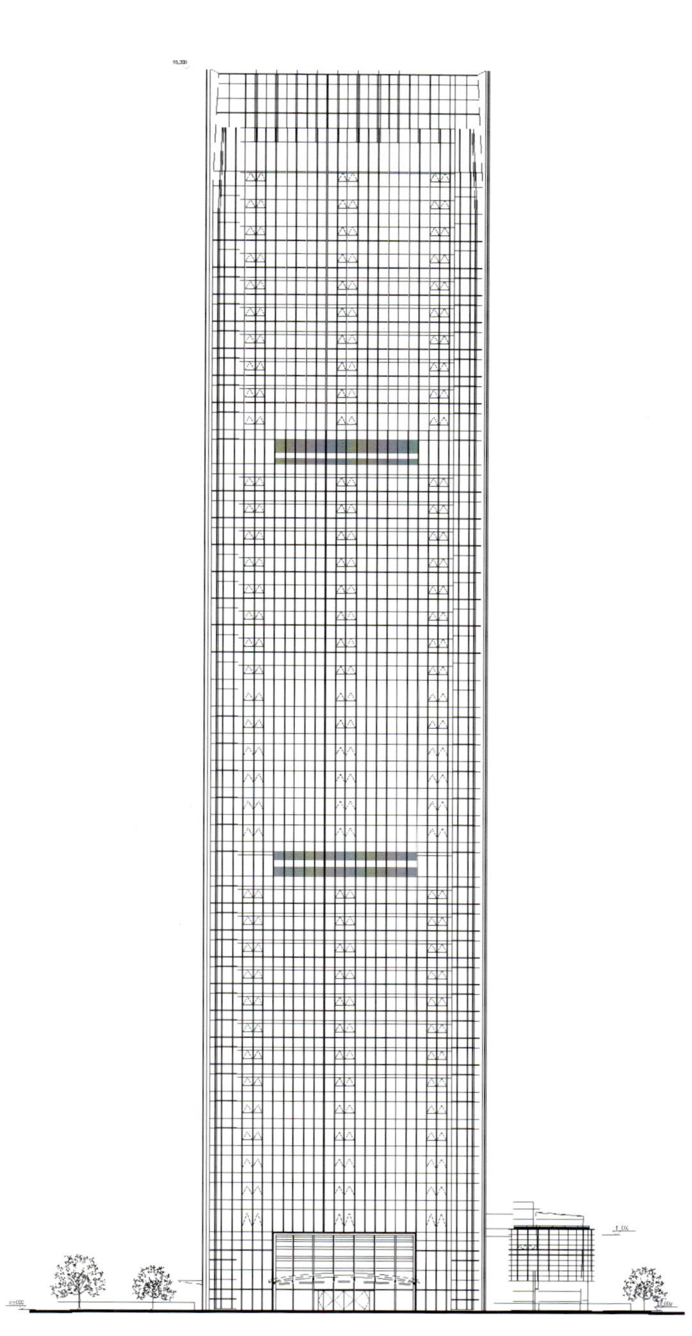

南立面图

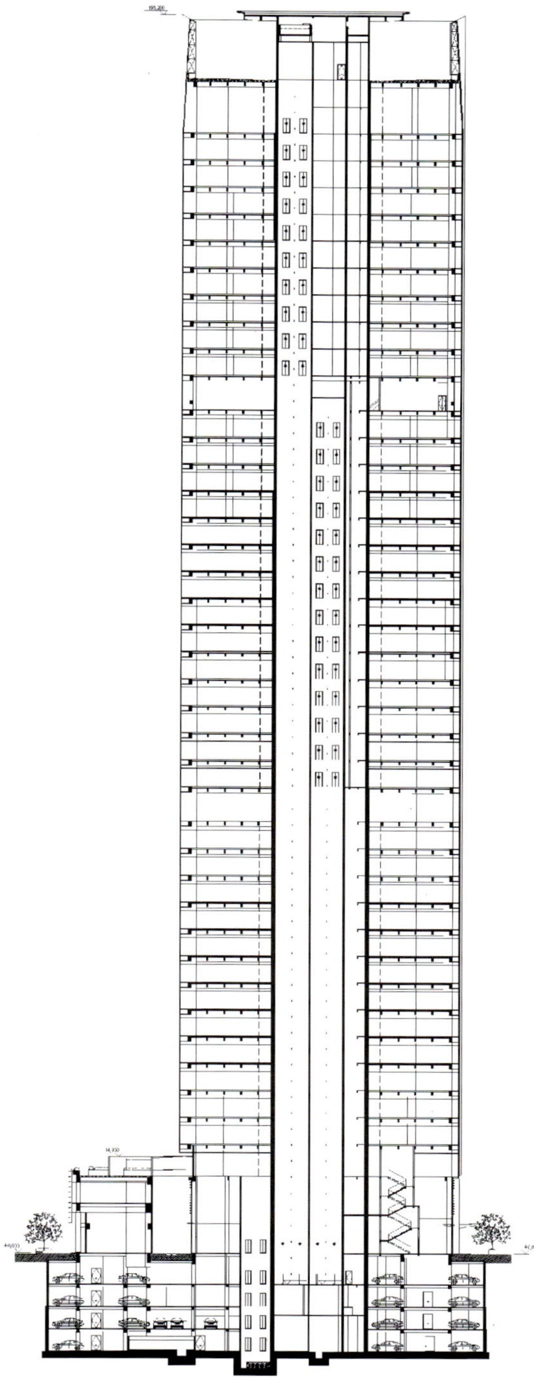

剖面图

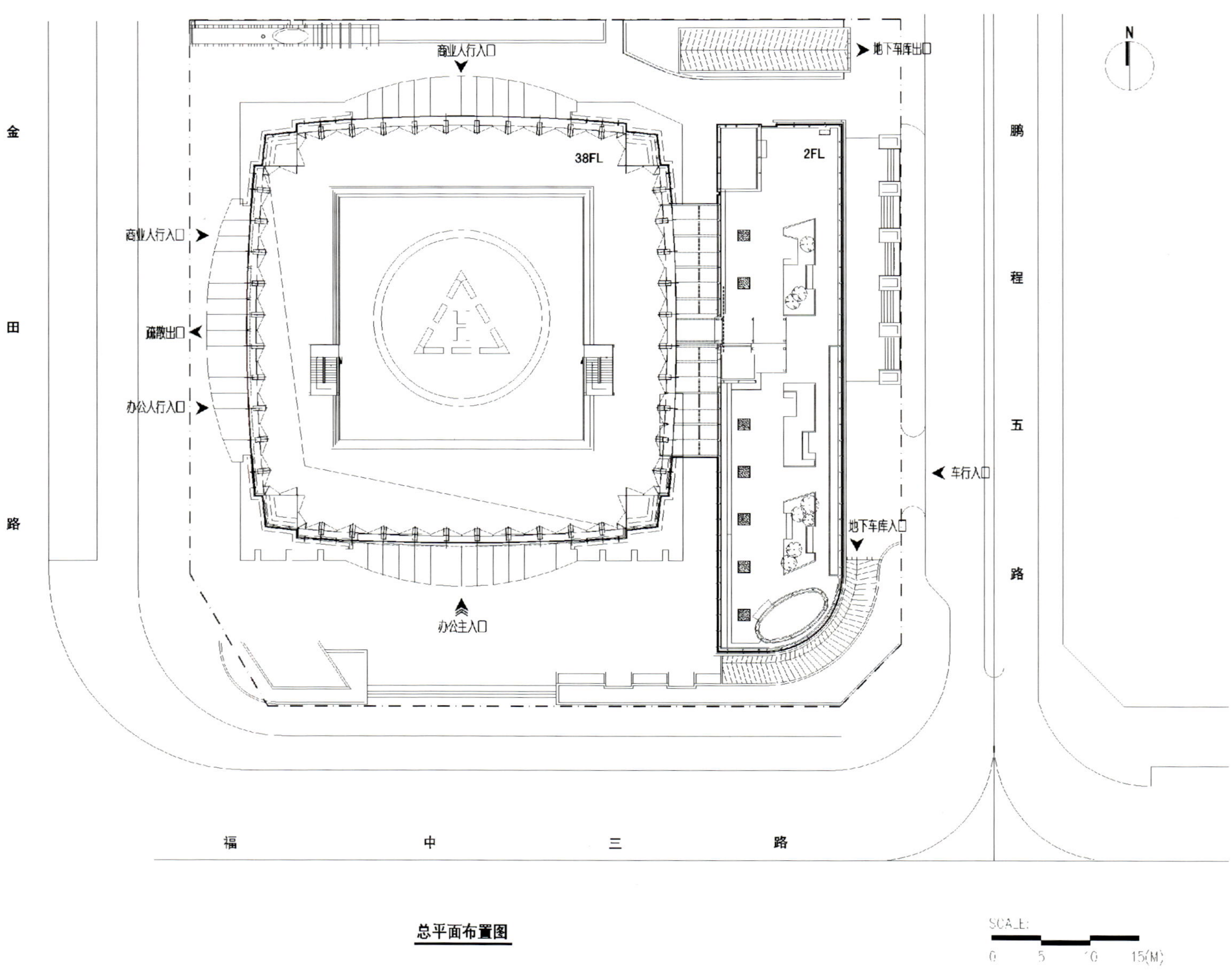
商业人行入口
地下车库出口
N
金田路
38FL
2FL
鹏程五路
商业人行入口
疏散出口
办公人行入口
车行入口
地下车库入口
办公主入口
福中三路
SCALE:
0 5 10 15(M)

总平面布置图

14
1

38
15

38
15
14
1

Yingu Mansion

银谷大厦

Real Estate Developer: Yingu Real Estate Co., Ltd.
Architecture Design: Dadi (International) Architecture Design Office
Project Location: Zhongguancun Science & Technology Park of Beijing City
Site Area: 12,182m²
Building Area: 80,540m²

开发商：银谷地产
建筑设计：大地（国际）建筑设计事务所
项目地址：北京中关村科技园区
用地面积：12,182m²
建筑面积：80,540m²

Located in the northeast cross corner of North Fourth Ring Road and Tsinghua South Road, Haidian District of Beijing City, also the core area of Zhongguancun Science & Technology Park. It is adjacent to urban greenbelt of North Fourth Ring Road in the south, urban planning road in the north, and Tsinghua South Road in the west. The mansion is connected to the main roads of urban traffic, and isolated by greenbelt, which makes a relatively quiet working environment with convenient traffic.

In order to solve the common issues among domestic medium and high grade office building, such as enclosed space, low storey height, limited public space, this project set the transportation core and internal aisle in the center, to increase the effective usable area and improve the flexibility. Meanwhile, a courtyard for ventilation and daylight is put in the center, to avoid defects of an internal courtyard. A 300m² sky garden is set every three floors to connect to the internal courtyard, forms a space sequence composed of outside space, lobby, group public space (i.e. sky garden) office. In this way, the internal corridors

银谷大厦位于北京市海淀区北四环路与清华南路交叉口的东北角，地处中关村科技园区的核心区，南临北四环路城市绿化带；北临城市规划路；西临清华南路。大厦与城市交通主干线相通，又有绿地相隔，营造了交通方便又相对安静的工作环境。

针对目前国内中高档写字楼内部交通空间封闭、建筑层高较低、公共活动空间少等问题，本项目居中设置了交通核及内走道，以提高建筑的有效使用面积和平面灵活性。同时，在大厦的平面中间设有一个通风采光天井。为避免以往内天井通风不畅、采光不亮的缺陷，又每隔三个楼层设置一个300余平方米的空中花园与内天井连接，形成了一个室外空间—楼门厅—组团公共空间（空中花园）—办公室的空间序列。这样不仅使大厦内部走道明亮并有很好的通风效果，也为楼内的工作人群提供了休闲交往空间，利于工作人员恢复疲劳，提高工作效率。同时，各个空中花园将楼内的楼层有机地连接为一个整体，使大厦内部既有相对安静独立的各楼层空间，又有浑然一体的立体空间，极大地增加了空间的识别性、趣味性及完整性。

look bright and well-ventilated, and staff working here is provided a leisure communication space, which reliefs them from exhaustion as well as improve their efficiency. Each sky garden integrates each floor organically, forms a quiet and separated space in each floor, as well as an entirely stereoscopic space, and enhances the identification, fun and integration of the space.

The mansion has 21 floors above the ground, is set along the east-west direction, and equips two relatively independent transportation cores inside. One core is composed of a guest panoramic elevator, 3 ordinary guest elevators, an internal guest-cargo and fire elevator, and a smoke-proofing stairs; the other is composed of 4 guest elevators, an internal guest-cargo and fire elevator, and a smoke-proofing stairs. Toilets, barrier-free toilets, pantries and special pipe shafts and ventilation shafts are equipped within the transportation cores.

A sky garden is set at the 4th, 7th, 10th, 14th, 17th, and 20th floor, whose height equals 3 floors, connects the courtyard to outside space, and solve the issues with ventilation and daylight. The 6 green shared spaces break the barrier of communication among staff, provide both staff and guests with communication and leisure space, and bring the serious business space with quiet and reserved beauty.

The 1st floor is 5m high, which has main entrance and exit in the south, and sub entrance and exit in the north, and both of them are connected to the lobby of the mansion. The first and second floors are connected to the internal courtyard through the lobby. An entrance and exit is set on both east and west side, which goes directly to the transportation core, and connects to the lobby through the corridor. The entrance and exit located in the west is for pedestrians, a duty room and business room are set there; the one located in the east is for internal staff, a fire monitoring room is set nearby; the other rooms closed to external walls on the first floor are used as commercial rooms for independent operation, whose doors can be open to the outside. The 2nd floor is 4.2m high, which is almost composed of commercial or exhibition rooms besides the transportation cores as well as a few rooms for necessary devices. The 3rd to 21st floors are 3.5m high,

大厦地面共21层，呈东西方向布置，在大厦平面内设有两个相对独立的交通核。其中一个交通核内各设有乘客观光电梯1部、普通乘客电梯3部、内部客货兼消防电梯1部、防烟剪刀楼梯1部。另一个交通核内设客梯4部、内部客货兼消防电梯1部、防烟楼梯1部。交通核内同时设有卫生间、无障碍卫生间、开水间及各专业管道井、通风井。

两个交通核之间为通风采光的内天井，天井贯通各层，在一、二层与大厦门厅连为一体。从四层开始，每三层与不同的空中花园连通一次。为避免天井高度与宽度比例失调，在十三层处布置会议室将内天井隔断，分为两个天井。

大厦的四层、七层、十层、十四层、十七层、二十层分别设有一个空中花园，花园贯通三层高度，将内天井与室外空间连通，有效地解决了通风和采光问题。六大绿色共享空间，突破了写字楼人际交往的壁垒，既为办公人士和到访者提供了自由交流和休息休闲的空间，也为严谨沉谧的商务空间注入了内敛宁静的理性之美。

大厦一层层高5m，南面设有大厦主出入口，北面中间设有副出入口，两个出入口均与大厦门厅大堂连通。大堂则贯通一、二两层并与其上空的内天井相连。东西侧各设有一个出入口直通交通核，并通过走廊与门厅大堂连接。在设计中将西侧的出入口定位为步行人员出入口，在这个出入口处设有值班、商务房间；将东侧的出入口定位为内部人员出入口，在其门厅附近设有消防控制室等；一层其他临外墙的房间定位为可独立经营的商业用房，并可以开设独立的直接对外的大门。大厦二层层高4.2m，除交通核和必要的设备用房面积外均考虑做为商业或展示用房。三层至二十一层层高3.5m，均考虑为大空间的灵活写字间，每隔两层在平面东、西端头设有两个小的新风机房；同时在每层东、西两端的南北面分别设置四个带卫生间的高级办公室。十三层平面中间部位设有三个会议室，将内天井断为两个。二十一层层高6.6米，既可以做为高空间会议室，又可以用轻钢楼板分隔成两层办公室。大厦的顶部共设3层空间，第一层为电梯缓冲层；第二层为电梯机房层；第三层为水箱层和部分设备用房。在两个交通核之间的

which all function as large and flexible office, and equips a tiny new fan room on both east and west ends every 2 floors; and sets 4 high grade offices with toilets in south and north side of the 2 ends. The 3 meeting rooms in the center of the 13th floor divide the internal courtyard into 2 parts. The 22nd floor is 6.6m high, which can be used as either a meeting room or 2floor offices divided by floor plate made of light steel. The top of the mansion is composed of 3 floors, the 1st one functions as buffering layer for the elevators, 2nd one functions as motor room for the elevators, and the 3rd one functions as both water tank layer and device rooms.

The buffering layer of elevators located between 2 transportation cores, are made a large meeting room of light thermal insulation materials; and the meeting room can hold about 300 people. The buffering and motor layers of elevators are nearby the external walls with 4 independent roof villas, whose internal area is 286m²; the 2 layers are connected through internal stairs, and provide separated roof platform. Roof water tanks and satellite TV machine rooms are set on the water tank layer of 2 transportation cores.

There are 3 floors of Yingu Mansion underground the ground; the B3 is 3.6m high, which function as 6grade defense storage warehouse during the war, as well as parking lot in peacetime. The external defense space between 2 transportation cores is used as water tank for fire extinguishing and daily life, pumping room and water system room. B2 is 5m high, equips with rooms to hold necessary devices, such as electricity distribution room, freezing room, staff restaurant, bathroom as well as leisure room. B1 is also 5m high, equips with service rooms, such as fitness club, sauna bathroom and restaurant besides the room for necessary devices. An independent parking lot is located below the south of mansion building, 3mdepth earth covering is above the top plate of the parking lot, which meets the needs of greening as well as extension of the pipes. Vehicle ramps are located on both side of the basement, and let vehicles get to the ground from B3.

Located in the center of Zhongguancun High-tech Park of Beijing City, the mansion is right on the intersection of 2 main urban roads of North Fourth Ring Road and Tsinghua South Road. Therefore, the stereoscopic effect really matters for urban landscape. The façade of the mansion focuses on both office image and modern style. A narrow-up glass curtain wall is used to form the south side, and various sky gardens make the façade serious and vigorous. Taking entire effect and the weather of North China into consideration, the north façade is designed as evenly separated windows and rhythmic stereoscopic shape. General evolution from virtue to actuality is made from south to north on the east and west façades. A 4mwidth ring road is around the mansion, which meets the needs of internal transportation as well as climbing solution for fire extinguishing. The collected urban greenbelt is made as a parking lot, which solves the problem of ground parking. 2 vehicle ramps getting to underground parking lot are set nearby the internal ring road, and lead some of the vehicles to underground parking lot. The existing roads in the south of this mansion connect the internal ring road to urban main road, i.e. North Fourth Ring Road; a sub entrance and exit for vehicles is set in the north, which leads to urban planning road. Convenient traffic network is made to integrate internal ring road and urban traffic organically. An entrance and exit for pedestrians is set in the west side, where leads to Tsinghua South Road, and is connected to the sub entrance of the mansion.

电梯缓冲层部位用轻质保温透光材料构筑一个大空间会议室，可同时容纳300人左右。在两个交通核的电梯缓冲层和电梯机房层的靠外墙部位，设有四个独立的屋顶别墅，每处别墅户内面积为286m²，用户内楼梯连通两层，并拥有各自独立的屋面平台空间。在两个交通核的水箱层，则分别设有屋顶水箱和卫星电视机房等。

银谷大厦地下共3层。地下三层层高3.6m，作为战时六级人防物资库，平时为停车库。在两个交通核之间为人防外空间，作为大楼的消防、生活水池、相关泵房及大楼中水系统用房。地下二层层高为5m，布置了变配电室、冷冻机房等大厦的主要设备用房，同时还布置职工餐厅、职工浴室及休息用房等。地下一层层高也为5m，除布置一些必要的设备用房外，并利用本层层高较高的特点，设有部分直接对外的服务用房，如健身、桑拿、餐厅等。在大厦主楼地下的南侧—代征绿地下设有一个独立的停车库，同时确保停车库顶板上有3m深的覆土，既可以满足绿化要求，又使地下出户管道得以出户。大厦主楼地下室的两侧分别设汽车坡道，坡道自地下三层通至地面。

大厦地处北京市中关村高科技园区的中心地带，位于北四环和清华南路两条城市主干道的交汇点上，因此立面效果的成败，对城市景观影响较大。在大厦立面设计过程中，兼顾庄重朴实的办公形象和活泼变化的现代化风格。大厦南面用向上弧线渐收的玻璃幕墙，并利用空中花园的变化，形成庄重且不失活泼的立面效果。北面考虑整体效果需要及北方天气的现实，将北立面设计为以均匀独立的小窗为主调的富有韵律的立面造型。东、西两面实现由南向北的立面变化，很自然地实现了由虚变实的渐变。

大厦四周设有一条4m净宽的环形道路，既能满足内部交通问题，又可作为消防环路解决消防登高要求。同时利用代征的城市绿地作成绿化停车场，解决地面停车的问题。在内部环路旁边，设两个通向地下车库的汽车坡道，将一部分停靠车辆引入地下车库。利用大厦南侧现有道路，将内部环路与城市主干线——北四环连接起来，并在北侧向城市规划道路开设一个汽车辅助出入口。这样就将大厦内部的汽车环路与城市交通有机地连为一体，形成一个便利的交通网络。考虑人车分流的问题，在大厦西侧通过清华南路的人行道，开通一个大厦的行人出入口，与大厦副入口相连接。

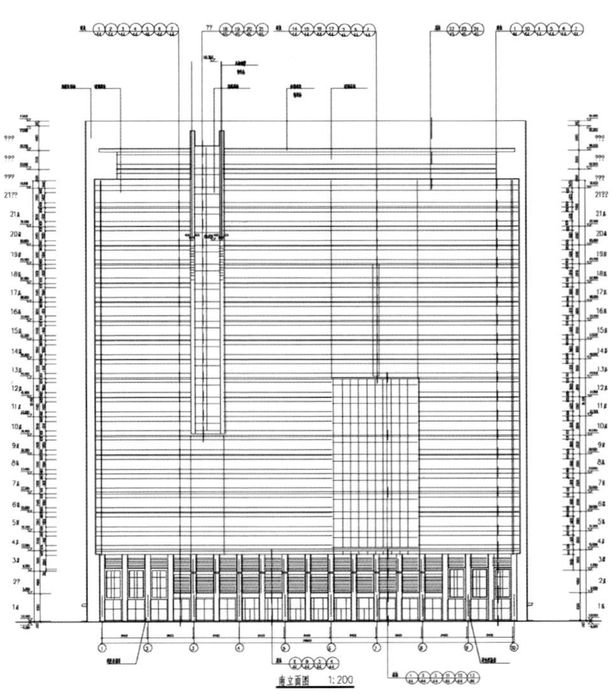

南立面图 1:200

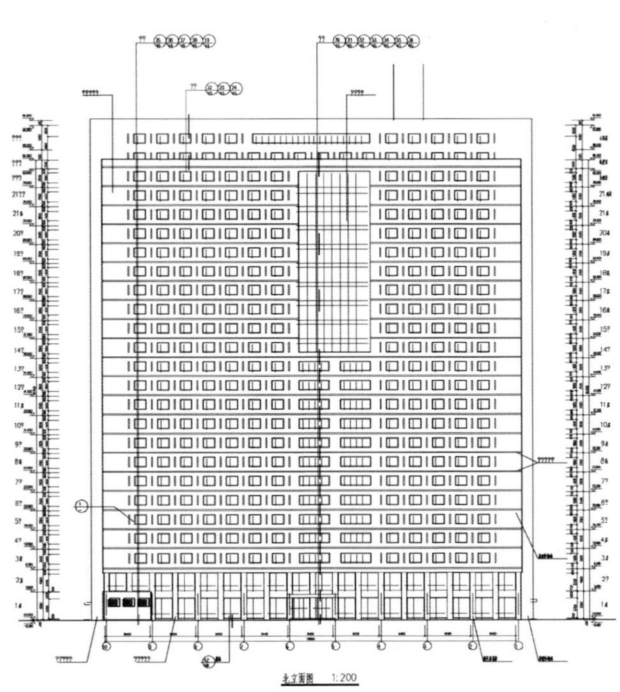

北立面图 1:200

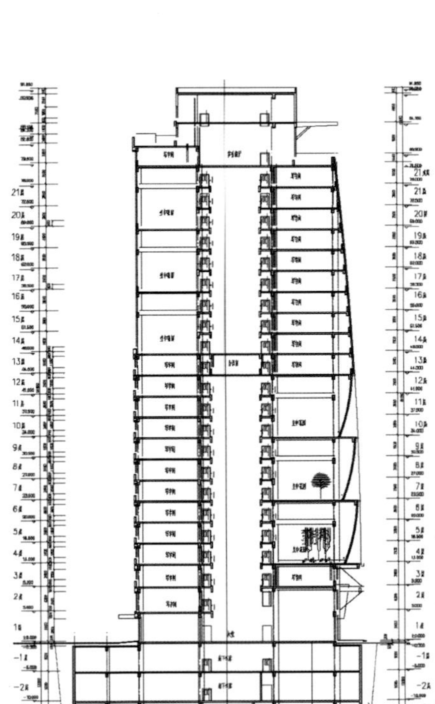

1-1剖面图 1:200

1 1:20

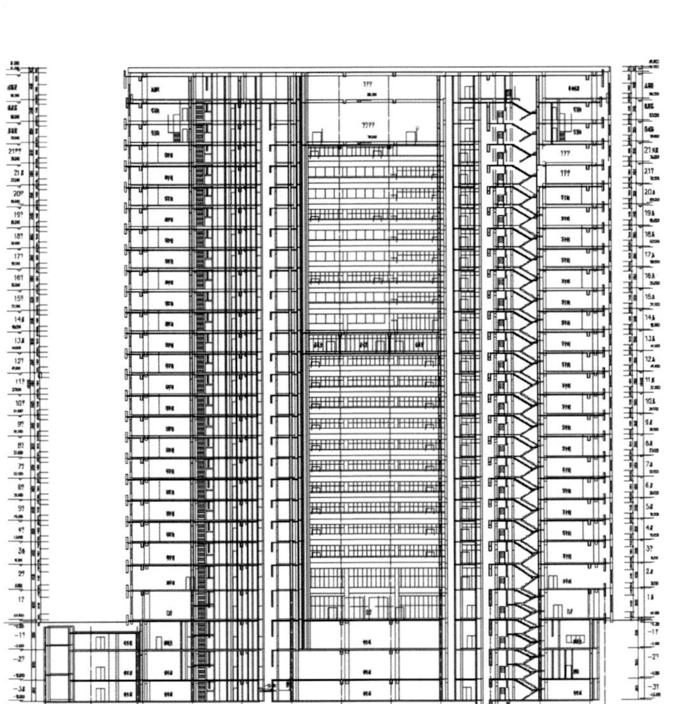

2-2剖面图 1:200

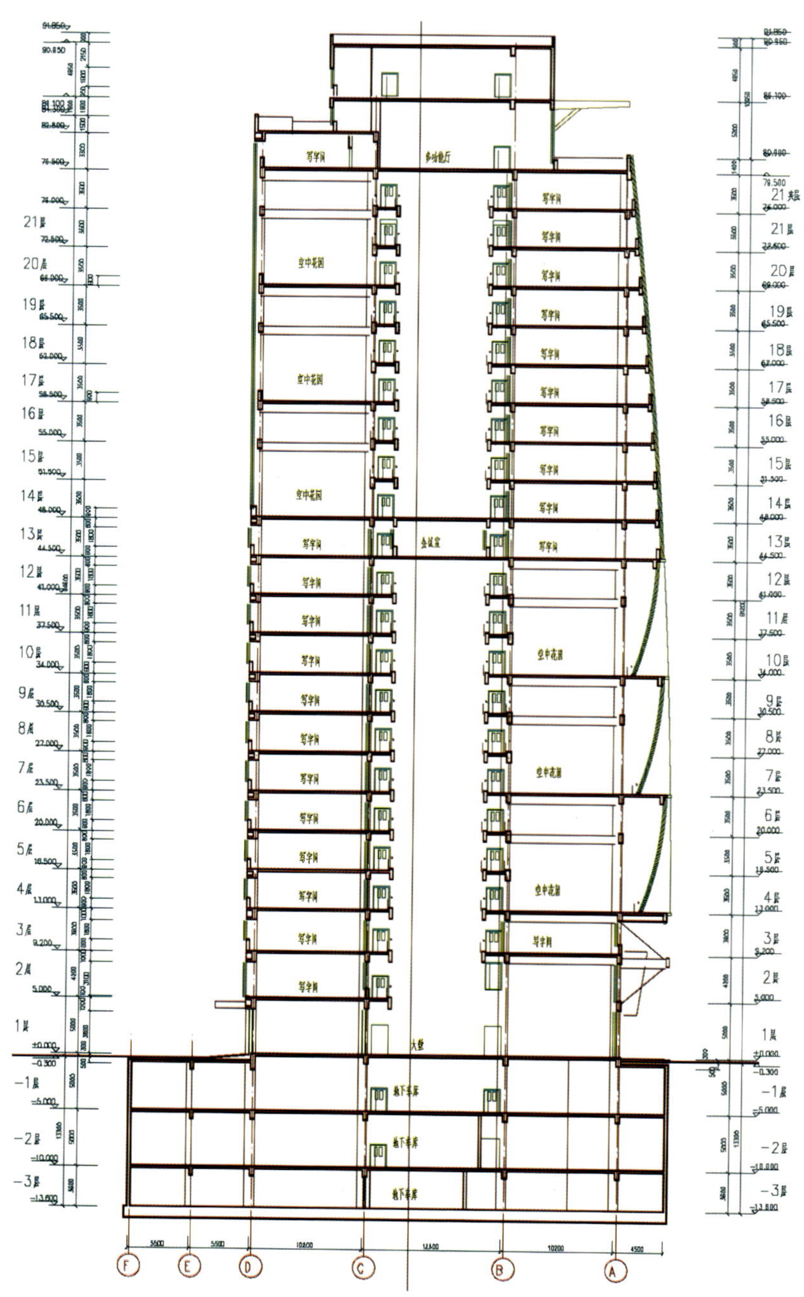

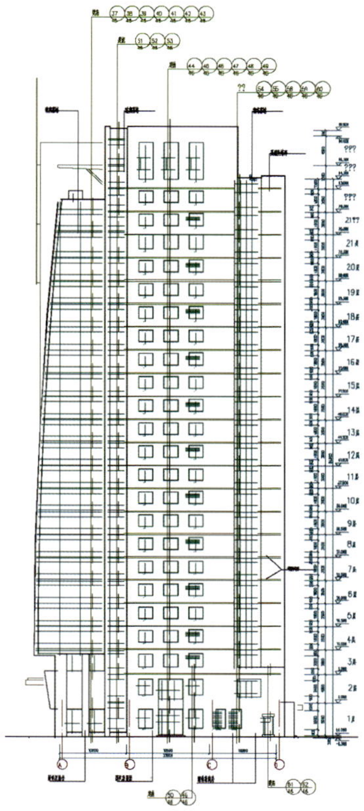

东立面图　1:200

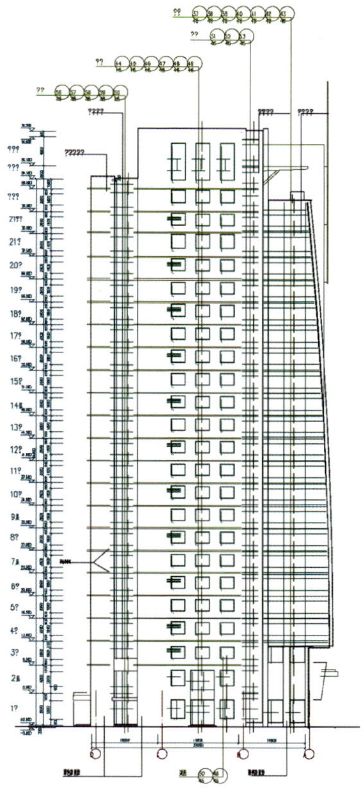

西立面图　1:200

China Mobile Mansion

中国移动大厦

Site Area: 12450.53m²
Building Area: 63474.67m²
Number of Floor: 3 underground, 16 above the ground
Building Height: 70m
Architecture Design: Orient Zhuzhong Construction Planning Design Co., Ltd.
Construction Operation: The 3rd Construction Company of the 2nd China Construction Bureau

占地面积：12,450.53m²
项目规模：63,474.67m²
楼层：地下三层，地上十六层
建筑高度：70m
设计公司：北京东方筑中建设规划设计有限公司
施工单位：中国建筑二局第三建筑公司

Beijing Mobile Communication comprehensive building is located in the northwest corner of East Second Ring Road and Haiyuncang Road, which is adjacent to the west sub road of East Second Ring in the west. The site is 135m long in east-west, and 92m wide in south-north, usable area is up to 12490.53m², building planning height is limited to 70m, and north of a planning public buildings group, south of 2 high-rise tower apartments.

The load of comprehensive and motor buildings of Beijing Mobile Communication is quite large, both architecture and devices demand more on structural beam section, stiffness, refractoriness and earthquake resistance performance, and shaped steel concrete frame as well as the shear wall structure achieve various targets. A great quantity of split long columns exists in office building, the 4 layers of plate at the bottom weaken down, and the transformation of inclined columns also suffers from difficulties. Shaped steel concrete structure enhances the security and extension, improves earthquake resistance performance; grouped beams relief self gravity and decrease the demand of steel.

北京移动通信综合楼位于东二环路与海运仓路的西北角，西邻东二环西辅路。基地东西平均长约135m，南北平均宽约92m，红线内用地面积12,490.53m²，规划建筑控高70m。基地北侧为规划中的公建群体，南侧为两幢高层塔式住宅楼。

北京移动通信综合楼机房楼楼面荷载大，建筑及设备对结构梁柱断面、刚度、耐火及抗震性能均有很高要求，型钢混凝土框架、剪力墙结构可以很好地满足各项指标要求。办公楼存在大量错层较多的长柱，底部四层楼板削弱严重，且存在斜柱的转换问题，采用型钢混凝土结构可以增加结构安全储备和延性，提高结构抗震性能、采用组合梁则有助于减轻自重，节约用钢量。

中国建设银行

Easy Home Tower

居然大厦

Real Estate Developer: Easy Home Investment Holding Group
Architecture Design: Zhongtian Wangdong International Design Consultant Co., Ltd.
Building Area: 620,000m^2

开发商：北京居然之家投资控股集团
建筑设计：中天王董国际工程设计顾问有限公司
建筑面积：6.2万m^2

Easy Home Tower is a 5A A-Grade office building invested by Beijing Easy Home Investment Holding Group, which located in golden area of Beijing City, is among the intersection of 6 CBDs in the east city. It provides convenient transportation, impressive popularity and distinctive location.

Easy Home Tower has 18 floors above the ground and 3 floors underground, whose site area is 8400m² and building area is 620,000m². The project was launched on March 26th, 2006, and structural capping was achieved on August 8th, 2007, then the entire building was completed on December 31st, 2008, finally is enabled to use today.

The followings are features of Easy Home Tower comparing with similar projects:

1. Architecture Design is novel, unique and modern. The sharp section and crossed lines show the dynamism and vigor of the building, the embracing and dancing image of 2 human bodies symbolizes harmony.

The principle part of Easy Home Tower uses steel structure, which realizes a

居然大厦是由北京居然之家投资控股集团投资兴建的5A甲级写字楼。大厦位于北京黄金地段，跻身于北京城东六大核心商圈交汇处，交通便利，人气十足，位置显赫。

居然大厦占地8400m²，建筑面积6.2万m²，地上18层，地下3层。工程于2006年3月26日正式破土动工，2007年8月8日实现了结构封顶，2008年12月31日整体竣工，即将正式启用。

居然大厦与同类项目相比具备以下几方面特点：

一、建筑设计新颖、独特，充分体现了现代与时尚。大厦利落的切面与相互交错的线条描绘出建筑的动感与活力，两个形体相拥起舞的形象寓意着“和谐”。

居然大厦主题采用了钢结构，实现了12m大跨度柱距，大大提高了楼层使用空间；办公楼层高3.9m， 精装修后净高可达到3m，使得办公环境通透、舒适。

huge 12m span between columns, and increases the usable area of each floor. Each floor is 3.9m high, and 3m high after fine decoration, which makes the office environment transparent and comfortable.

2. Construction materials are environment friendly, and energy saving. The façade is made of low radiation LOWE hollow laminated glass curtain wall, and shows the environment friendly and energy saving concept completely.

3. Devices and equipments are advanced and high-grade. 8 international top branded elevators imported from Finland KONE are used to serve vertical transportation; in addition, 2 cargo elevators as well as 4 escalators are equipped. Direct-fired units from Japan Ebara are provided for cooling and heating during the whole year. Perfect 5 A smart control system, composed of high-speed fiber, networking, remote controlling, fire extinguishing, security smart monitoring shows extreme intelligence of the mansion.

二、材料的选用环保、节能。大厦主体采用纯钢结构，外立面选用低辐射LOW－E中空夹胶玻璃幕墙，充分体现了环保、节能的理念。

三、设备、设施先进、高档。大厦选用了8部国际顶级品牌纯进口芬兰通力电梯，用于大厦垂直客流运行服务，另外还配有2部货梯和4部自动扶梯；大厦配备的日本荏原品牌直燃机组，可全年自主提供冷热源；大厦拥有完善的5A智能化控制系统，从高速光纤、网络接入，到楼宇远程控制，消防、安防智能控制等，处处体现了大厦的高度智能化。

克缇生活馆
Kelti 克缇
克缇生活馆

New Poly Plaza

新保利大厦

Architecture Design: American SOM Company
Site Area: 10,600m²
Building Area: 103,308m²

建筑设计：美国SOM公司
占地面积：10,600m²
总建筑面积：103,308m²

New Poly Plaza is a master piece of work invested by China Poly Group, a top class, A-grade office building which has 27 floors in total, 23 floors above the ground and 4 floors underground. The project is located in the southwest corner of east fortieth bridge, center of East City District of Beijing, intersection of East Second Ring Road and Pingan Street. It is endowed with excellent position as well as convenient transpiration. Besides international standard offices, the mansion also provides boutiques, high-grade restaurants, fitness clubs, etc; in addition, Poly Art Museum, which is famous for purchasing national treasures from the Old Summer Palace, is going to move to is mansion. New Poly Plaza enjoys the convenience of peripheral transportation, and business equipments, it seeks peace and quiet in noisy surroundings and meets the need of high-grade enterprises offices. The project is designed by American SOM Company, shows the spectacular style and is called "a masterpiece handed down for generations" through the largest flexible glass curtain wall of the world, a suspension villa, the first vertical shielding stone blinds in domestic, and the top

新保利大厦是中国保利集团在北京地产的倾心力作，为高档甲级写字楼，总27层，地上23层，地下4层。项目位于北京市中心东城区东四十条桥西南角，地处东二环与平安大街交界处，地理位置优越，交通极为便利。大厦除了国际标准开间的写字楼外，还配套有精品店、高档餐饮、健身中心等，以收购圆明园国宝而蜚声海内外的保利艺术博物馆将迁至大厦内。新保利大厦既享受了周边交通、商务配套的便利，又闹中取静，很适合高档企业的办公需求。新保利大厦由美国SOM公司精心设计，在建筑上体现了大气磅礴的丘城气质，拥有世界上最大的柔索玻璃幕墙、悬浮式吊楼、国内首例竖向遮光石材百叶、北京第一中庭等建筑特点，被誉为"传世之作"。大厦在内部功能上达到了世界级写字楼水准，同时强调了自然生态、高效节能的人性化标准。充分体现现代化人文办公、绿色办公的理念。通过灯光语言对建筑的二次诠释，使建筑的恢弘气宇、民族精神、历史文化等特点得到了完美展现。

HARVEST

one courtyard of Beijing. The internal functions achieve international standard office building, emphasis nature, ecology, efficiency and energy saving, and show the concept of modernized cultural and environmental friendly office. The spectacular image, national spirit as well as historical civilization is exhibited perfectly through the interpretation of lightening language.

CB

杭州银行

Shanghai Rainbow Bridge Silver City Far East Internati

上海虹桥银城远东国际

Real Estate Developer: Shanghai Lianda Real Estate Co., Ltd.
Architecture Design: Hualei & Maiji Associated Architecture Office
Extension and Construction Drawing Design: Shanghai Second Light Industry Architecture Design Institute & Shanghai Environmental Engineering Institute
Project Location: No. 933, Zhongshan West Road of Shanghai City
Site Area: 3,626m^2
Building Area: 29,181m^2

开 发 商：上海联大房地产有限公司
建筑设计：华雷迈基联合建筑事务所
扩初及施工图设计：上海二轻建筑设计研究院
上海环境工程设计研究院
地理位置：上海市中山西路933号
占地面积：3,626m^2
总建筑面积：29,181m^2

The site is narrow and irregular, adjacent to raised highway of Shanghai Zhongshan West Road in the west. General plan takes the advantage of underground space, makes various sub systems, such as power supply, communication, parking lots into position based on the decision of building position, environmental greening, internal & external transportation. Among hundreds of worldwide cases, it is important that the design of image and façade of the high-rise office building should be novel, which is achieved by investigated the internal & external architectural environment completely. Rainbow Development Zone has been established and operated for 20 years, where locates various style of high-rise buildings, and the site is adjacent to Rainbow and Silver River Hotels, it is not easy to be creative on Architecture Design.

Through repeated comparison and validation, it is decided to extend underground of the building, and to form a triangle architecture surface, whose bottom is on the north of this site, and to make the main façade face to Zhongshan West Road,

基地西临上海市中山西路高架路，占地狭小，地形不规则。总体规划设计在周密推敲地面的大楼定位、环境绿化、内外交通流线的基础上，充分利用地下空间，使各种相应的辅助系统如电源、通信、停车位等能各适其位。

综观世界各国案例，高层办公楼设计最主要的是造型和立面设计要创新，而创新的基础是对内外建筑环境的充分研究。由于虹桥开发区已建设经营20年，高楼林立、风格各异，基地又与虹桥、银河两大宾馆相邻，建筑创新颇有难度。

经多方案反复比较和论证，最后确定建筑地下外扩，地上以北为底的三角形建筑平面，主立面面向中山西路，与虹桥、银河宾馆隔路相望。规划设计强调个性、简洁、新颖、挺拔，充分体现时代特点，并在现代风格上有所突破。在此共识的基础上，选定为比例适度的“格子布”式、粗细搭配的全明框玻璃幕墙。实践证明，这样的立面和体型新颖，而且是本地区独一无二的属于较为成功的案例。

and Rainbow Bridge and Silver River Hotels can be viewed from the opposite side of the road. Planning design emphasizes individuality, simplicity, novelty and straightness, shows features of times, and breaks through the modern styles to some extent. According to the agreement stated above, transparent glass curtain walls shaped as appropriate grids are used. It is proved that its façade and shape is quite novel and unique successful case among local office buildings.

During the graphic designation of the triangle high-rise building, first of all the positions of office rooms and public transportation are settled; the core vertical transportation system and auxiliary part are set in the rear, and office spaces are set in the front. It features appropriate entrance and exit, convenient access, suitable appearance, efficient usable areas; the entrance is centered, main façade is symmetrical, and the rear is flexible; auxiliary stairs are hidden behind. In summary, the triangle graphic designation is integrated with the layout of basement organically, which makes various functions are arranged reasonably.

The lobby is 30m long, centered entrance is 12m deep, both sides are 7.5m deep, and the ceiling is 6m high. Business centers are set at both ends of the lobby; external walls are composed of of glass curtains, dark brown polished granite columns and low windowsills. The transitional business center entrances on both ends are decorated with glass curtains and frame-free glass doors. The internal walls, which face to the main entrance, are decorated with black-stripped natural marbles, and ten impressive decorative pieces are put on the left and right sides; the subject in the middle is made of white marbles, and asymmetric black-stripped blocks are put on the top and bottom. The floor of lobby is basically made of grey and white-striped polished granites, and few irregular black spots are arranged diagonally. The ceiling is basically decorated with white parallel beams, and dotted with striped wind gaps on the edge. The entire lobby is bright, contrastive, gentle, exquisite, and shows elegant, novel and high-grade style of business lobby.

The elevator hall at the bottom is the flow distribution center, which is connected to the lobby as well as forms a separate room itself. Design of the walls is consistent with the lobby, and decorated with fine aluminum alloy strips, makes the style of basically consistency and diversity in details. The design of floor is quite similar with the lobby, except a few variations in size, pattern and color, in order to achieve appropriate ratio as well as comfort. The ceiling of elevator hall is about 5m high, whose subject is composed of two curved hanging plates with several down lamps, and the lid is made of aluminum plates with dots; alarming and shower equipments are set inside. The slots on both sides function as ventilation blinds, the entire space is interesting and harmonious, provides the similar experience with the decoration of the lobby.

The sample storey is basically decorated with light plastic floors and white plasterboard ceiling; both sides of the corridor are decorated with plasterboard walls and frame-free glass doors. People could feel simple and bright as soon as entering the corridor from elevator hall. Rest rooms are decorated with rice polished tiles, light grey aluminum plastic panel wall and silver aluminum reticular ceiling, whose standard, style and toner are consistent with the subject. The design of sample rooms implements two different styles and decorations for office units, including glass partition, stainless steel accessories as well as fashion layout of furniture, which are all extremely impressive.

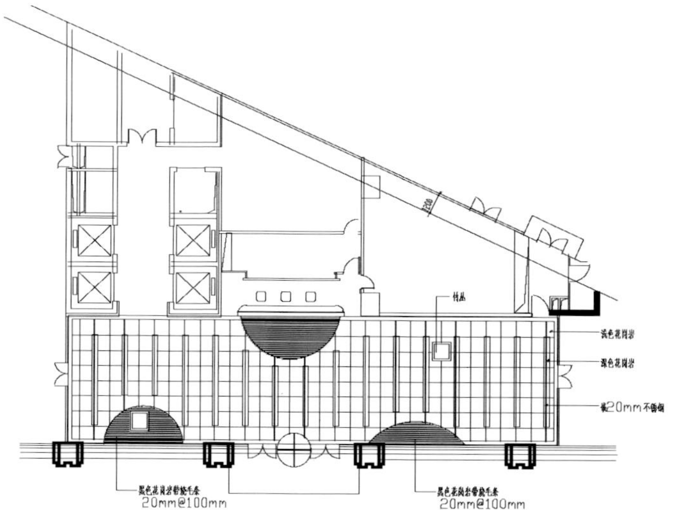

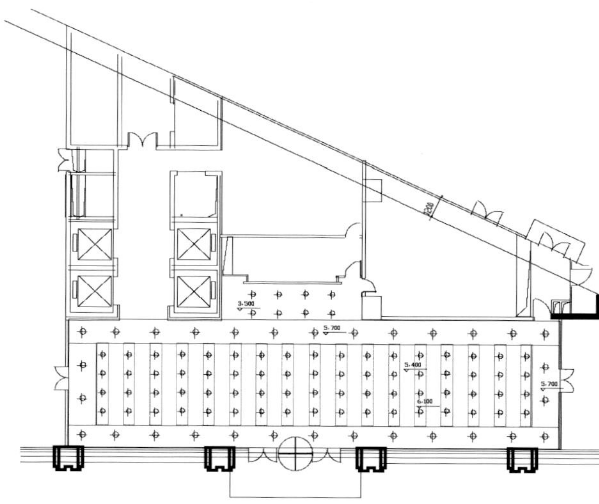

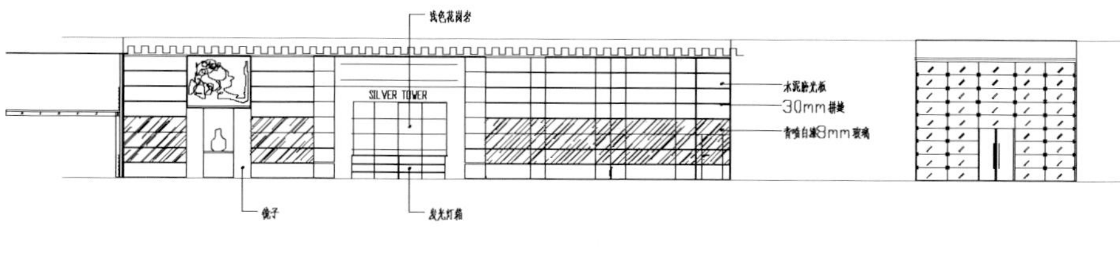

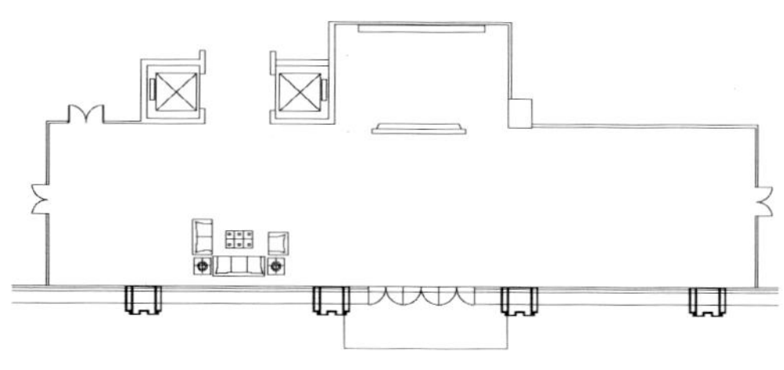

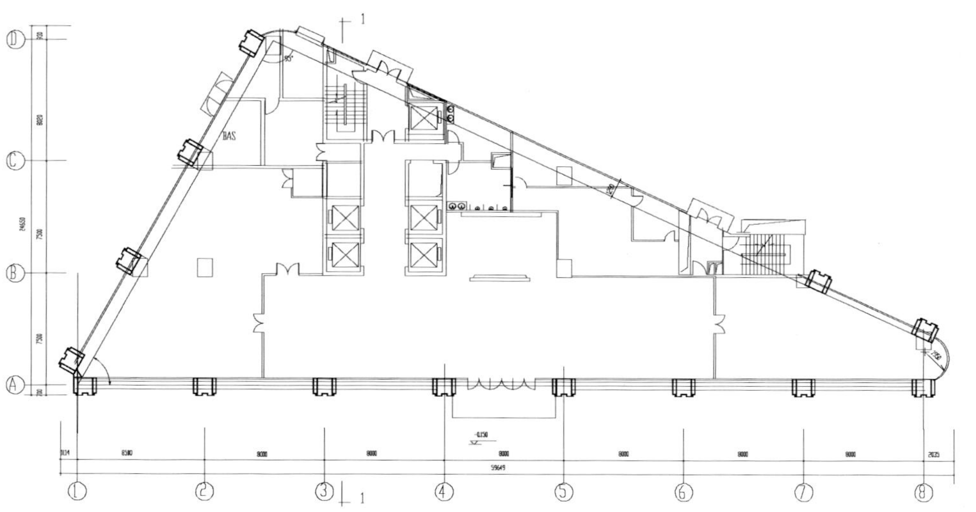

在三角形高层建筑平面设计中，首先确立办公房间和公共交通的定位，将最核心的垂直交通系统筒体和辅助部分设在后腰部位，前部主要部位是办公区域。这样，就显示出入口适中、进出便捷、外型得体、平面利用系数较高等优越性：底层中部设垂直交通系统，方便地上、地下联系，有利内外进出；入口居中，主立面可对称处理，后立面能区别对待；辅助楼梯间在背后比较隐蔽。总之，三角形平面的设计与地下室布置有机结合，使各项功能得到巧妙安排和合理布局。

大堂设计长30m，中部入口处进深12m，两侧进深 7.5m，吊顶高6m。大堂两端为商务中心，外墙为全玻璃幕墙、深褐色磨光花岗石柱子和低窗台。作为两端商务中心的过渡入口处，均用上凸下收的全玻璃墙和无框玻璃门；在主入口对面的内侧墙面，以折线型浅底黑纹天然大理石为背景，其左右墙面则以大块实体作大手笔饰块；中部主体为外凸白色大理石面，上下为不对称的收进式黑色横细条墙砖作衬底。大堂地面以灰色和白色条形磨光花岗石相间为基调，辅以不规则的若干黑色小方块对角布置作点缀；天棚主体为白色平行出梁一通到底，四周由条形风口兜圈镶边。整个大堂通透明亮、虚实对比强烈、色调柔和、材料名贵、做工精细，呈现出高雅、大方、新颖的高级商务大堂气派。

底层电梯厅是人流集散中心，它与大堂相连又自成空间。墙面设计上采用大堂统一的实墙处理手法，并在电梯门口用精细的铝合金条作竖条细部处理，形成“大体一致，细部有变”的格调；地面也与大堂“大同小异”，仅在尺度、图案、色彩上有所变化，以求比例协调，更显亲切感。电梯厅吊顶约5m高，主体为两片带若干筒灯的长弧形吊板，中间盖以带小圆孔的铝板作为主带，内置报警、喷淋设施；两边凹槽内则为送、排风百叶带，勾画出空间丰富而又和谐的一方天穹，使人体会到电梯厅与大堂似同非同的装饰效果。

样板楼层在总体上全部选用浅色塑料地板和白色石膏板吊顶，走廊两侧是石膏板墙和无框玻璃门。从电梯厅进入走廊，就有一种简洁、明亮的感觉。卫生间选用了米色抛光地砖、浅灰色铝塑板墙面和银灰色铝板网吊顶，其标准、格调、色彩与主体部分相称。样板房设计实施了两套不同平面形式和装修手法的办公单元，包括房间的全玻璃分格、不锈钢饰件和时尚家具布置等等，均让人眼前一亮。

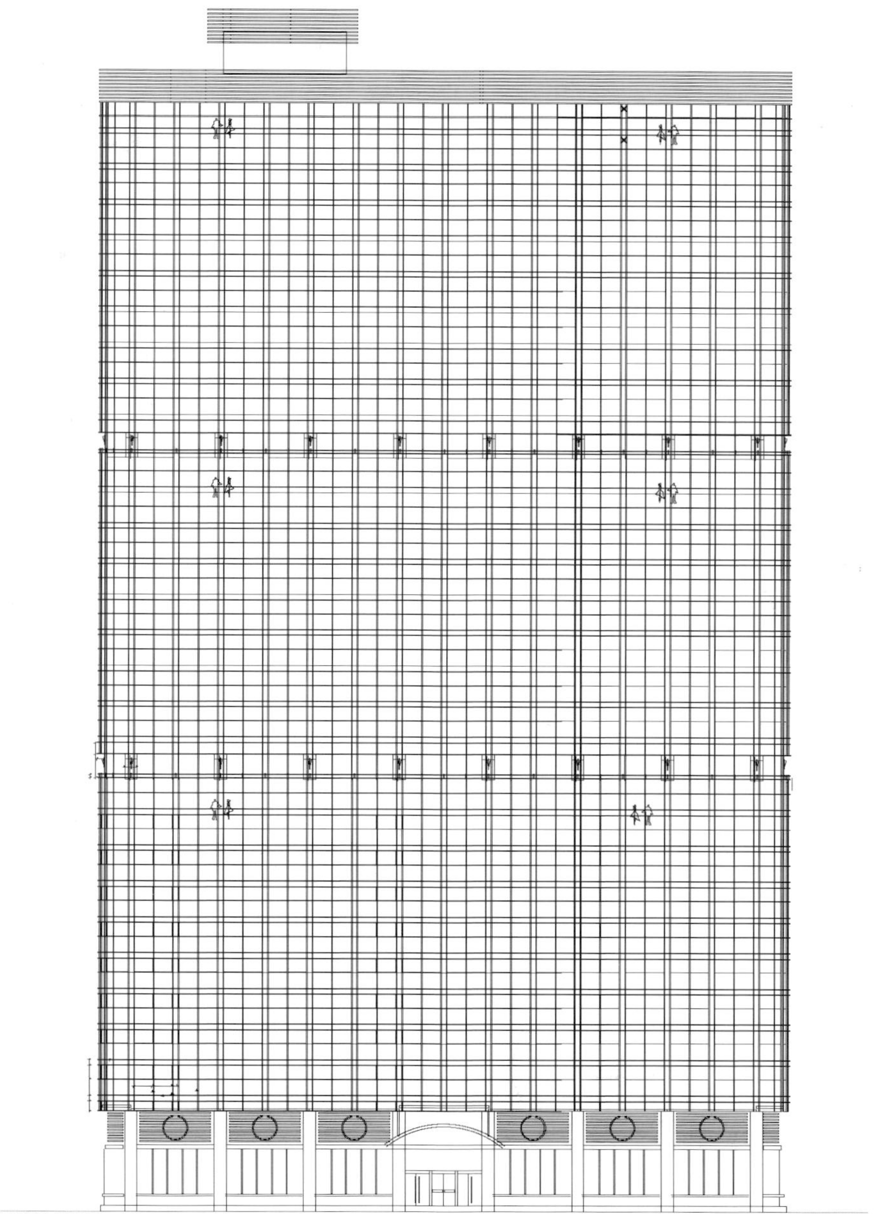

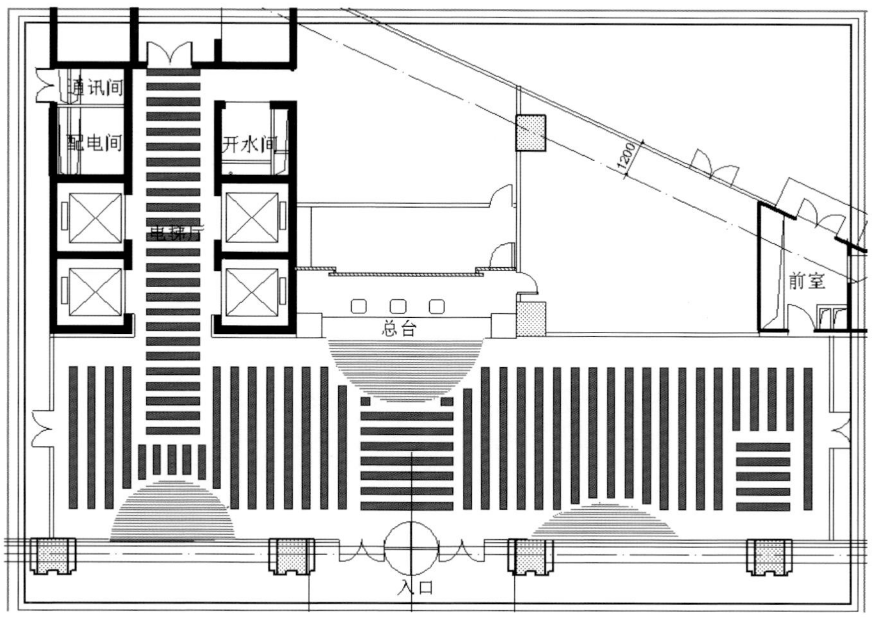
讯间
电间
开水间
1200
前室
总台
入口

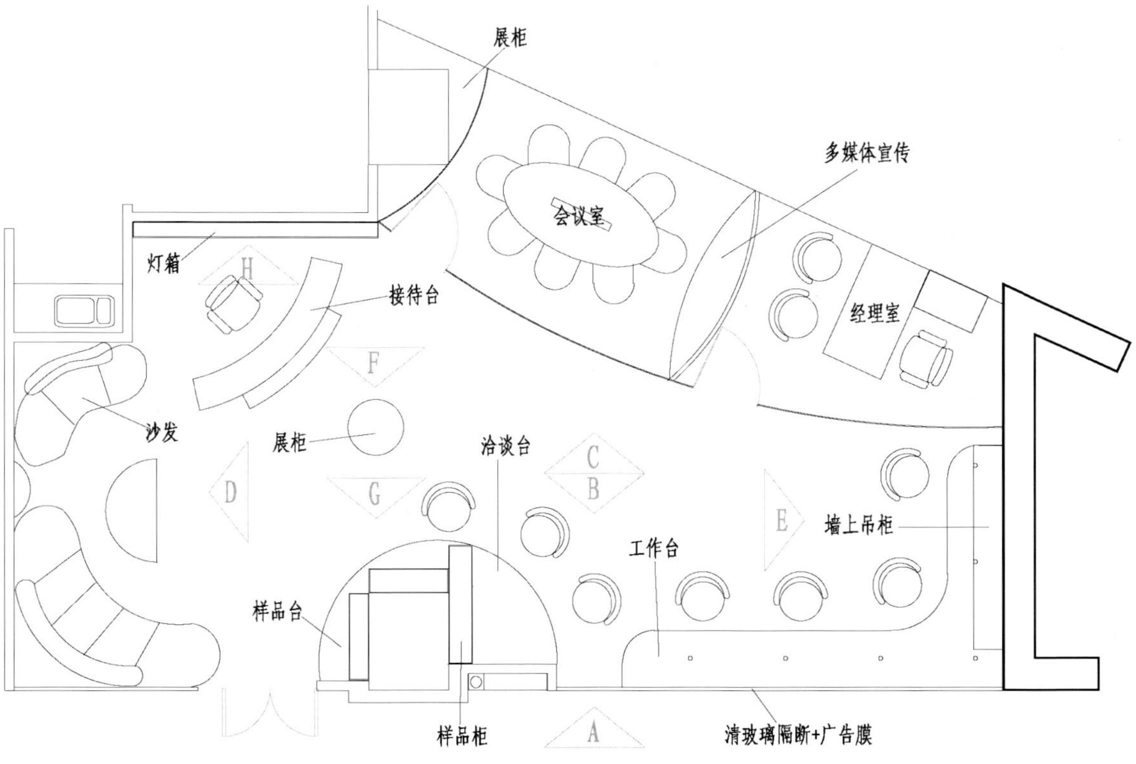
展柜
多媒体宣传
会议室
灯箱
接待台
经理室
沙发
展柜
洽谈台
墙上吊柜
工作台
样品台
样品柜
清玻璃隔断+广告膜
A
B
C
D
E
F
G
H

平面布置图 1:50

安全出口
EXIT

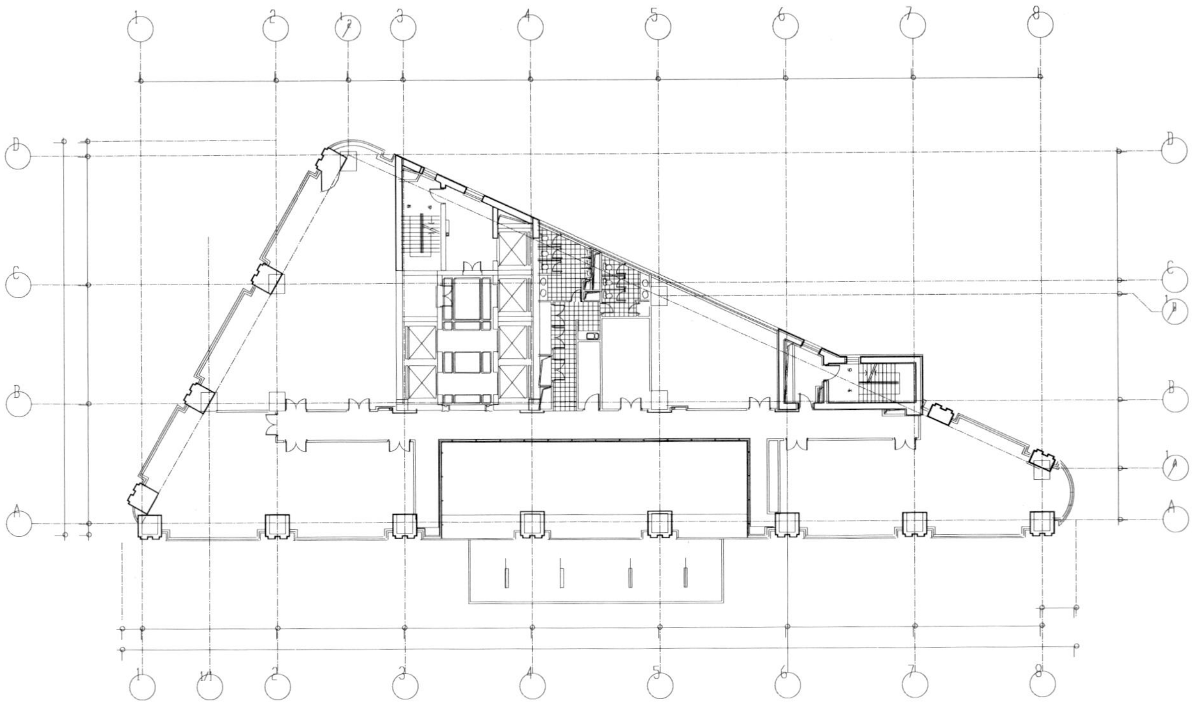

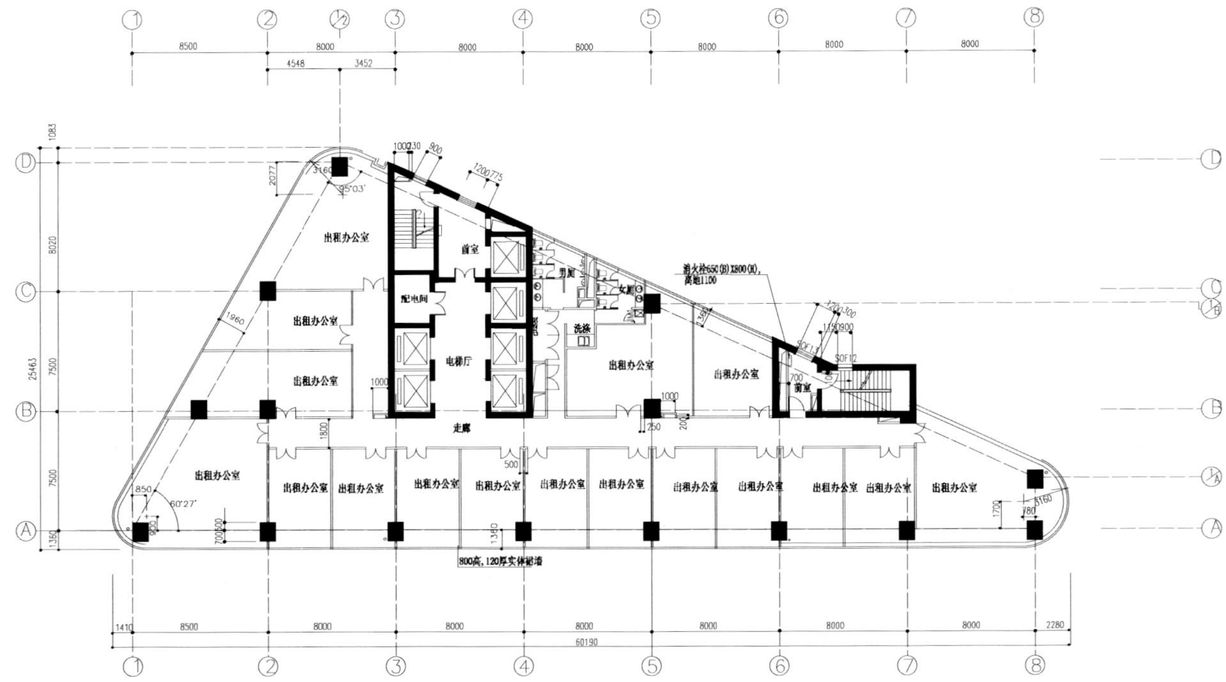

图例

钢筋混凝土墙，柱

轻钢龙骨，MC板

泰柏板

四层至十一层平面图
十三层至二十一层平面图
二十三层至三十层平面图

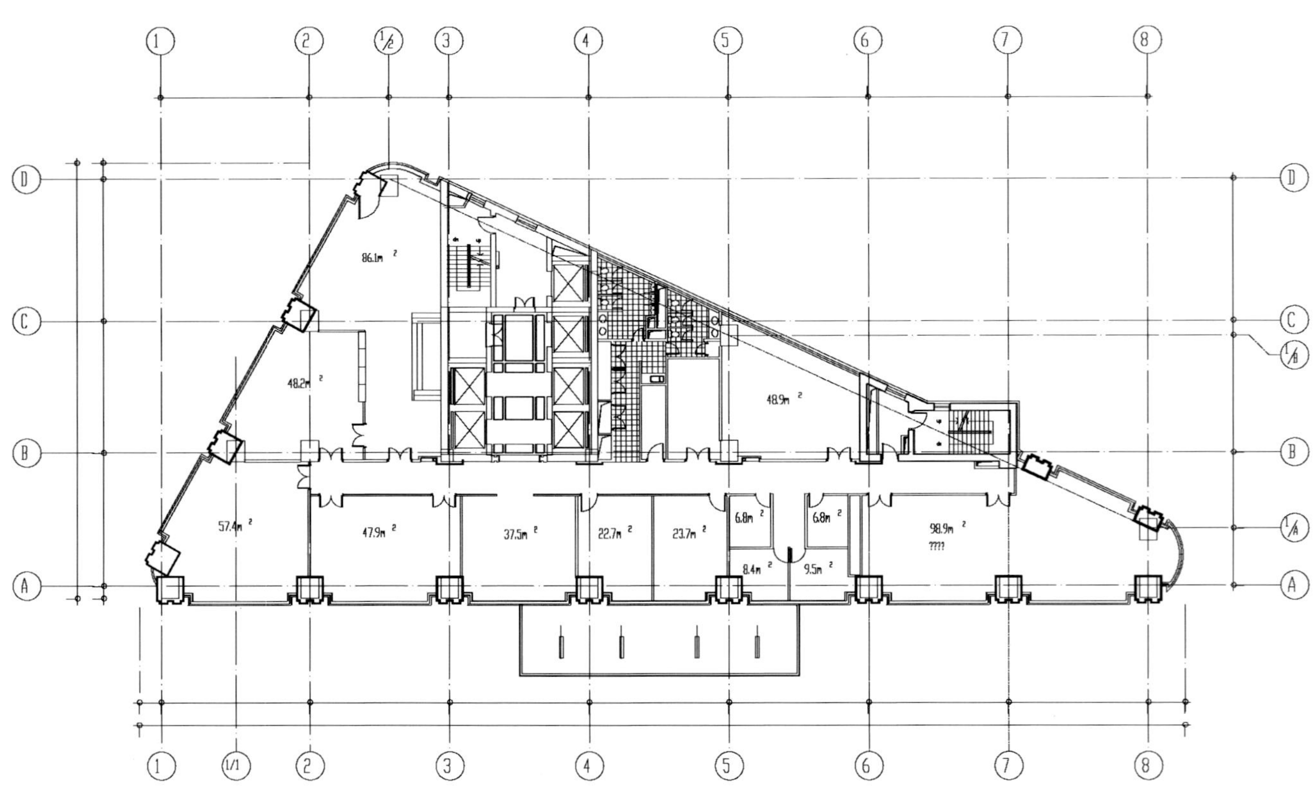

Christian Dior

Shenzhen Airport Information Building

深圳机场信息大楼

Building Area: 53,920.31m²
Site Area: 5,307.49m²
Greenbelt Area: 4,082.48m²
Building Height: 39.9m
Floor Area Ratio: 2.26
Green Ratio: 25.66%

总建筑面积：53,920.31m²
建筑基底面积：5,307.49m²
绿地面积：4,082.48m²
建筑高度：39.9m
容积率：2.26
绿化率：25.66%

Airport Information Building is located in the north of Shenzhen Airport Staff Entertainment Center, adjacent to the entrance path of the airport. It has 8 floors above the ground, and 2 floors underground, the site area is oval shaped. The 8 floors above the ground function as business rooms, such as security management, operation management, emergency direction center, and networking management; while the floors underground the ground function as parking lot, equipment rooms and defense spaces, whose land area is 15904.7m², total building area is 53,920.31m², and coverage ratio is 33.37%.

Each side of the site is neither parallel nor regular shaped; in addition, the buildings surrounded are all established in early ages, whose styles are not consistent, therefore, it is quite difficult to perform designation for this building. However, a building should be harmonious, integrated with the surroundings, we are trying to figure out an architectural form, which can be merged into the environment. The oval shape is composed of curved lines in each direction, whose orientation is not indicated, and can be suitable to complicated

机场信息大楼位于深圳机场职工活动中心北侧，紧临机场进场大道，建筑地上八层，地下二层，平面呈椭圆形，地上八层分别为安全管理、运营管理、应急指挥中心、网络管理等业务用房，地下层主要为车库和设备用房及人防，用地面积15,904.7m²，总建筑面积53,920.31m²，覆盖率33.37%。

由于建筑用地各边均不平行，地块呈不规则形状，加之周边的建筑均为早期的建筑，建筑形态各异，没有统一的风格，因此给设计上带来了一定的困难。而建筑是依附于环境而存在的，建筑与环境是有机的整体，所以我们力图寻找一种建筑形式，能与现有的环境融于一体。由于椭圆形各个方向均由圆滑的曲线构成，方向性不明确，从而能够适应复杂的周边环境和地形地貌，其次椭圆形体型完整、形态优美，容易给人留下深刻的印象，因此本方案平面形式选择了椭圆。在总图布置上，椭圆的长轴平行进场大道，为东西向，短轴为南北向，建筑退让进场大道63.50m，使建筑与进场大道之间形成了40多米宽的绿化带，在美化城市环境的同时，也缓解

surroundings and terrains; in addition, the oval shape is integrated, and elegant, which impress people easily. In the general layout, long axis of the oval shape is parallel to the entrance path, which is east-west orientation; while the short axis is south-north orientation. The building is 63.50m away from the entrance path, which forms a 40m greenbelt between them, optimizes urban environment, reliefs the pressure of traffic, as well as eliminates noises and pollution to direction mansion caused by traffic. The other 3 sides are far from the red lines; certain distance makes it possible to perform construction due to its relation with the subway underground tunnel of Baoan Avenue in the west, and meets the requirements of fire extinguishing regulations due to the gas station in the east.

Main and sub entrances are set on both west and east side of the site, whose neighborhood is a greenbelt with a ground parking lot as well as entrances to underground parking lot, that makes the transportation simple and convenient. A outside crossover bridge located in the south is connected to staff restaurant, which makes it convenient to contact each other.

The plan of this building is a long oval shape; it has 8 floors above the ground and 2 floors underground. The 1st floor functions as security management rooms, and large, medium, small meeting rooms; the 2nd floor functions as operation management center and information service rooms; the 3rd to 7th floors function as emergency direction center and auxiliary rooms; and the 8th floor functions as networking management, main motor rooms, data backup rooms; a ring shaped corridor is set on the roof to provide space for outside activities. The middle of this building is a rectangle shaped block, whose ends are used for vertical transportation, and whose middle part is a meeting room. In addition, the enclosure of the building forms 3 outside courtyards, which solves the problems of daylight and ventilation in corridor, stairs and rest rooms. 2 floors underground function as parking lot, equipment rooms and underground defense space during both war and peacetime.

The Architecture Design achieves simple, generous and clear, whose façade is composed of horizontal stripes, makes the entire building smooth and emphasizes the integrity of ovalshape. The overhead board on the top and partial blinds on the wall makes the entire building more graceful. Constructive materials of this building include glass, aluminum boards and stone, the diversity of textures and colors shows modernism and beauty in visual; and the mansion is definitely a fresh highlight in this region after it is completed.

了对道路的压迫感，并降低了交通带给指挥大厦的噪声和污染。建筑的其它三边均对周边红线有所退让，西侧考虑到与宝安大道地下隧道的关系，退让距离满足建筑建造的可行性，东侧退让现有加油站，从而满足防火规范的要求。

在场地的西东侧分别设置了主、次出入口，在入口附近结合绿化设有地上停车场及地下车库出入口，交通流线简洁、便利，建筑南侧设有一个室外天桥与职工食堂相通，以方便两者的联系。

本建筑平面呈长椭圆形，地上八层，地下二层，建筑一层为安全管理用房及大中小型会议室，二层为运营管理中心用房及信息服务用房，三至七层为应急指挥中心及辅助用房，八层为网络管理、主机房和数据备份用房，在屋顶上部设有一个环形架空廊，便于室外活动，大楼的中部为一个矩形的体块，两端为竖向交通，中部为会议室，另外，通过建筑的围合形成了三个室外天井以解决走廊、楼梯间及卫生间的采光、通风。大楼地下二层为停车场、设备用房和平战结合的地下人防。

建筑造型设计力求简洁、大方、明快，立面设计采用了横线条处理，使整个大楼圆润流畅，强化了椭圆形体的完整性，建筑顶部的架空飞板以及墙面上局部百页，使整个建筑更显飘逸，大楼的主要建筑材料为玻璃、铝板和石材，通过材质的变化和色彩搭配，体现了建筑的时代感，给人以美的视觉享受，建成后的大厦将成为该地区的一个新的亮点。

北立面实景

南立面实景

深圳机场信息大楼

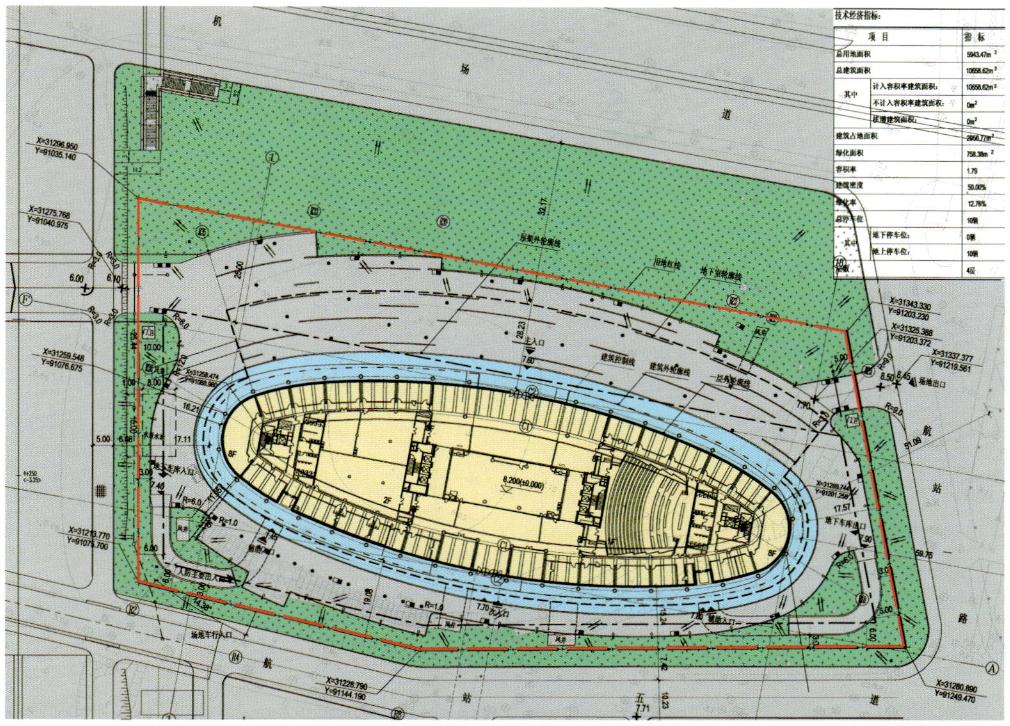

总平面图

深圳机场信息大楼

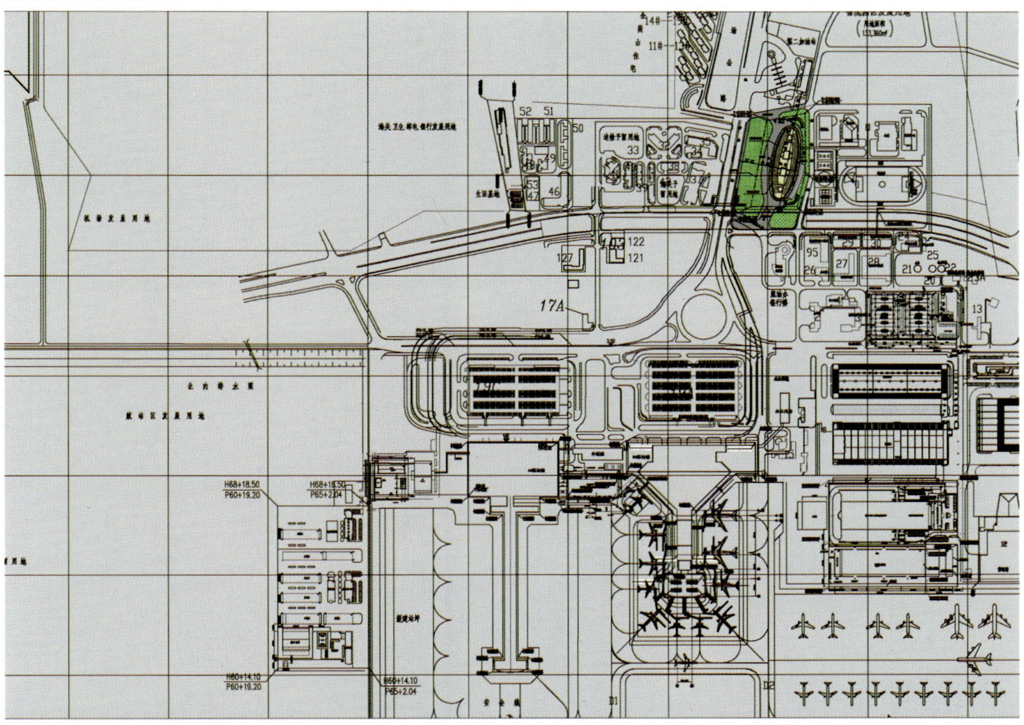

区域位置图

深圳机场信息大楼

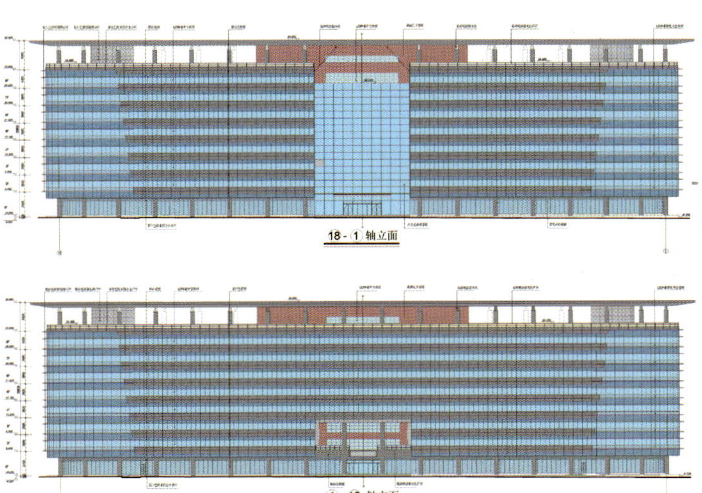

立面图

深圳机场信息大楼

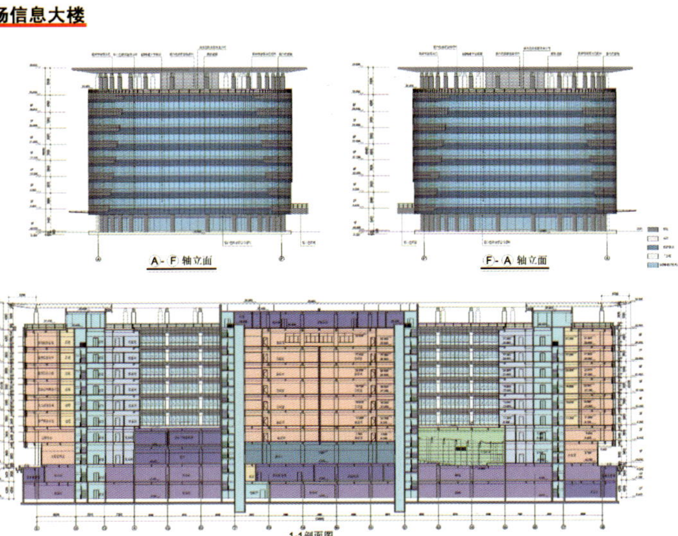

立面图

深圳机场信息大楼

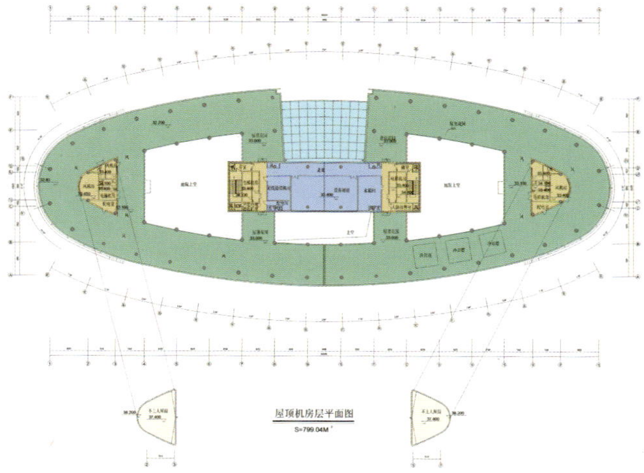

屋顶机房层平面图
S=799.04M²

屋顶机房层平面图

深圳机场信息大楼

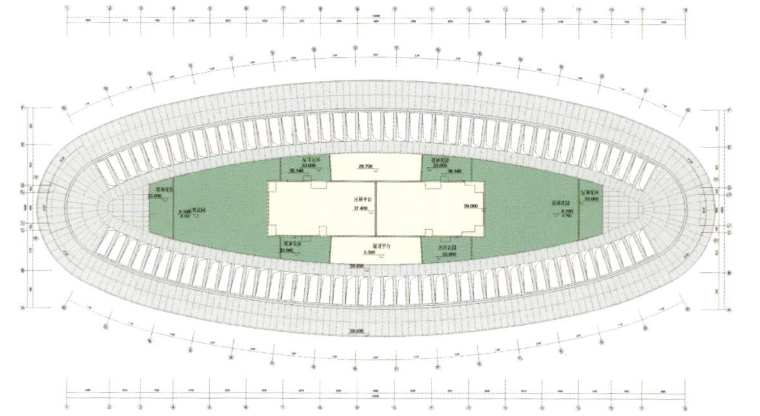

屋顶平面图

屋顶平面图

深圳机场信息大楼

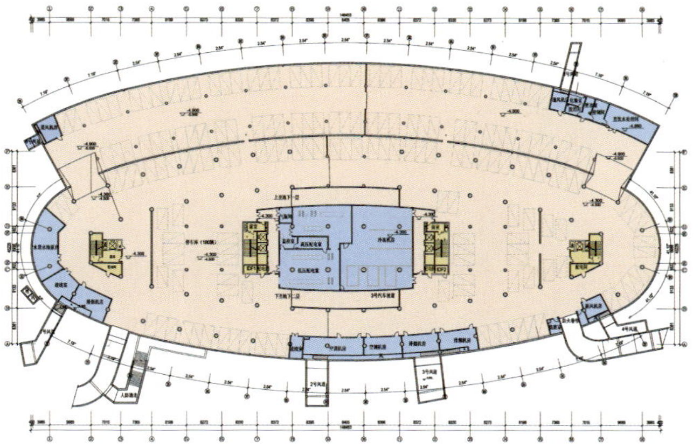

地下一层平面图 1:150
S=9112.85M²
停车位：180辆

地下一层平面图

深圳机场信息大楼

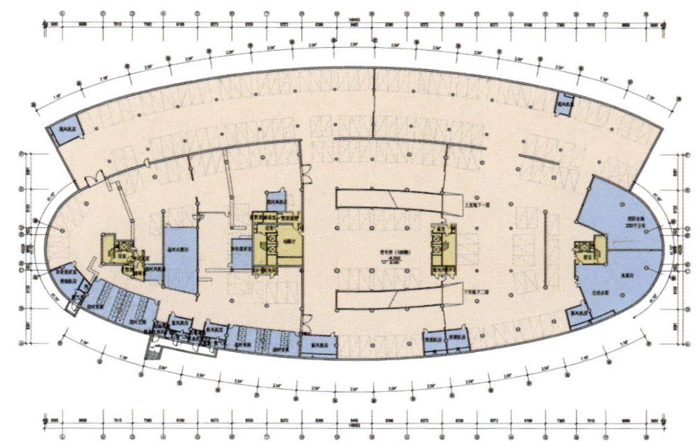

地下二层平面图 1:150
S=8807.01M²
停车位：188辆

地下二层平面图

深圳机场信息大楼

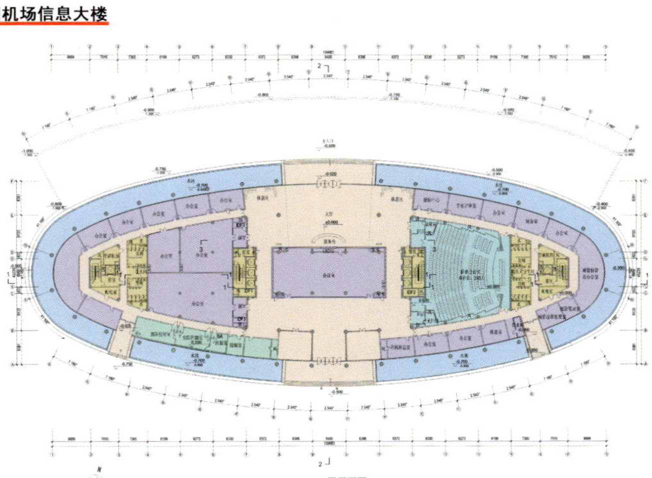

一层平面图
S=5307.49 M²

一层平面图

深圳机场信息大楼

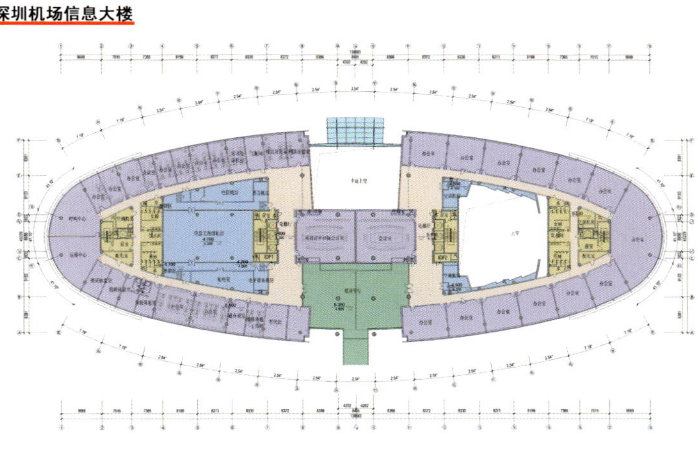

二层平面图
S2=4706.32M²

二层平面图

深圳机场信息大楼职工食堂

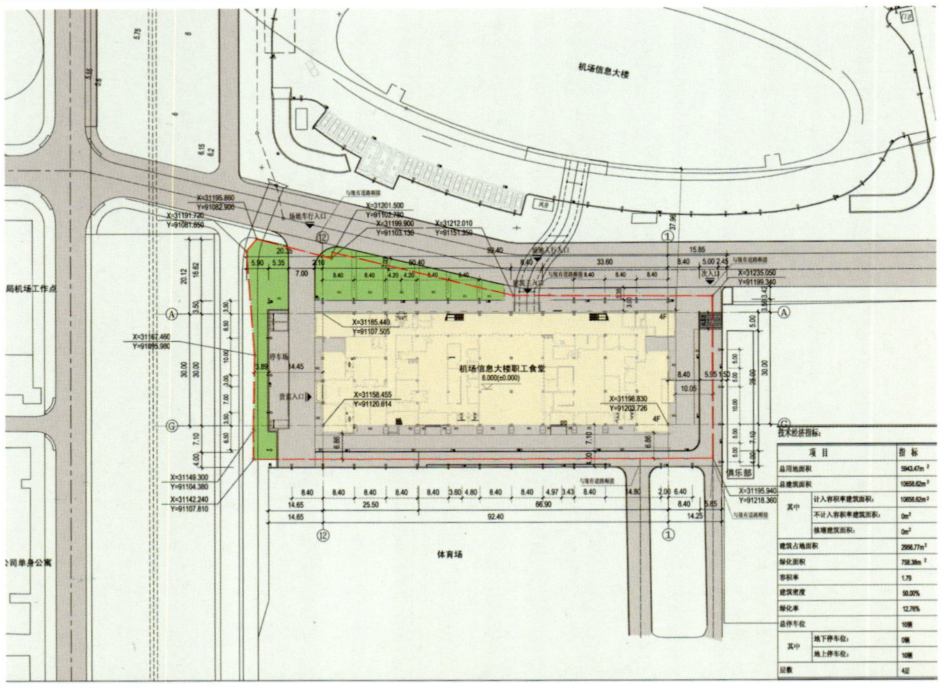

总平面图

深圳机场信息大楼职工食堂

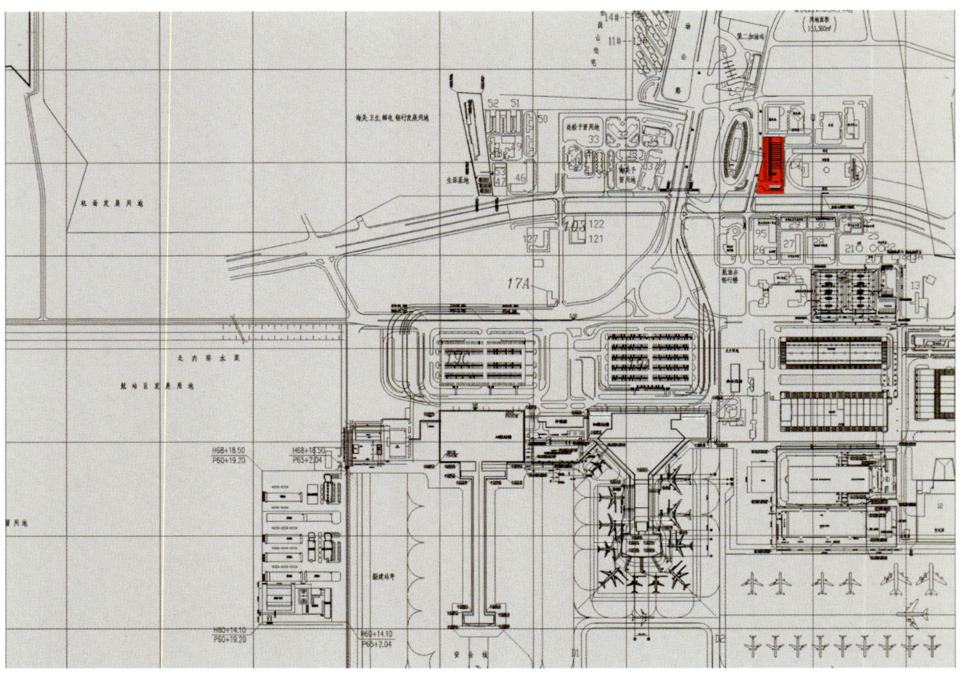

区域位置图

深圳机场信息大楼职工食堂

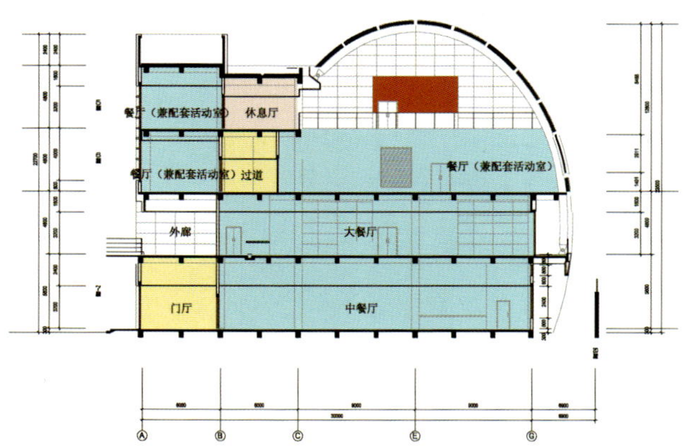

2-2 剖面图 1:100

剖面图

深圳机场信息大楼职工食堂

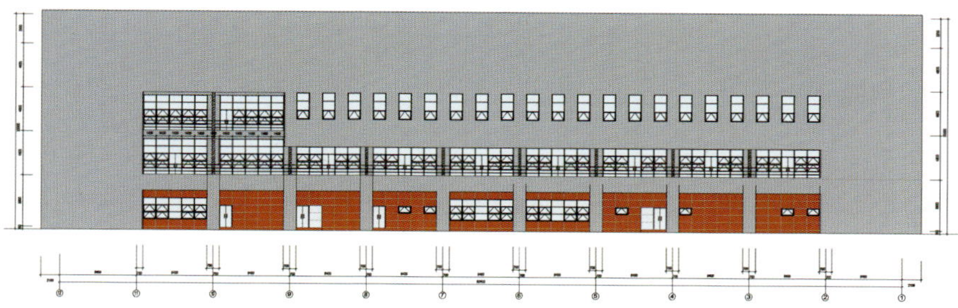

⑫—① 轴立面图 1:100

立面图

深圳机场信息大楼职工食堂

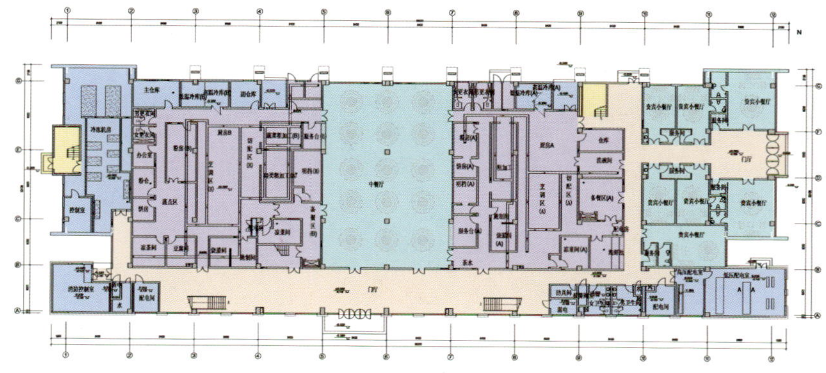

一层平面图 1:100

S=2956.77 M²

一层平面图

深圳机场信息大楼职工食堂

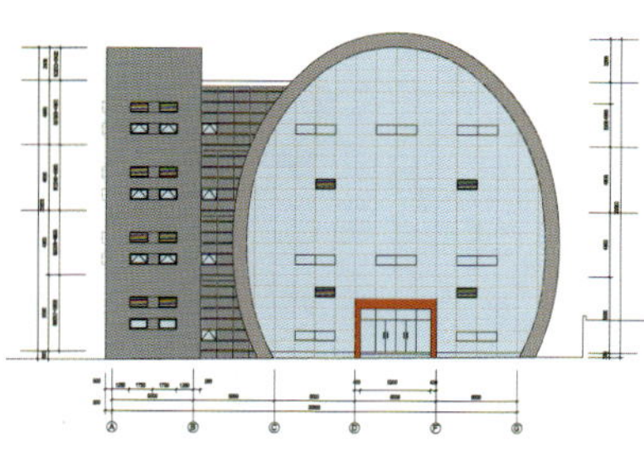

Ⓐ—Ⓖ 轴立面图 1:100

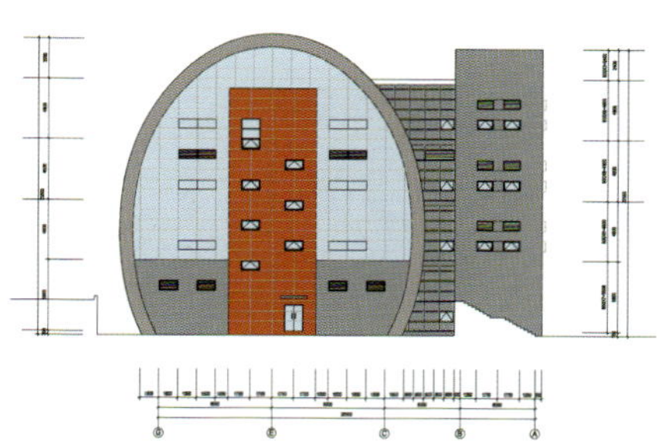

Ⓖ—Ⓐ 轴立面图 1:100

立面图

深圳机场信息大楼职工食堂

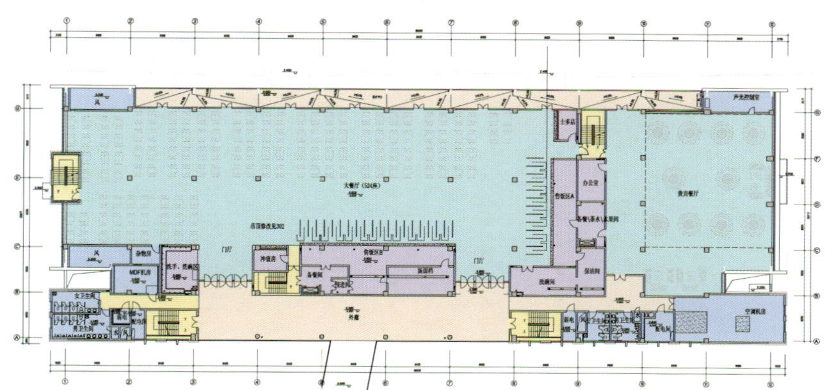

二层平面图 1:100

S=3115.43 M²

二层平面图

深圳机场信息大楼职工食堂

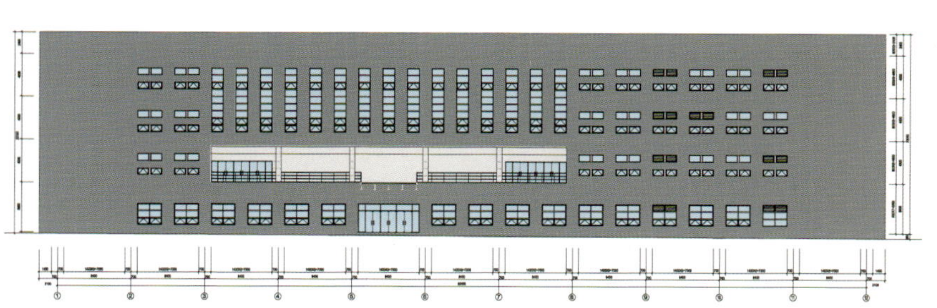

①—⑫ 轴立面图 1:100

立面图

深圳机场信息大楼职工食堂

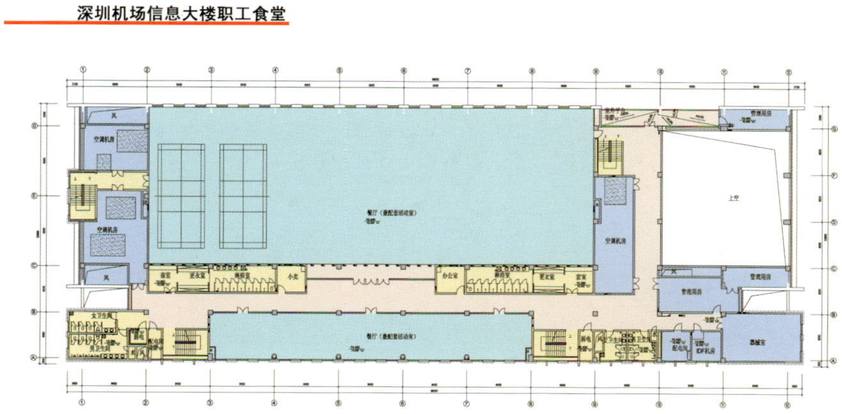

三层平面图 1:100

S=2800.30 M²

三层平面图

Shenzhen Era Science & Technology Mansion

深圳时代科技大厦

Project Location: Bamboo Forest Block, Futian District of Shenzhen City
Site Area: 10,695.8m^2
Building Area: 117,330.77m^2
Floor Area Ratio: 8.0
Real Estate Developer: Shenzhen Agricultural Science Group

建设地点：深圳市福田区竹子林片区
总用地面积：10,695.8m^2
总建筑面积：117,330.77m^2
建筑容积率：8.0
开发商：深圳市农科集团

Shenzhen Era Science & Technology Mansion is located at the intersection of main road of Shenzhen, i.e. Bamboo Forest Block and Agriculture Park Road, both in the north of Shenan Avenue, is adjacent to the 256mhigh Shenzhen World Trade Center. Architecture Design of this mansion follows "Urban Design Guidelines", and meets the demand to form blocks on both sides of Shennan Avenue; 170mhigh tower appearance optimizes the original skyline of the city. To avoid shading caused by building volume, setback appearance is used to the north side of the building, which forms a semi-enclosure outside square and provides a comfortable space for the citizens. The façade of the building is made of several kinds of textural glass curtain walls, and creates various shaped effect.

深圳时代科技大厦位于深圳主干道深南大道北侧竹子林片区、农园路和深南大道交会处，毗邻 256m高的深圳世界贸易中心。时代科技大厦的设计遵循了《城市设计指引》所强调之深南大道本地段两侧形成街墙的要求，170m 高的端塔式造型优化了原有的城市天际线。为避免建筑体量对北侧产生日照遮挡，建筑北部采用退台式处理，同时在北侧形成了具有半围合感的室外广场，为市民提供了一处宜人的城市空间。建筑立面以几种肌理不同的玻璃幕墙穿插对比，形成变化丰富的体形效果。

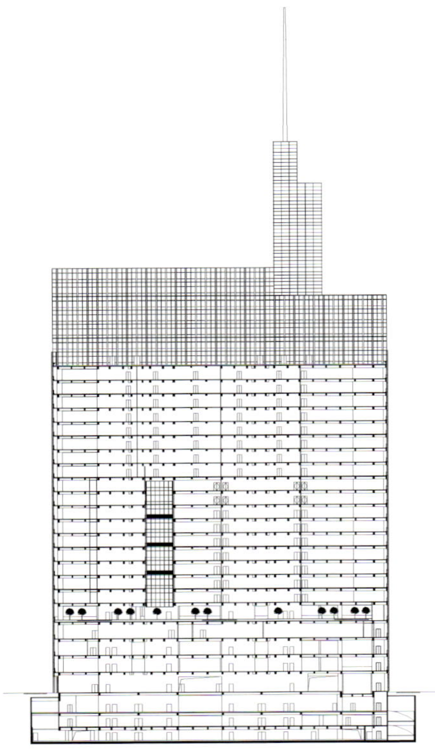

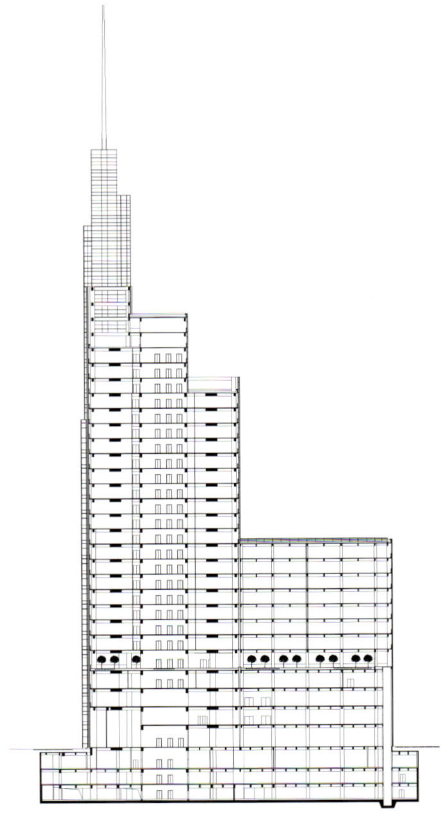

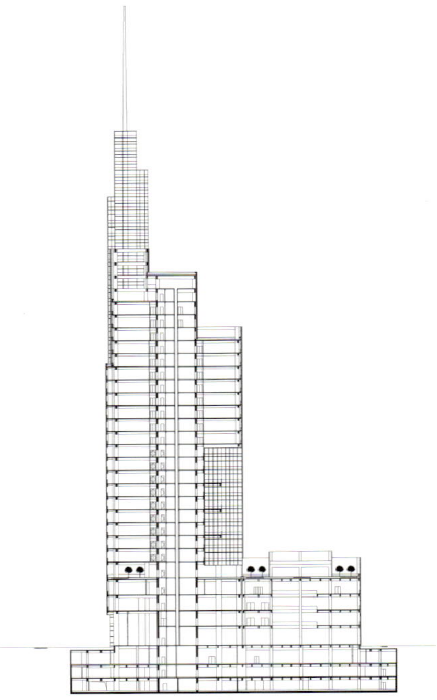

3-3剖面图　1:250

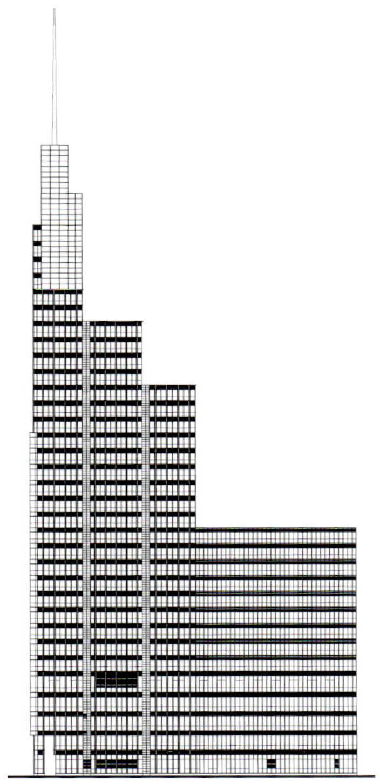
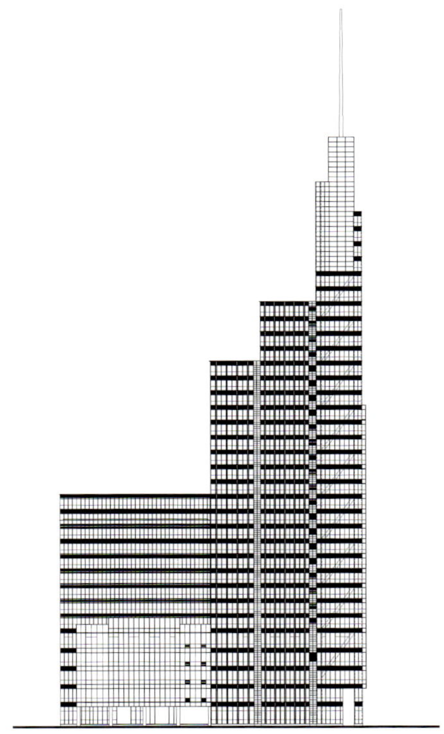
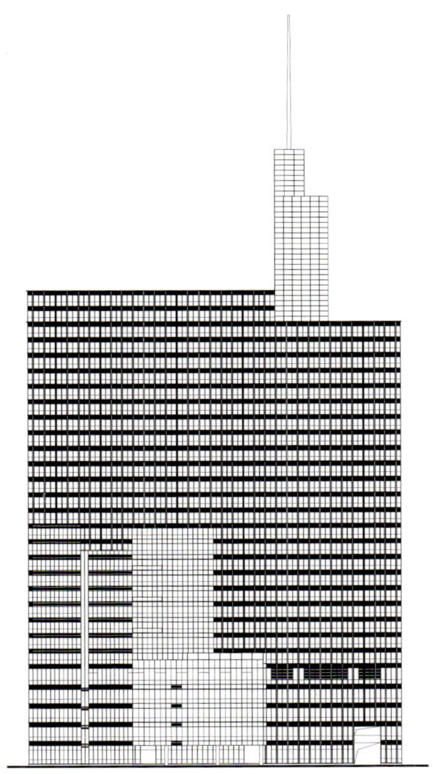
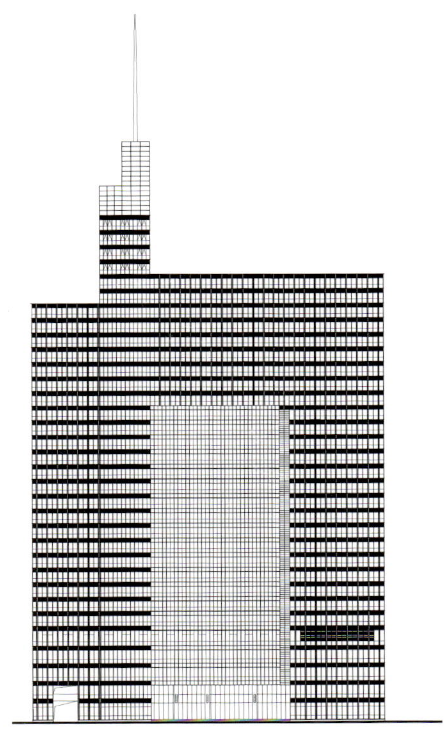

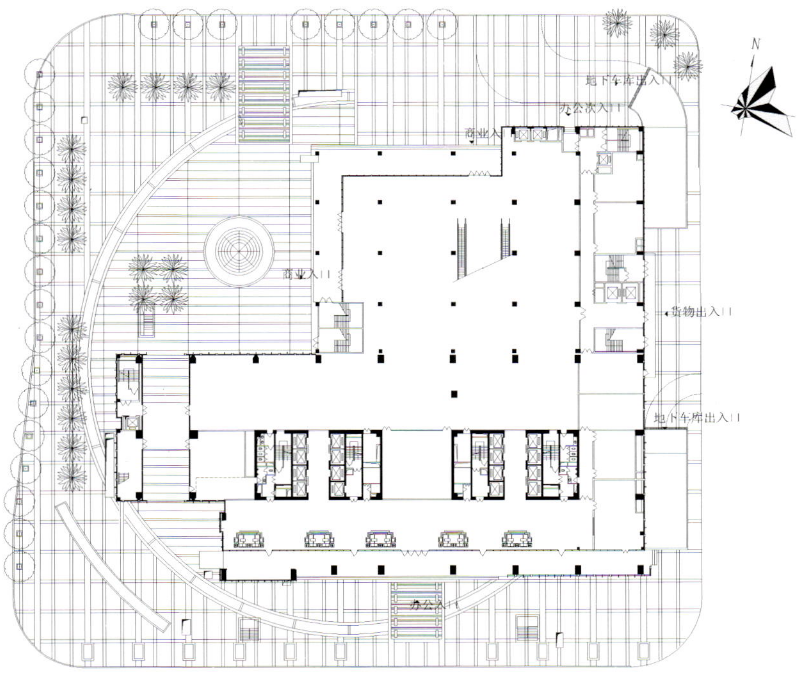

一层平面图

0 30m

十六~二十层平面图 1:150

注:本层建筑面积:3168.21M²

0 30m

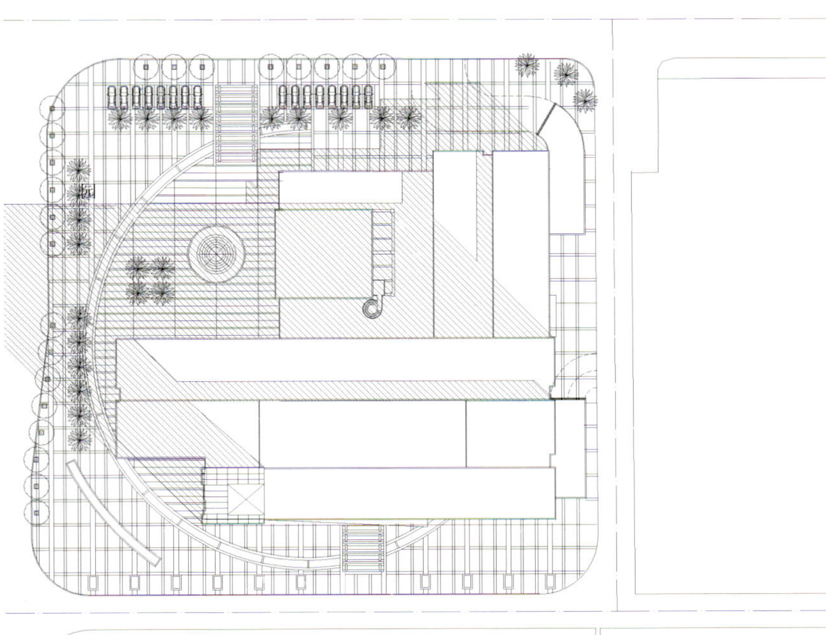

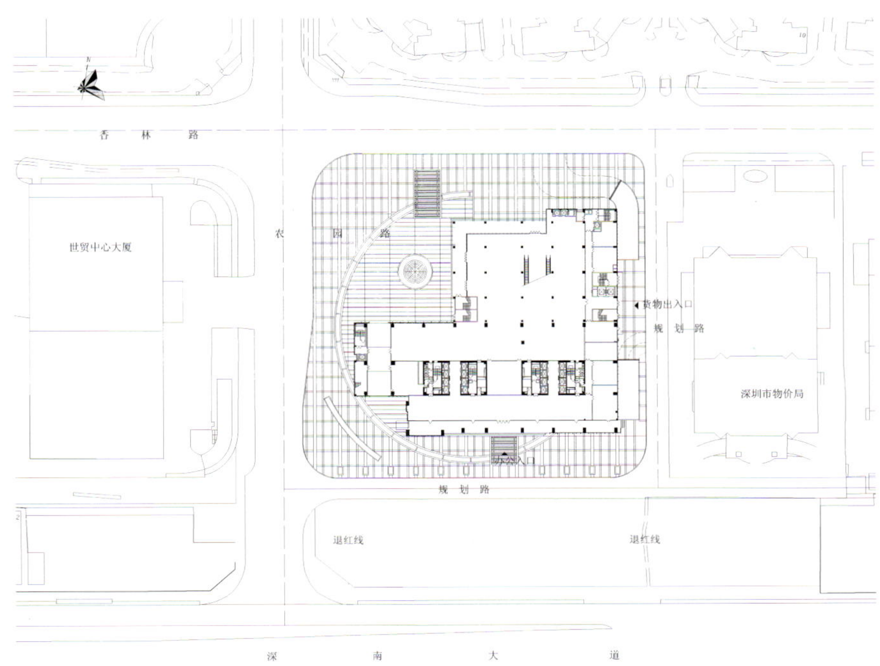

(一层)总平面图 1:400

0 100m

Shenzhen Galaxy Century Plaza

深圳星河世纪广场

Project Location: CBD of Shenzhen City
Site Area: 11,700.7m^2
Building Area: 156,626.21m^2
Floor Area Ratio: 10.4
Real Estate Developer: Galaxy Real Estate (Shenzhen) Co., Ltd.

建设地点：深圳市中心商务区（CBD）
总用地面积：11,700.7m^2
总建筑面积：156,626.21m^2
建筑容积率：10.4
开发商：星河实业（深圳）有限公司

Galaxy Century Plaza is located at the intersection of Caitian Road and Shennan Avenue in the CBD of Shenzhen and is the main landmark building at the east end of CBD. The complex is formed by a 39floor super high-rise office building with a total height of 170m, and two apartment buildings connected by a four floor shopping podium.

The door shaped office building forms a vivid icon in a simple way. The "two floors in one grid" module applied to the inherent characters of the elevation creates an expression of openness and elegance. Though slightly different from the conventional office building in texture, it is consistent to the design style of the podium in overall scale. The large scaled frame of the podium elevation is harmonious with the main tower and defines the vast city square along the street which enhances the integrity of the mixed-use building.

The exterior of the office tower and the commercial podium are glass and stone curtain wall while the apartment tower is finished with brick veneer and paint.

Thus achieves an overall sense of high quality.

星河世纪广场位于深圳中心商务区（CBD）彩田路和深南大道交会处，是深圳CBD东端主要标志性建筑物。建筑综合体由超高层办公塔楼、两栋公寓塔楼和连接它们的4层商场裙楼组成，办公塔楼总高度为170m，39层。

办公塔楼的"门式"造型，手法简单却对比强烈，具有很强的标志性；公寓塔楼立面结合自身的特点，通过两层一格的处理，获得开朗的表情，并不介意在肌理上与办公楼有所差别，但在较大尺度层面上与办公裙楼设计手法保持一致。裙楼大尺度的框式造型与主题塔楼协调，并在沿街勾勒出恢宏的城市广场，强化了综合体的完整性。

办公塔楼及商业裙楼建筑外墙材料主要为玻璃及石材幕墙，公寓塔楼主要为面砖及涂料，但整体却保持了较高的品质。

星河世纪

8372 8888

星河世纪

JOC
Simmons
Better Sleep Through Science.
Villeroy & Boch
1748
KOHLER
Artemide

ROYAL DOULTON
一站式豪华家居主题购物中心
第三空间 五月绽放!

TOTO
Artemide
ipe CAVALLI
THE LORD

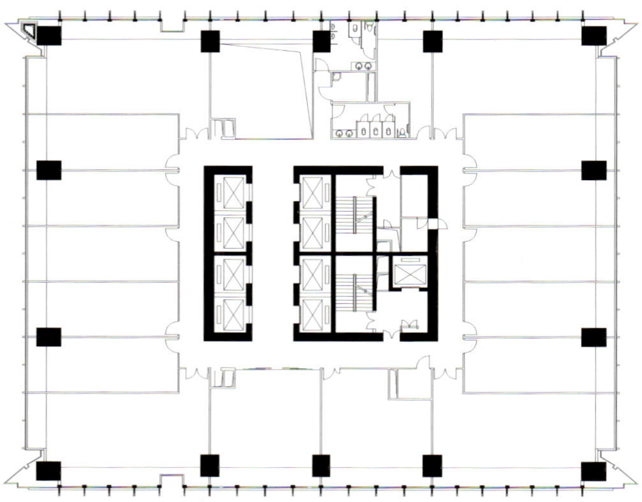

A栋：18层平面图 1:100

0 20m

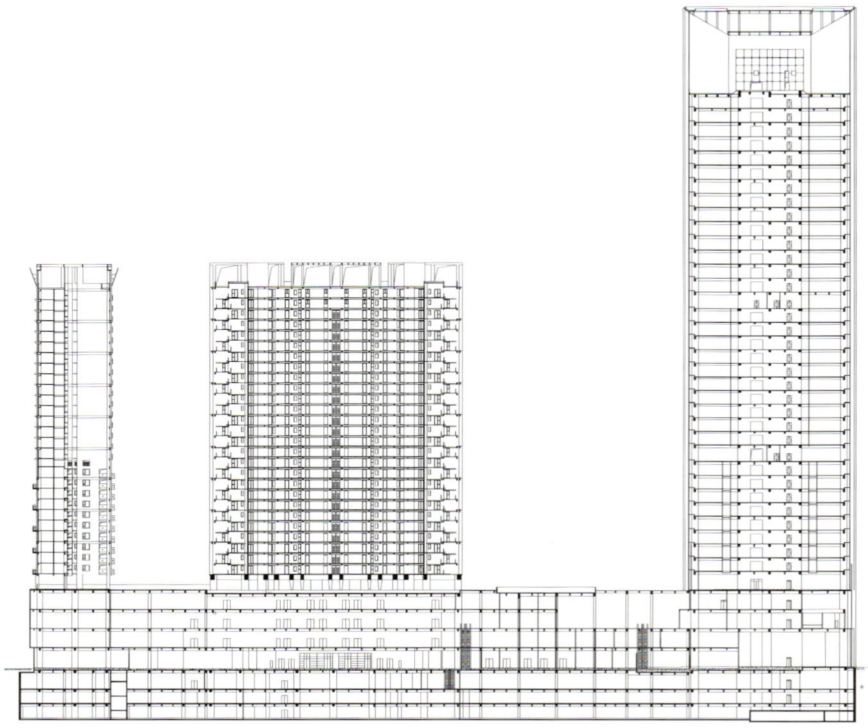

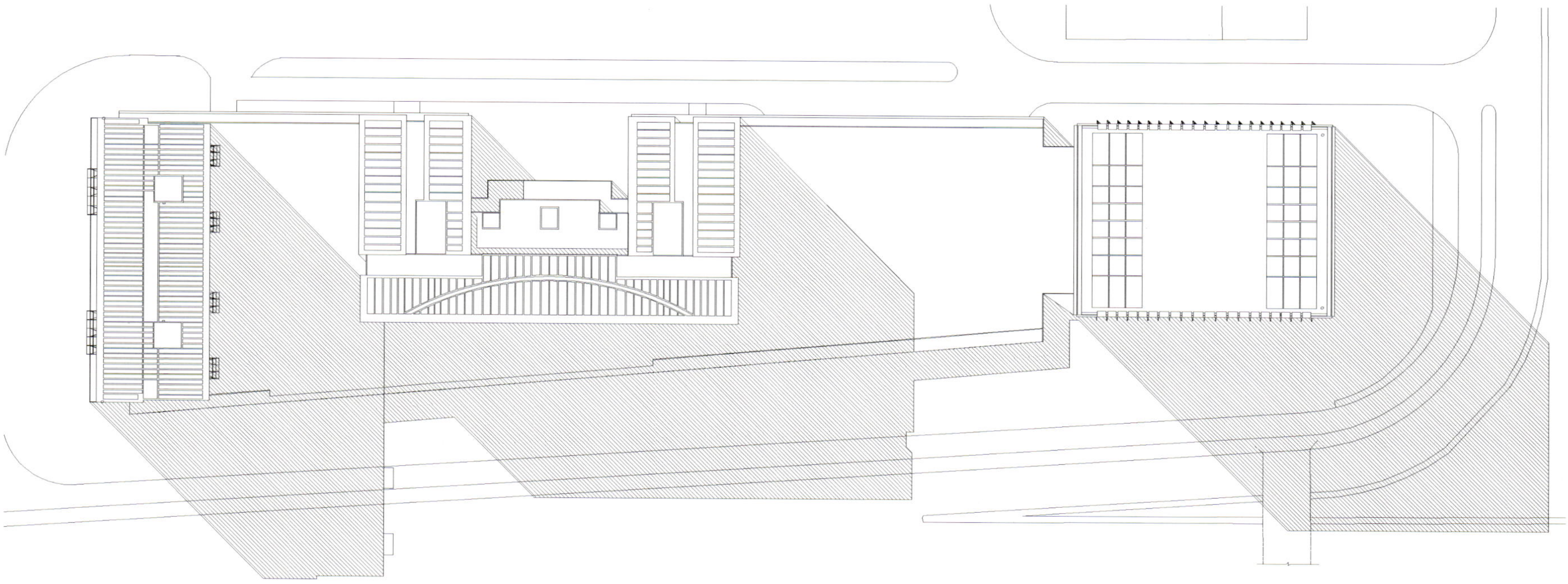

建筑红线
用地红线
住宅出入口
住宅出入口
住宅出入口
车库出入口
消防登高面
消
防
登
高
面
住宅出入口
C栋
B栋
A栋
32F
29F
32F
办公出入口
裙房屋顶花园
商场出入口
商场出入口
商场出入口
彩　田　路

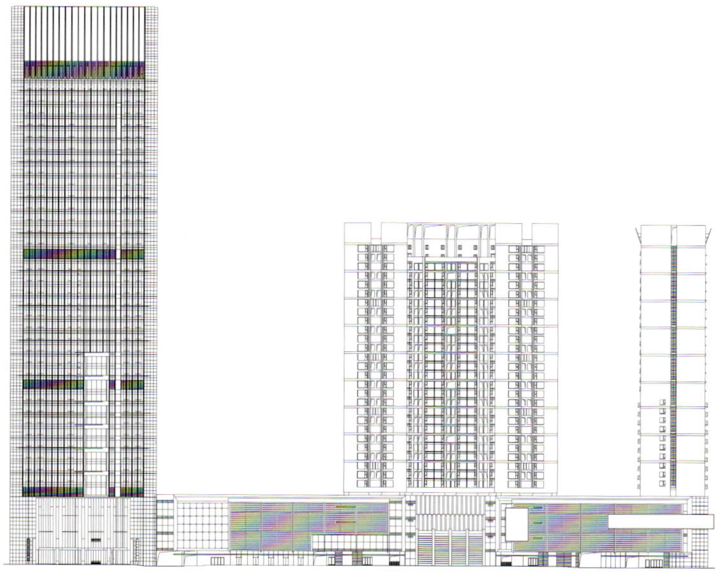

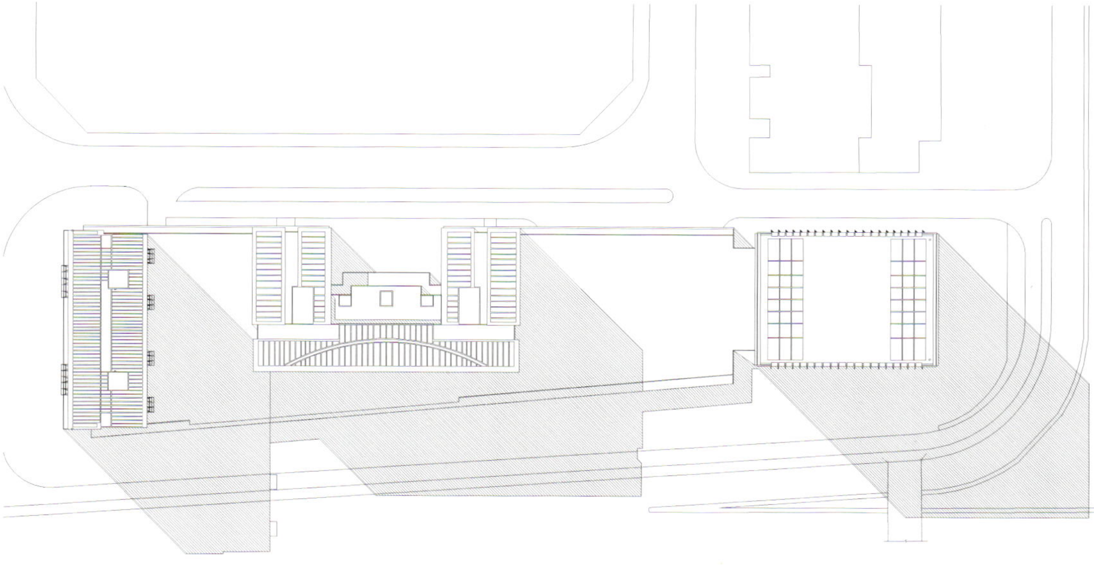

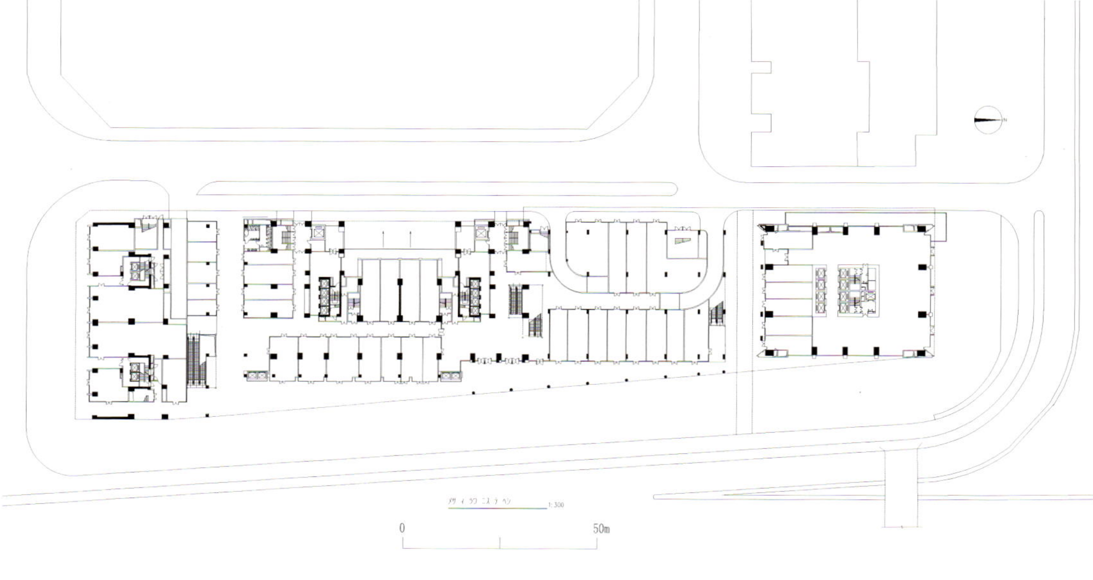

Tianli Central Business Plaza

天利中央商务广场

Real Estate Developer: Shenzhen Tianli Real Estate Group Co., Ltd.
Architecture Design: DDB International Co., Ltd.
Site Area: 17,233.80m^2
Building Area: 147,193.90m^2

开发商：深圳天利地产集团有限公司
建筑设计：DDB INTERNATIONAL，LTD
占地面积：17,233.80m^2
建筑面积：147,193.90m^2

The appearance of this building is modern, spectacular and simple, duration and usable space is absolutely increased with exposed concrete design and techniques, as well as strict control on reinforced protective layers. Repeated optimization of structural layout makes the building much more practical, impressive and convenient. Office space is disciplined and effective, whose internal space is flexible without any columns, and easy to separate. A 7.2m high sky garden is set on every 2 floors of the tower, which creates leisure, comfortable and intelligent working environment. A setback roof garden is in picturesque disorder and clearly demarcated, which achieves visually extensive effect.

External walls: Nanbo low-e glass, which is energy saving and environment friendly, is used on the whole office building; and hanging stone materials, which are high-grade and textural, are used on podium.

Internal walls and floors: Granite is used on office and elevator lobbies, which are 5A, Agrade standard and make coastal business space with international standard.

建筑外观现代、大气，线条简洁明快，清水砼设计和工艺，严格的钢筋保护层控制，不但大幅提高了建筑耐久性，同时有效地增加了空间的使用面积。结构布局的反复优化，使建筑具有了实用、美观、通道便捷等多项优势。办公房间方正，使用效率高，内采用无柱设计，空间灵活多变，可自由分隔。塔楼隔层设置7.2m的空中花园，创建轻松舒适的生态智能办公环境。退台式屋顶花园高低错落、层次分明，视觉空间的有效延伸。

外墙：写字楼全部采用南玻low-e玻璃，节能环保；裙楼采用干挂石材，高档质感；

内墙及楼地面：办公大堂及电梯厅——花岗石；

5A甲级设计标准，打造国际水准的滨海商务空间。

MSUNG
Whirlpool 惠而浦
LG

SUNING

天利中央广场
全球招商
招商中心

天利名城
TILEY FAME CITY

中国银行
BANK OF CHINA

NBA

中国银行 BANK OF CHINA
中国银行 BANK OF CHINA

ROBAM
SUNING 苏宁电器
SAMSUNG

SAMSUNG

天利名城
COASTAL CITY

苏宁电器 2F
数码城
40

得福轩
精品潮州菜
老西安
经典陕西民间风味
美容 SPA 纤体 美甲
Sexy Queen

900 9600 9300 9300 9300 9600 900
8800 8400 8400 8100 750
09 08 07
10 06
12 11
01 05
02 03 04

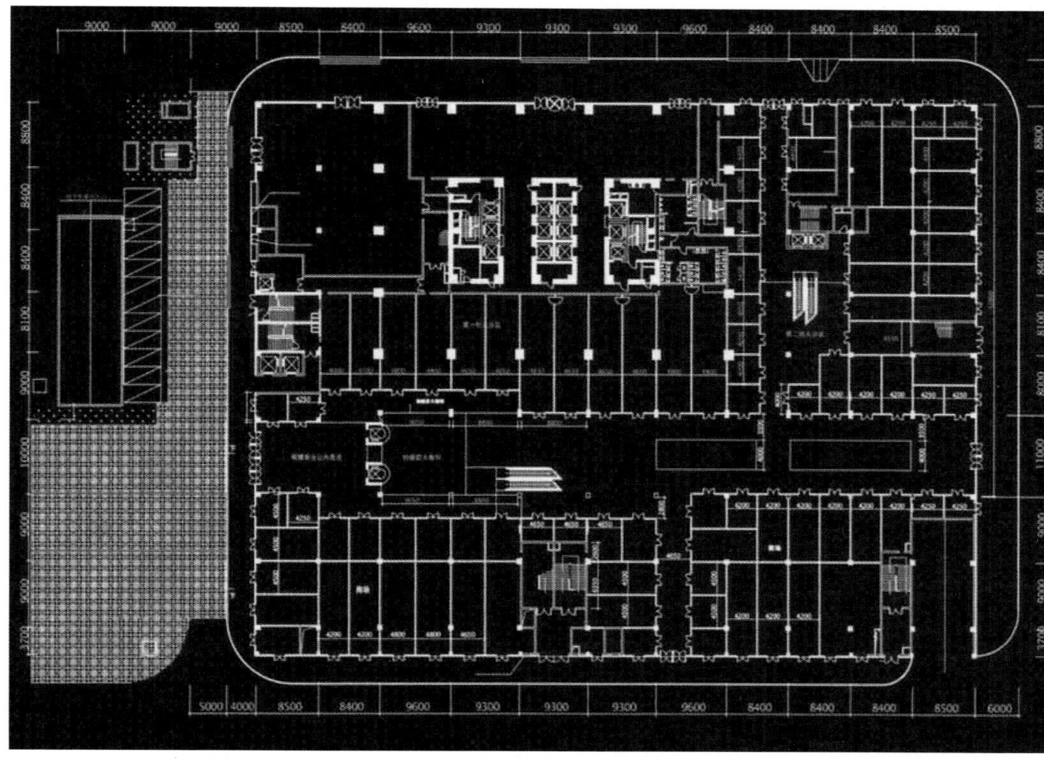

900 9600 9300 9300 9300 9600 900
8800 8400 8400 8100 750
09 08 07
10 06
01 05
02 03 04

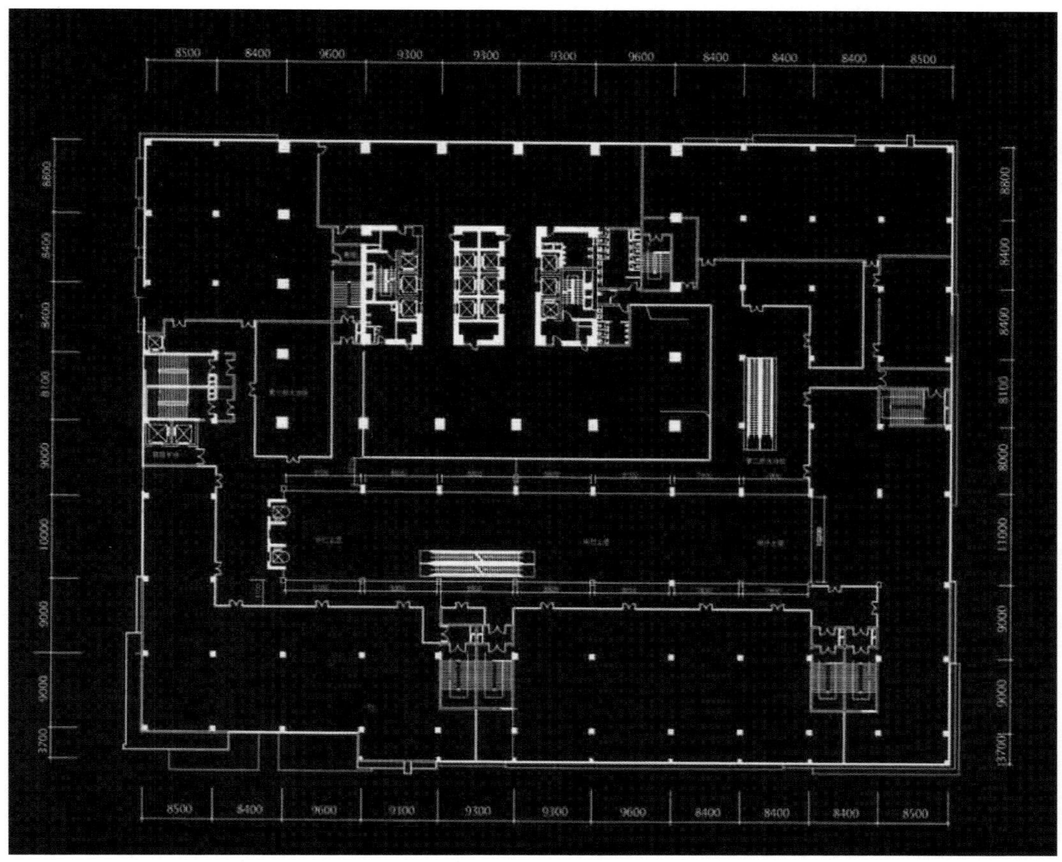

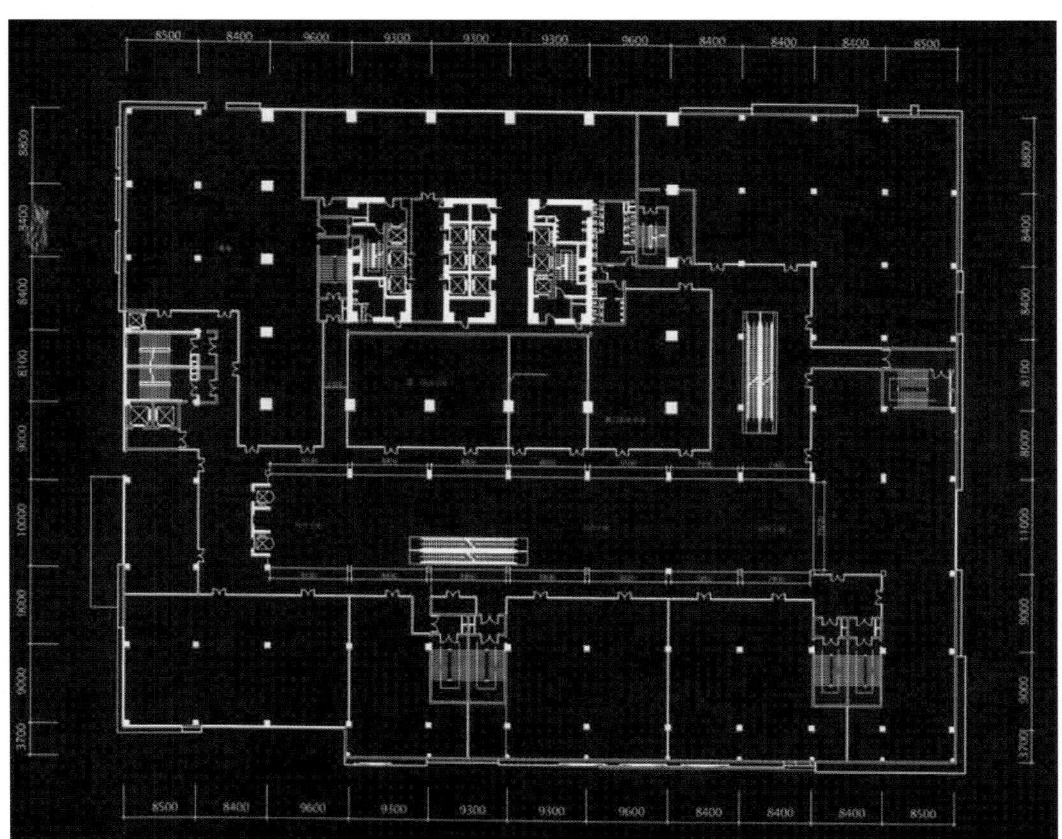

Shenzhen Subway Bamboo Forest

深圳地铁竹子林

Shennan Avenue, Shenzhen Overseas Chinese Town
Gross Building Area: 53,395m²
Height: 24m
Storey: 6
Building Area with Floor Area Ratio: 33,744m²
Basement Building Area: 21,812m²
Storey: 6 floors above the ground, and 1 floor underground
Earthquake Intensity: 7 Degree, implement 8 Degree Earthquake resistance structure Actions
Site Category: Ⅱ Class
Construction Limited Height: <24m

深圳华侨城深南大道；
总建筑面积：53,395m²
高度：24m
层数：6层
计容积率建筑面积：33,744m²
地下室建筑面积：21,812m²
建筑层数：地上6层，地下1层
抗震设防裂度：七度，按八度采取抗震构造措施
场地类别：Ⅱ类。
建筑控制高度：<24m

Vehicle portion of Bamboo Forest is located in the south of Shennan Avenue, east of Overseas Chinese Town, which looks like a rectangle area. Architecture design follows the significant features of "City in the Park" in Overseas Chinese Town, avoids using units in large volume, and provides high-grade, rhythmic landscape.

Vehicle portion of Bamboo Forest extends along the south of Shennan Avenue, distance is up to 1m from east to west, within which continual and single style architectural volumes blocked the landscapes on both sides, and it is hard to have something impressive among the boring architectural interfaces. This case adopts a collection of solutions to divide and group, in order to achieve the landscape with rhythm.

The plan of operation comprehensive building is a triangle divided into 3 parts, which are connected each other with 2 glass bridges. The lobby of each unit looks like a glass box inserted into the light yellow curved marble walls, the shape of building groups is simple and elegant.

竹子林车辆段位于深南大道南侧，华侨城东路以东，近似一个长方形的区域。设计延续了华侨城的“花园中建城市”的城市肌理的重要特征，建筑单体避免较大的体量，具备较高的景观质量并具有很强的节奏感。

竹子林车辆段基地沿深南大道南侧展开，东西距离达1km长，在如此长的距离中，连续而单调的建筑体量不仅封闭了道路两侧景观，而且由于超长的建筑硬质界面的枯燥使东西来往视线中难以形成令人振奋的视觉焦点。因此设计采用一种分解、群化的群体方案来达到产生富有节奏韵律感街景的目的。

运营综合楼的平面近似一个被截成三段的三角形，这3个分而不离的体量之间分别以两座玻璃桥相连。每个形体的门厅如同一个斜插入淡黄色弧形大理石墙内的玻璃盒子，建筑群形态简洁优雅。

运营综合楼的东侧为公安局和培训中心，便于紧急行动。运营办公楼和调度中心靠西侧布置。食堂设于运营办公楼下层，位置居中，便于两侧使用。运营综合楼设计为智能办公建筑，大空间办公室，易于灵活布置和

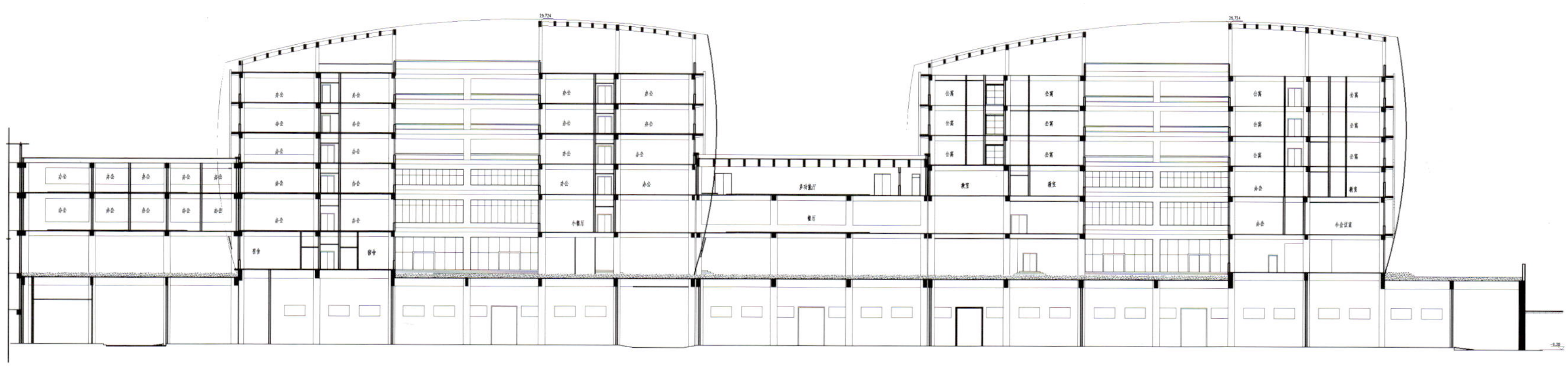

剖面图

Police station and training center are located in the east of operation comprehensive building, which is convenient for emergency actions. Operation office building and dispatch center are located in the west. Staff restaurant is under the office building, and centered for convenience from both sides. Operation comprehensive building is an intelligent office building provides large spaces flexible to layout and split. 3 office buildings are separated and connected to each other, each of them equips with independent and convenient transportation system. Division and characters of vehicles and people flows are taken into consideration during designation. A semi-island parking lot is located in the north of the site, to meet parking needs of office buildings.

The site is quite long and narrow, so the landscape is designed as a theme park. A forest and fountain square, an ecological park, a sight-viewing courtyard and a spiral greenbelt are all extended along Shennan Avenue. A garden within site, a 500m long sight-viewing corridor, an open greenbelt square is dotted among the building groups, which follows the planning concept of "City in the Park" in Overseas Chinese Town.

Pre-stressed tubular pole foundation, framed structure, girder structure, and 1860 grade pre-stressed steel strand wires are inside of the girder. B-grade water-proofing roof, whose duration is up to 15 years, the polymer water-proofing coats are used in rest rooms, pantries and kitchens, except the reinforced concrete structure, self water-proofing level of the underground is S≥0.8mpa. Polymer concrete and sand water-proofing is used for external walls.

The project also adopts several new technologies, such as solar energy, which makes the buildings much more environment friendly, energy saving and power conservative.

分割。三栋办公楼既功能独立又便于相互联系。每栋楼都设有独立便捷的交通系统。设计充分考虑车流、人流的分流及不同性质人流的组织及分流。基地北侧设半岛式停车场地，满足办公大楼停车的需要。

由于基地地段狭长，景观的设计采用主题公园的形式。林泉广场、生态植物园、怡景中庭、螺旋式台地绿化沿深南大道依次展开。厂区花园、500m景观绿廊、开敞式绿地广场等穿插于建筑群体之中，成为建筑的花园，真正延续了华侨城“花园中建城市”的规划思想。

预应力管桩基础，框架结构、梁板结构，梁板内设1860级预应力钢绞线。屋面防水等级为二级，耐用年限15年。卫生间、开水间、厨房聚合物防水涂层，地下室防水除钢筋砼结构自防水S≥0.8mpa。外墙防水采用聚合物水泥砂浆防水。

该工程还使用了太阳能等多项技术，使建筑更加环保、节能、节电，效果良好。

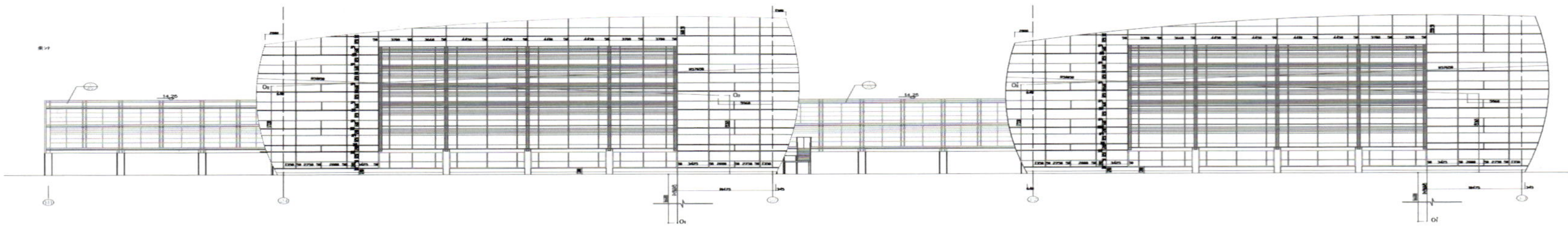

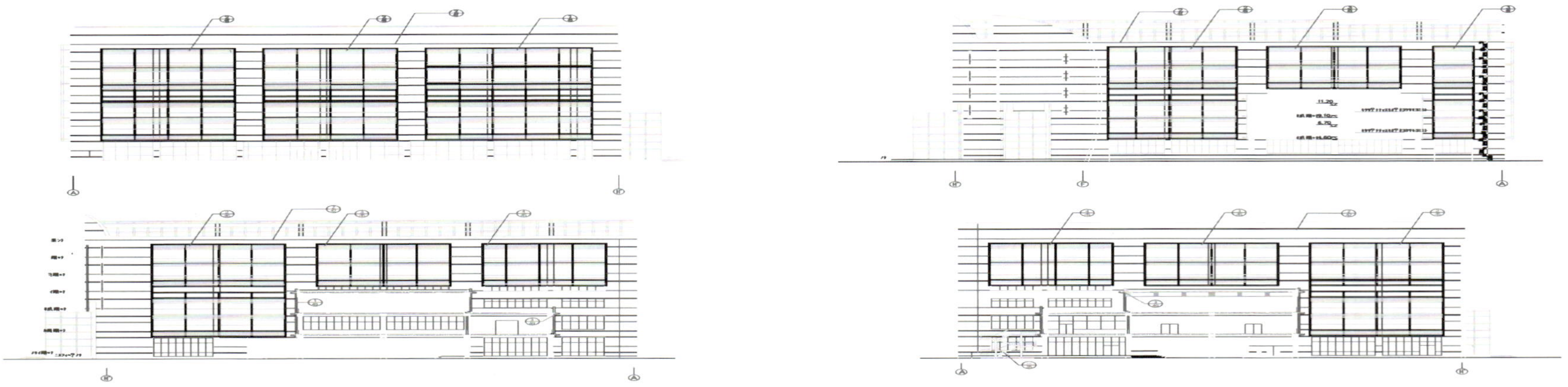

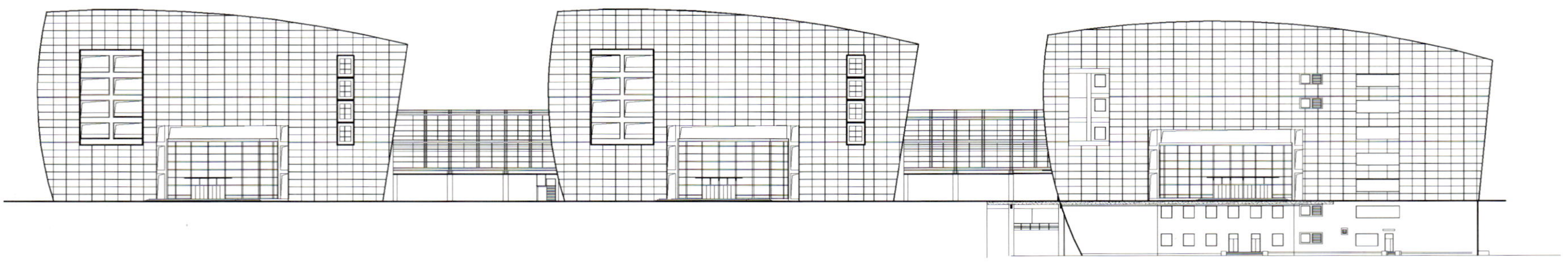

立面 1　　立面 1

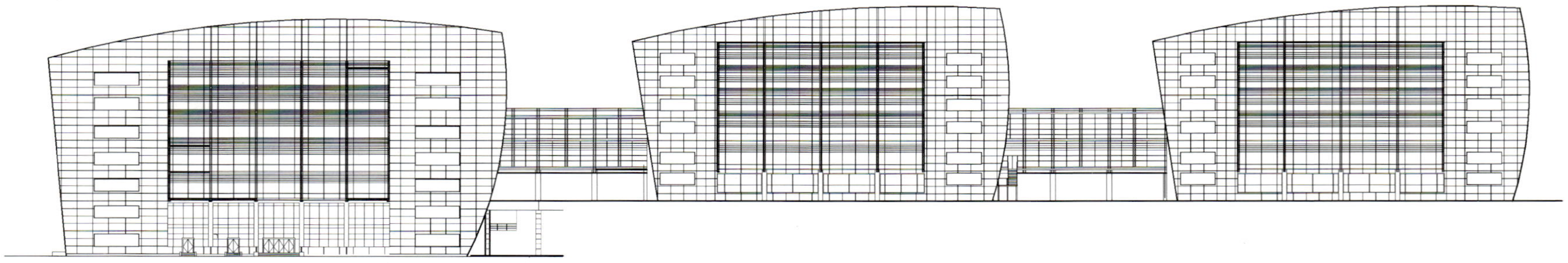

立面 2　　立面 2

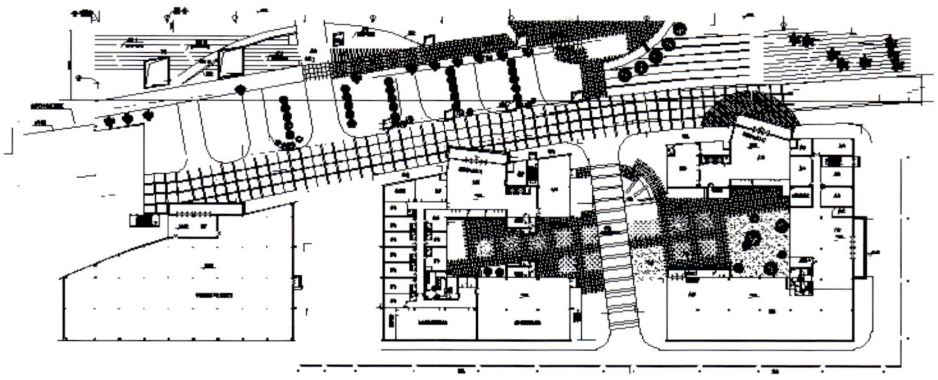

Han's Laser Science & Technology State Center

大族激光科技州中心

Site Area: 77,000m^2
Building Area: 230,000m^2
Number of Floor/Height: 14 Floors/64.8m

占地面积：7.7万m^2
建筑面积：23万m^2
层数高度：14层/64.8米

The project is located in High-Tech Development Zone of Shenzhen City, the site is evenly divided by Han's Innovation Mansion and the other project with the axis from south to north; the 2 projects form a symmetrical and enclosed building group. The intense contrast between glass and stone adopted by architecture design divides certain volume ratio again, and create a human and unique architectural style, whose external part is quite suitable for vehicles and internal part is comfortable for pedestrians, shows the feature of tight outside and loose inside.

该建筑位于深圳市高新区，地块由大族创新大厦与另一项目沿南北向平分，整体形成对称、围合的建筑群组细胞。项目设计运用玻璃与石材的虚实对比，将既定的体块比例进行二次分割，塑造出建筑外侧适应车行尺度、庭院内侧适宜人行惊讶的“外紧内松”型的人性化研发建筑独特气质。

24h自助银行服务

电子监控禁停路段
违法停车罚款200元
拖走

宝安、蛇口
深南大道

Chongqing Citic Plaza

重庆中信广场

Site Area: 3,672m²
Building Area: 55,165m²

占地面积：3,672m²
建筑面积：55,165m²

Chongqing Citic Plaza is located at the entrance of Guan Yin Bridge Pedestrian, a famous commercial street in China, north of Jialing Park, south of Yuao Bridge. Its gross building area is 55,165m², the building is 99m high, has 26 floors in total, 4 floors underground. Floor 1-4 are commercial set, 5-6 are business set, 7-26 are international (5A) A-grade offices.

The mansion is invested by Citic Real Estate Company, an expert in top class office buildings in China, and the total investment amount is up to 0.23 billion, the quality of the building is guaranteed. The appearance is made of brown glass and stone curtain walls, the combination shows a retro style. The office building equips with intelligent system inside, keeps pace with international technological intelligence, and matches the business concepts of customers.

重庆中信大厦坐落于获誉“中国著名商业街”的观音桥步行街街口，北眺嘉陵公园，南接渝澳大桥。中信大厦总建筑面积为55,165m²楼高99m共26层，地下四层，其中1~4层为商业配套、5~6层为商务配套、7~26层为国际（5A）甲级写字楼。

大厦由中国顶级写字楼专家中信置业倾力打造，项目总投资达2.3亿，写字楼品质保障毋庸置疑，楼体外观由玻璃幕墙与石材幕墙合奏出棕褐色调，并肩演绎着一种现实源自历史的情怀，大厦内全面装备写字楼智能化系统，以同步世界的科技智能，匹配您谋划全球的商务智慧。

方案一

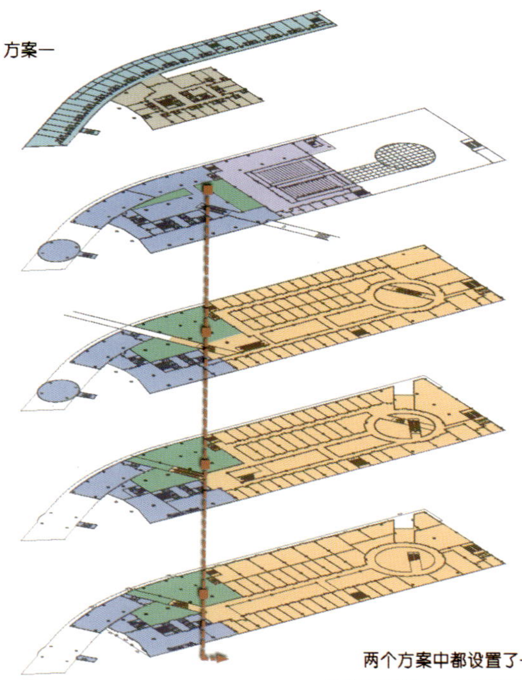

方案二

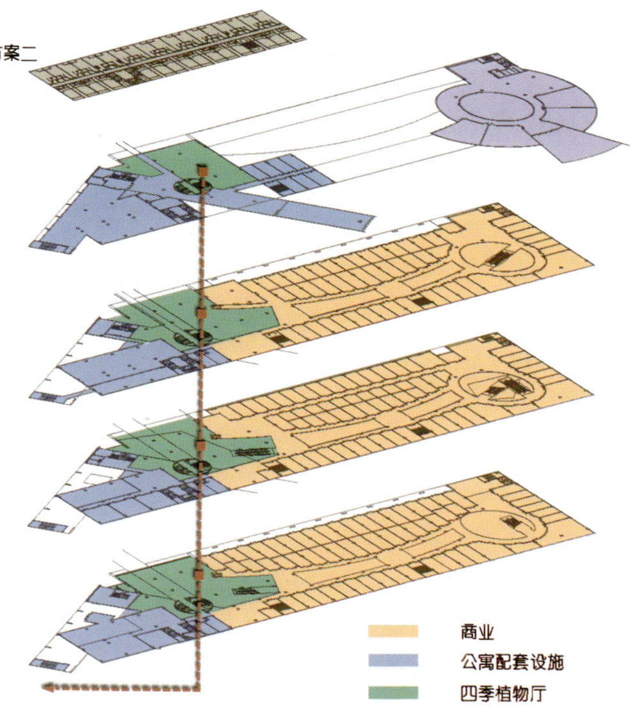

两个方案中都设置了一个前文所述的虚体空间——四季植物厅，它是各种功能联系的一个枢纽，同时也是各功能区间互相冲突的缓冲。此外，它还是自然之间的联系。

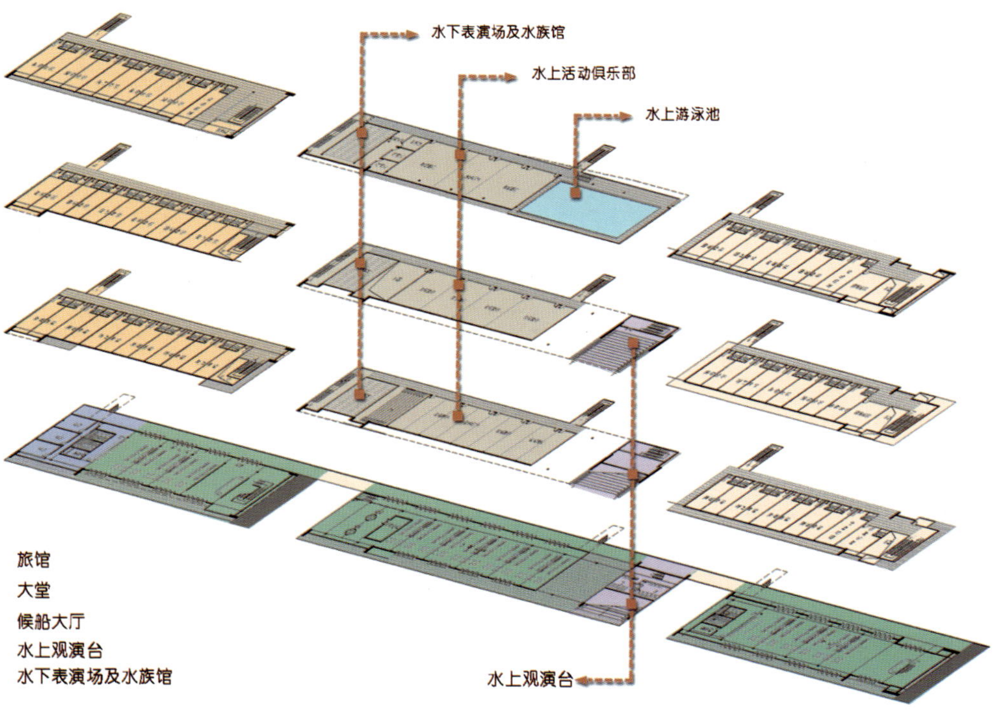

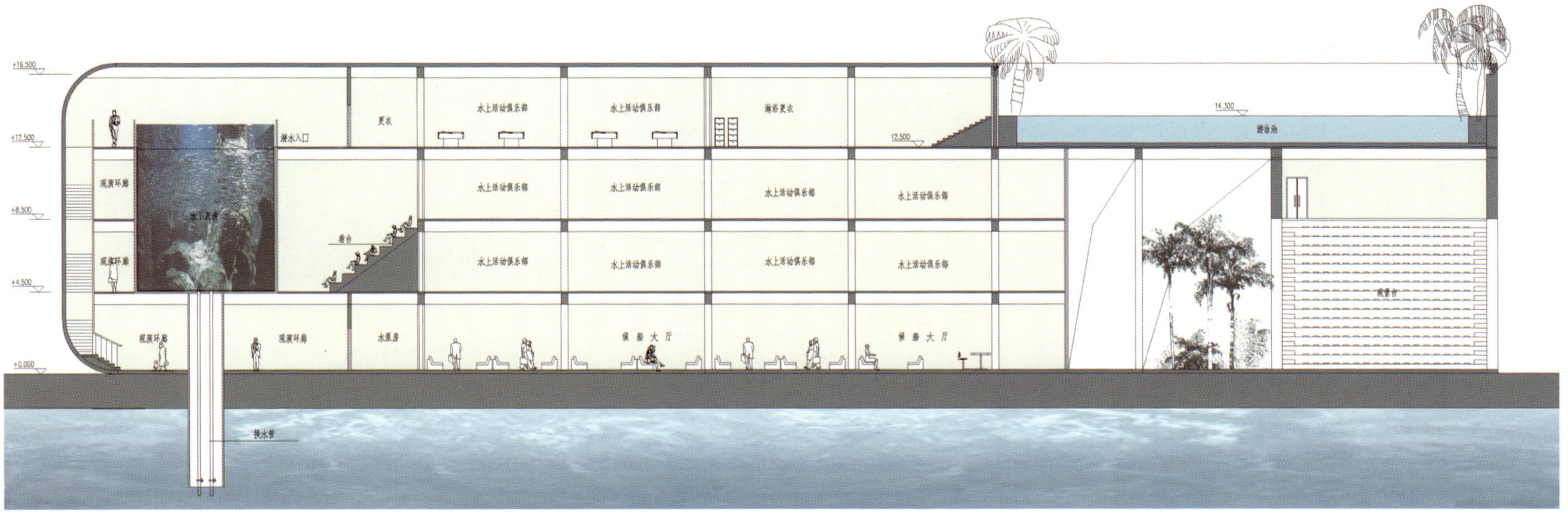

SOHO
SOHO

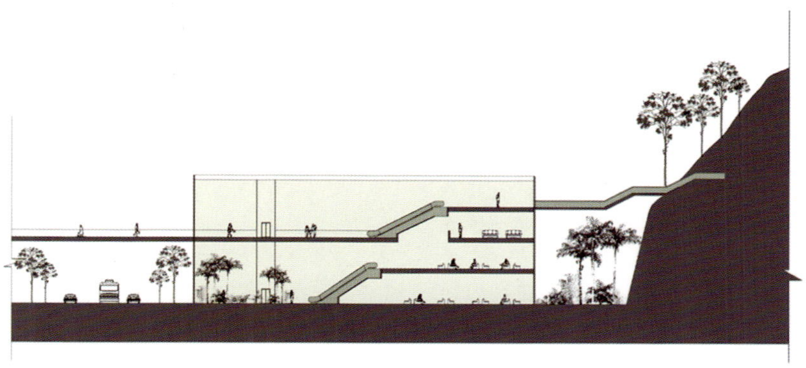

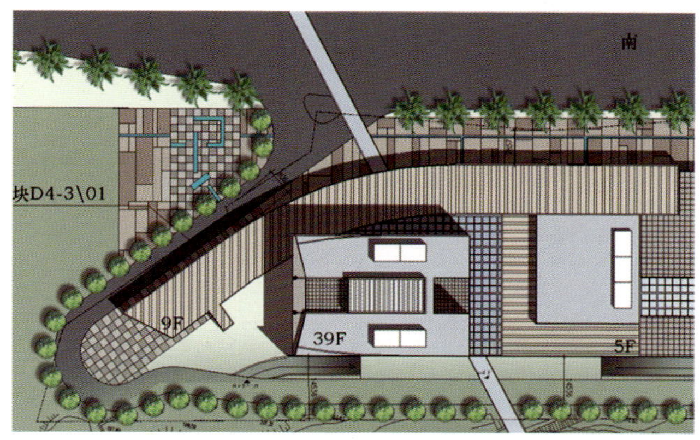
南
块D4-3\01
9F
39F
5F

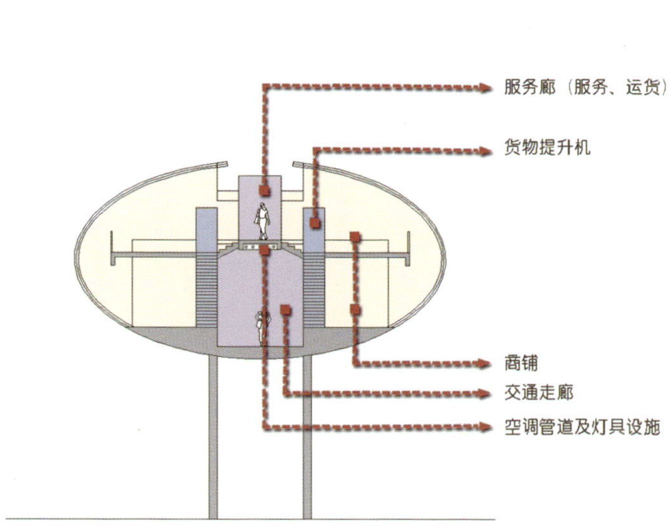
服务廊（服务、运货）
货物提升机
商铺
交通走廊
空调管道及灯具设施

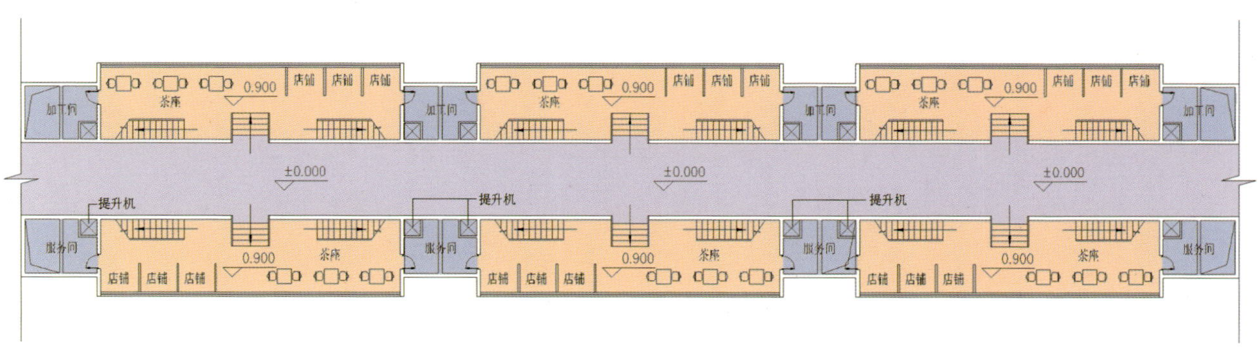
店铺
茶座
0.900
加工间
±0.000
提升机
服务间

Shenzhen Coastal Town

深圳海岸城

Gross Building Area: approximate 120,000m²
Gross Height: 30m, 5 floors above the ground, and 2 floors underground
Parking Space: more than 1,500

总建筑面积：约12万m²
总高度：30m，地上5层、地下2层
停车位：1,500余个

Coastal Town is located in Back Lake, South Mountain District of Shenzhen City, south-east of Back Coast Road, Innovation Road, north-west of Coast Avenue and Back Lake Avenue. It acts as a commercial core, leading Kempinski Hotel, Poly Cultural Plaza, Tianli Central Plaza to establish a commercial and cultural center in South Mountain District.

Coastal Shopping Mall combines famous department stores, top-class retailers, characteristic restaurants, fashion entertainments, high-grade cinemas, Skating Rink, etc., promotes various modern lifestyle, such as shopping, dining, leisure, sports, cultures, fitness and sight viewing. It gathers 50 more restaurants with great popularity worldwide, about 300 branded stores with various scales and features, becomes the most huge, functional, popular, typical and fashionable urban complex with coastal style.

海岸城坐落于深圳南山后海，东南毗邻后海滨路、创业路，西北接壤滨海大道、后海大道，以核心商业的姿态，携领凯宾斯基酒店、保利文化广场、天利中央广场，共同构建出南山商业文化中心区。

海岸城购物中心集知名百货、高级零售、特色餐饮、潮流娱乐、高档影院、真冰溜冰场等多功能休闲服务于一体，荟萃了购物、美食、娱乐、运动、文化、康体、观光等多种现代生活方式，云集了50家全球风行的特色美食，近300个规模不一、功能各异品牌店铺，是迄今为止深圳西部经营面积最大、功能最齐全、服务人群最广泛、最具代表性、时尚气息最浓郁的滨海风情多功能城市复合体。

海岸城

江南厨子
bossini
STARBUCKS COFFEE

AEON
COASTAL
CITY
海岸影城

恒生银行

海岸城
COASTAL
CITY

天使湾
铁板烧
亚马逊巴西烤肉
3F
4F

Corridor and Water-feature in Coastal Town

海岸城连廊及水景

Gross Building Area: approximate 120,000m²
Gross Height: 30m, 5 floors above the ground, and 2 floors underground
Parking Space: more than 1,500

总建筑面积：约12万m²
总高度：30m，地上5层、地下2层
停车位：1,500余个

South Mountain Bay area provides complete independent non-motorized transportation system, i.e. natural bay pedestrian promenade system, including Red Forest Sight Viewing Path (Hang Bridge Path), Coastal Pedestrian, Bicycle Path and Coastal Commercial Pedestrian, which function as public space, and connect the non-motorized transportation system with coastal leisure zone within Shenzhen Bay as well as the former Back Lake Pedestrian, realize connection of urban coastal space with the central portion. Urban integration into nature is achieved through directing designation of main flow of people, and taking advantage of urban public space, which leads to vast Shenzhen Bay over Shahe West Road.

South of the plaza is harmonious with water feature through simple lines and plain materials, shows modern, dynamic and leisure theme. North of the plaza takes 10,000m², equips with multi-functional stage, and super-large LCD, which provides opportunities for various shows, parties, commercial activities as well as public entertainments for citizens.

作为环境优良的公共开放空间，南山内湾地区将提供完善独立的非机动交通体系——自然海湾的步行系统，包括沿内湾的红树林观赏线路（浮桥步道）、滨水步行道、自行车道及滨海商业步行街等，令非机动交通体系与深圳湾滨海休闲带及原南山后海片区步行体系的衔接，将实现城市滨海空间与城市纵深地区的便捷联系，并通过对主要人行流线的引导性设计，利用城市的公共空间平台跨越沙河西路引向广阔的深圳湾，实现城市向自然运动的“无缝”连接。

南广场与水景完美结合，通过简单的线条和朴素的材质，展现现代、动感的休闲主题；北广场面积约10,000平方米，设有多功能大舞台和超大电子屏幕，为各类演出、聚会、商业活动和市民的公共休闲娱乐活动提供了广阔空间。

SUNING
创富
8633 1499
东方创富
东方创富

海德信数码港
招商业态
品牌电脑、笔记本电脑、数码产品、
数码通讯、数码摄像机、DIY装机商、
电脑配件、多媒体及网络产品。
招商电话：86356899
(0755)：86356399
天利中央广场
全球招商
招商中心
魅力女人

远邦教育
浦发银行

深圳公证处
SHENZHEN NOTARY PUBLIC OFFICE

美胸 纤体
享受好身材
海岸城
COASTALCITY

华尔街英语·中国
10年
您心中信赖的品牌
400 650 0661
本座二楼

得福轩 精品潮州菜
招商进行中
SUNING 苏宁电器

SUNING 苏宁电器

海岸城 COASTAL CITY

East Block of Shenzhen Coastal Mansion

深圳海岸大厦东座

Project Location: Nanshan Business and Cultural Center Zone of Shenzhen City
Gross Site Area: 68,377.9m^2
Gross Building Area: 295,819.2m^2
Floor Area Ratio: 3.07
Real Estate Developer: Shenzhen Coastal Real Estate Development Co., Ltd.
Partial Co-operation Partner: American Collison Office

建设地点：深圳市南山商业文化中心区
总用地面积：68,377.9m^2
总建筑面积：295,819.2m^2
建筑容积率：3.07
开发商：深圳市海岸房地产开发有限公司
局部合作设计单位：美国凯里森事务所

Located in the Nanshan Business and Cultural Center Zone of Shenzhen City, it is a component of the Coastal Town and composed of business podium buildings and high-rise office towers. It is adjacent to Kempinski Hotel in the east, and Poly Cultural Plaza in the south, the gross building area is 85,700 m^2, and the area for commercial operations is 20,000 m^2. The main body of the building is composed of a main building which has 24 floors, and an attached building which has 14 floors. It is 100m in height, and a 7m high sky garden is set for each floor, in order to create ecological and healthy concept, as well as the unique coastal style.

The shopping pedestrian promenade in the podium connects the street level and upper floor city corridor into the inside of the mall creating a pedestrian friendly shopping promenade. The office tower building is of 100m and 120m in height and the standard floor adopts internal-tube and external-frame design. The glass barrels form simplified and characteristic features; air garden provides rest place to staff and the stereoscopic greening system brings the building with a rich ecological sense.

海岸城东座位于深圳南山商业文化中心区，是海岸城的组成部分，由商业裙房和高层办公塔楼组成。东侧紧邻凯宾斯基酒店，南侧是保利文化广场，项目总建筑面积约85,700m^2，其中商业面积约20,000m^2。主体建筑包括24层高的主楼和14层高的附楼，主体建筑高度100m，每层均设有约7m高空中花园，营造生态、健康的概念，具有独特的滨海气质。

位于裙房部分的商业内街将地面和架空的城市走廊空间引进建筑内部，形成吸引人流的步行商业街。办公塔楼高1000m及120m，标准层采用内筒外框式布置；通过跌落的体形加上相互咬合的玻璃筒体形成建筑简洁大方而个性鲜明的特征；穿插于平面不同位置的空中花园，给办公人员提供休憩放松的场所，而立体的绿化系统使建筑带有浓郁的生态气息。

海岸地产

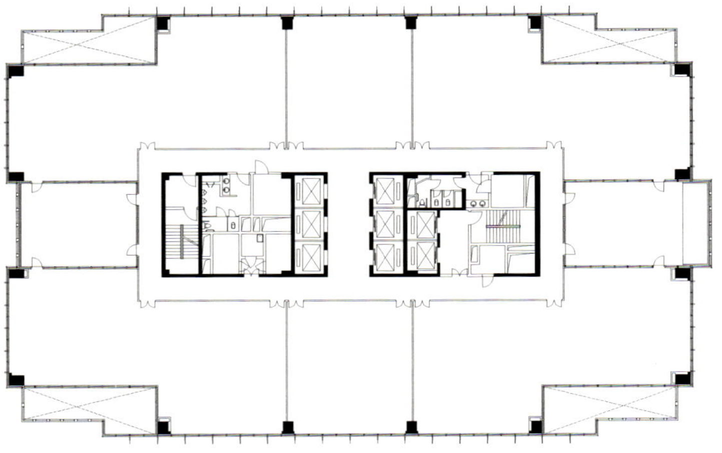

十九、二十一、二十三层平面图

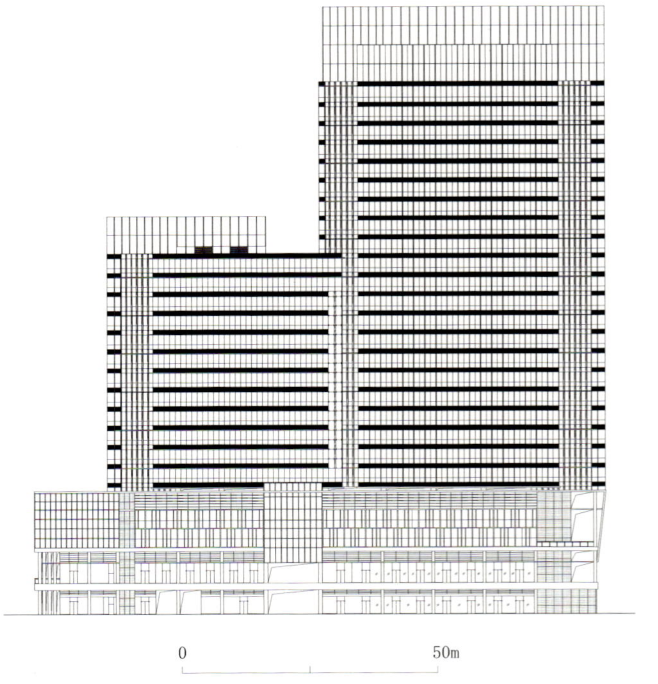

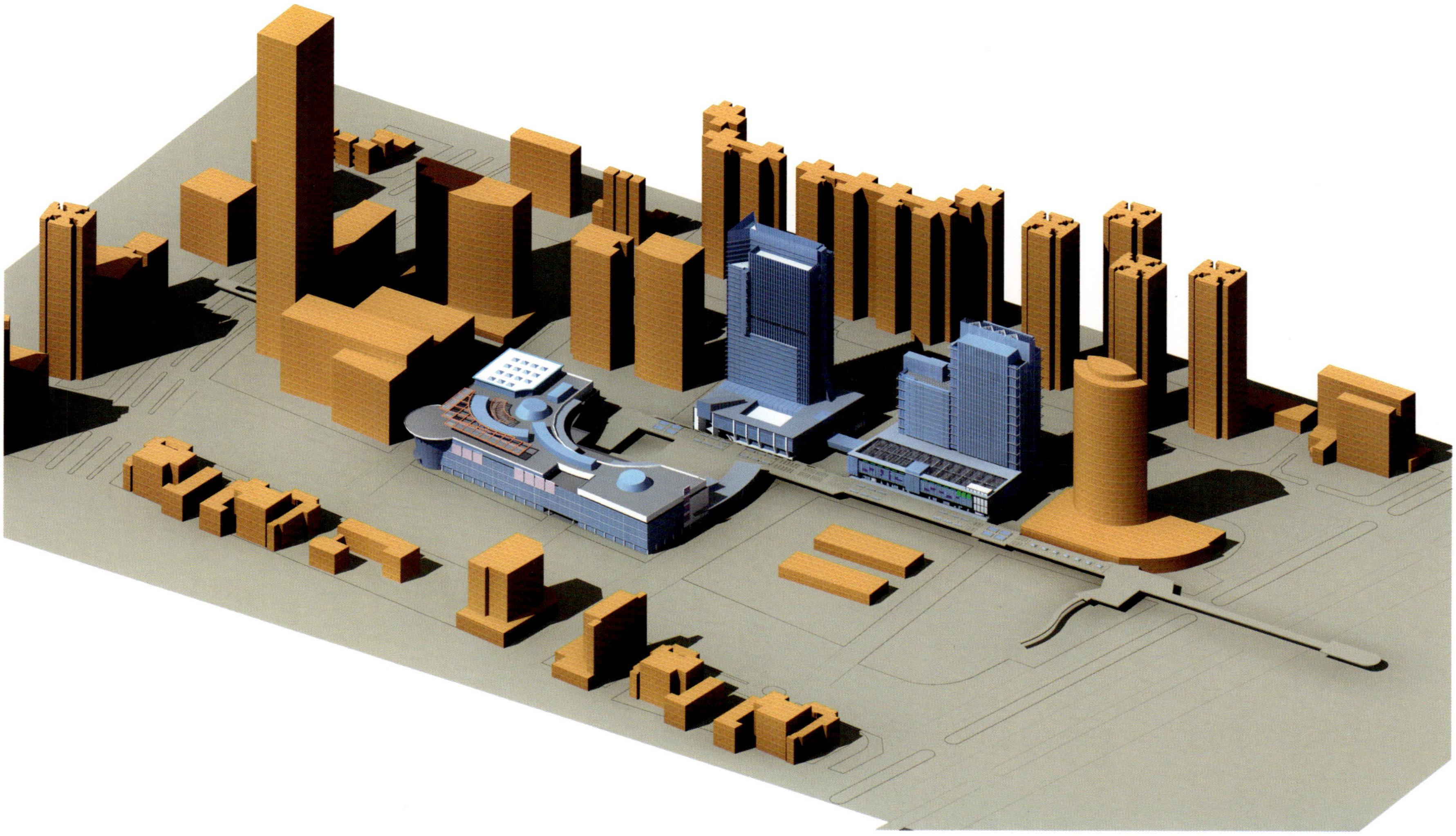

旅游
88875222

West Block of Shenzhen Coastal Mansion

深圳海岸大厦西座

Site Area: 14,600m²
Building Area: 85,000m²
Building Height: 155m, 32 floors above the ground and 2 floors underground
Architecture Design: American Collison Office & Shenzhen Chenmai Architecture Design Co., Ltd.

项目占地面积：1.46万m²
建筑面积：8.5万m²
建筑高度：155m；地上32层，地下2层
设计单位：美国凯里森建筑事务所、深圳市城脉建筑设计有限公司

It is located in the Nanshan Business and Cultural Center Zone of Shenzhen City. Standard height for each floor is 3.75m, the super huge luxurious lobby is 13m in height and takes 700m² area; tower building provides 130300m², which is flexible to divide and combine, the largest combination area can achieve 2000m², meets customers' demands for space. The private garden style sun & green office is original, 2 private gardens are set for each floor of the tower. The unique and luxurious elevator halls and CEO rest rooms are set for floors above 20. Getting rid of the straight image of office buildings, this project creates a "Multi-D Polylines" façade, remodels the fashion and business image of Shenzhen office buildings.

海岸城西座位于深圳南山商业文化中心区的核心区。标准层层高为3.75m；13m高、700m²的超大豪华大堂；塔楼空间由130~300m²,可灵活分割、组合，最大组合面积可达整层面积2000m²，完善满足客户的空间需求；原创私家花园式阳光绿色办公空间，塔楼每层均设有2个私家花园；20层以上独备阔绰豪华候梯厅及总裁卫生间；区隔写字楼惯用的直线形象，独创"多维折线"的外立面，重塑深圳写字楼的时尚商务感。

Nanshan District and is located in the sub - center of the belted structure city. The Center Zone plays an important role in the development strategy of "1 line connects 3 dots" for Nanshan District Government. As future commercial, business and cultural center in the west of Shenzhen, Nanshan Center Zone aims at top-class modern coastal urban, the master piece of Nanshan District construction; the long-term planning of it makes people feel more confident with the development potential in the future.

B东

B东

保利国际影城
VERASACE

B东
订座热线

Raffliers Plaza

来福士广场

Real Estate Developer: Capitaland Real Estate
Architecture\Landscape\ Interior Design: Sparch Architecture Design Company
Building Area: 150,000m²
Site Area: 97,000m²

开发商：凯德置地
建筑\景观\室内设计：思邦建筑设计公司
建筑面积：15万m²
占地面积：9.7万m²

Raffilers Plaza is located at the intersection of East Second Ring Road and Dongzhimennei Street of Beijing City, adjacent to Guijie Street famous of 24-hour dining service. Its site area is 150,000m², connects to Dongzhimen underground transportation center, i.e. the largest transportation harbor in Asia, which is in the north-east corner of the plaza.

The entire project is composed of office, retailing, resident and service apartment functions, which are associated and independent to each other. The association allows each commercial body supplementing to each other, while the independence allows them operating separately. It is challenging for the architects to meet the need of higher floor area ratio and independent unique entrance for each functional zone. And it is more complicated that vehicles are forbidden to enter service area and underground of west and south of the plaza by urban regulations.

The entire project is composed of 4 independent buildings, which are a 90m high office building having 21 floors, a department store podium having 5 floors, a

来福士广场选址于北京东二环与东直门内大街的连接要塞，毗邻以全天候餐饮营业而闻名的簋街，占地15万m²。地下与广场东北角的亚洲最大的交通港——东直门地下交通枢纽相连。

整个项目包含办公、零售、住宅及服务公寓4个功能，他们之间相互连接又彼此独立。相互连接可以使各商业体互补，而其独立性又可以令各商业体独立营运。较高的容积率及业主对每个功能区须具备独立个性及独立入口的要求，使建筑师面临挑战；而城市规范限制车辆进入落客区及广场西面和南面的地下层，使情况看起来更加复杂。

整个项目由4幢独立建筑组成：东北角是一幢90m高的21层办公楼、一幢5层的商场裙楼，裙楼之上东面是一幢17层的服务公寓；西面则是一幢75m高的15层住宅板楼。思邦选择将塔楼与裙楼分离，这样既降低建筑的体量，又能独立调整个体至最佳朝向。

我们将设计重点放在基地的东北角，办公楼及商场裙楼被转向45°以

17-floor service apartment building located on the east-top of podium, and a 75m high resident building having 15 floors located in the west. Sparch makes the tower separated from the podium, in order to decrease the architecture volume, as well as optimize the direction of each building respectively.

Designation focuses on the north-east corner of the site, the office building and department store podium are turned at 45° to match the diagonal of transportation center, which extends to external space, forms a narrow opening between the office building and department store podium, and provides a public corridor connecting urban and the resident area in the rear of this project. A glass crystal body connects the inside of department store with office building, and forms an entrance lobby of office building. The resident building and service apartment located above the department store podium are interlaced with each other, and set inclined a little to obtain more daylight as well as keep distance from the office building nearby.

2 resident buildings are connected through the grid-pattern club located on the top of podium, whose functions include apartment lobby service, swimming pool, fitness club and restaurants.

The basic concept of architects is to surprise and impress each customer who enter and walk around Raffilers Plaza. Therefore, Sparch desires to provide a comfortable space allowing customers go through Second Ring Road and landscape belt. Varying landscapes forms a leisure space as well as leads flow of people to the entrance of department store in the north-east corner. Fountain and huge LED on the curtain wall located at the entrance of department store, makes the plaza become a fashionable landmark.

The entrances of department store podium are set at the north-east and south-east corner, also the main windows open to Second Ring Road. The core of the courtyard, which is 5-floor in height, is a crystal lotus, i.e. a crystal patterned hanging structure. It rises from the Food Zone underground, and holds the curved glass roof on the top. The edges extends while rising, make it possible to set stalls with large area on high floors, and allow Sparch designing activity space below the crystal lotus hanging structure. 1500m² spaces for retailing inside of the crystal lotus, decreases pressure and adjusts the depth of peripheral stalls.

对应对角点的交通枢纽。此视角对应延伸至空间，形成办公楼与商场裙楼间的狭缝，提供了一条连接城市与项目背面住宅区的公共走廊。其中，一个玻璃水晶体提供商场与办公楼的室内连接，并形成办公楼的入口大堂。

位于商场裙楼上方的住宅楼和服务公寓俯视呈交错状，并且略微倾斜以摄取更多光线及与邻近的办公楼之间保持最大的距离。

两幢住宅板楼由位于裙楼顶部的带有格子状花纹的会所连接，会所功能包括服务公寓大堂、游泳池、健身房及餐厅。

建筑师的基本理念是带给每一位步入并留连于来福士广场的顾客以惊喜、难忘和享受其中。由此，思邦渴望提供给人们一个愉悦的空间，让人们自由穿越于二环路及景观带间。变化的景观造型提供休憩空间，并引领人流至东北角的商场入口。商场入口的喷泉及幕墙上的巨型液晶显示屏，使广场成为这一地带的流行地标。

商场裙楼的入口设于基地的东北角和东南角，成为商场面向二环路的主要窗口。5层高的商场中庭的核心是一个水晶状的悬臂结构——水晶莲。它由地下食品区拔地而起，向上支撑起顶部的弯曲状的玻璃屋面。由于其周边往上延伸时扩张，这样大面积的商铺可置于高层，使思邦得以在位于水晶莲悬臂结构的下方设计活动空间。水晶莲的内部可提供约1500m²的零售空间，使周边的店铺面积压力减少，以调整周边店铺的纵深范围。

业主对于商场流线的重点设计参数之一是吸引消费者前往上层。水晶莲螺旋状的设计，能带给人不断变化的空间体验，并引领人们的视线往上移，可吸引人们往上层探索。

在顶部，水晶莲穿过中庭泛光的玻璃顶面，令视线可以穿越裙楼6层顶部的空中花园。由裙楼顶部看来，水晶莲呈现出如雕刻般生动的表面，与花园景观相得益彰。

对于天花，思邦建议采用背后发光的灯带设计，连绵的灯带在天花形成动向，引领人们游走于商场内。

办公楼大厅光滑的几何造型，由一系列的分段式拱形结构构成。门架

Owners value one of the important parameters of department store streamlines, which is to attract consumers to the upper floors. Spiral designation of the crystal lotus provides varying experiences of each space, and leads the consumers to explore towards upper floors with their sight of view.

The crystal lotus goes through the blinking glass roof of courtyard on the top, makes it possible to view the sky garden located on the top of podium; viewing from it, vivid, sculptural appearance of the crystal lotus is in harmony with the landscape in the garden.

As to the ceiling, Sparch suggests to adopt lamp belts blinking from the rear, and the continual lamp belts makes the ceiling dynamic, as well as leads consumers walk around the department store.Smooth and geometrical shape in the office building lobby, is composed of a series of segmented arch structures. The mast crosses between the office building and 2nd floor of department store lobby. The height differences between them makes the mast look like a crank, which rises from the entrance towards the center gradually, and the top of it is 12m high. Over-viewing the lobby through the podium, there is a dent area functioning as a café. The wall and ceiling of the café and elevator hall are covered with large spiral crossing wooden blinds, which allow filtering and directing light, as well as shelter the installation of motors.

The club located on the top of the podium can be accessed through the special lobby in the south of Raffliers Plaza, then the elevators transport customers to the 2-floor high lobby on the top of the podium. The space of lobby extend to the lift swimming pools face to the elevator doors, the unique connection pattern of swimming pools makes the tiny hotel lobby more spacious.

The club is covered by glass and gradient black-white striped aluminum plates, whose color and size match the glass curtain walls of the 2 resident buildings perfectly. The swimming pool is located between the 2 resident buildings, held with a series parallel crank mast. A strip of transparent glass belt goes through at an angle, makes the outside scenery, landscape on the top of podium and the crystal lotus visible. The free pattern of the crystal lotus, the club and the office building lobby is used to connect functional zones next to each other, and makes an attractive public space. A 3D space composed of triangle surfaces provides several functions, the curtain walls of each building are connected, and the variability of triangle surface combination can be changed with the requirements of interior functions.

The designation of the crystal lotus takes tiny space of bottom floor into consideration, makes an extremely huge mirroring space under the hanging arm based on the tiny base. A lobby with high ceiling is formed by extension of club external layer, which connects to the swimming pool, wraps and hides the elevator shaft and motor room. Similarly, an irregular space located in office building lobby is extended with the connection of dent rooms with various areas.

The club and office building are surrounded by sculptural glass roofs, which form a roof platform eventually. Sparch desires to make this area an urban garden, another vigorous public space in the city. People could join all kinds of activities held in Raffliers Plaza, and enjoy the crystal lotus in the center of the department store, as well as the continuous mountains in the west of Beijing City with a special angle.

The shopping mall is an independent area, whose stalls face inside, while the walls face outside. Therefore, Sparch desires to provide the original boring spaces with a dynamic external layer. The surface of curtain wall is composed of pixel black-white alternative glass units, tiny black-white spots covers the surface; the area of spots are increased on purpose, in order to be visible from distances. The aluminum plate with identical spots is placed behind the glass, makes the façade with feeling of depth, and the overlapped spots are changed with the movement of people. The pattern on office building façade functions as the boundary between transparency and non-transparency, the black spots extend from the non-transparent area covers floor to the transparent area above and below themselves.

Pixel spots extend from the department store podium to the club and the 2 resident buildings, and the density of pixels decreases with the rising height of the office building.

In the evening, colorful LED effect of Raffliers Plaza responds to the visible effect in daytime, looks much more dynamic.

横跨于办公楼和商场大厅的二层楼板间。由于办公楼和商场间的高度差，使门架呈曲柄状。这一空间从入口处逐渐往中央升高，最高12m。从裙楼一侧俯视大厅，有一块类似凹室的区域，未来可作为咖啡厅。咖啡厅和电梯间的墙壁及天花板由大面积的呈螺旋形交叉延伸的木质百叶覆盖，百叶不仅具有过滤和导向光线的作用，同时也遮挡了机电安置。

位于裙楼顶部的会所由来福士广场南侧一层的专有大厅进入，然后乘电梯到位于裙楼顶的两层高大堂。大堂的空间延伸到电梯门对面的提升式游泳池，这种与游泳池并列的独特连接方式使原本面积不大的酒店大堂看起来更加宽阔。

会所由玻璃及渐变的黑白相间的铝制夹心板覆盖，夹心板的颜色和尺寸与上方两幢板楼的玻璃幕墙浑然一体。游泳池位于两幢板楼中间，就像办公楼大厅一样，由一系列平行的曲柄式的门架支撑着。其中一条透明玻璃带斜向穿越，使人可看见室外的景色及裙楼顶的景观与水晶莲。

项目中的自由形态结构水晶莲、会所、办公楼大厅都可用来连接相邻的功能区，并营造瞩目的公共空间。由三角面组成的三维空间有多项功能：首先他们将建筑物间的幕墙连接起来，其次三角形面的组合可变性则能根据不同的室内功能要求而变化。

水晶莲的设计，考虑到底层的空间较小，因此底座很小，但在悬臂下方却形成很大的镜面空间。会所外层的延伸形成一个挑高的大厅，连接游泳池，然后包裹并隐藏大楼的电梯井及机电空间。同样，办公大厅形成一个不规则的空间，不同面积的凹室连接形成一个延展性的区域。

会所和办公楼被雕塑式的琉璃屋顶环绕着，最后形成屋顶天台。思邦的理念是将此区域设计成城市花园——城市中又一个充满活力的公共空间。从这里人们可以参与来福士广场的所有活动，并且从一个独特的角度来欣赏商场中庭水晶莲，当然还有北京西面连绵的群山。

购物中心是个独立的世界，店铺的门面均朝里，而墙体朝外。所以，思邦对商场幕墙的设计理念是将其原本看似毫无生气的空间打造成动感外层。幕墙表面由像素化的黑白相间的玻璃单元组成，小块的黑白斑点布满幕墙表面，斑点的面积故意加大，以便处于远距离也清晰可见。有同样斑点的铝板被置于离玻璃150m²的背后，这样使立面看上去有纵深感，并且重叠的斑点在人移动时会产生变化。办公楼立面的图案被用作透明与不透明区域的界线，其黑色斑点会从遮盖楼板的不透明区域延伸至其上和其下的透明区域。

像素斑点从商场裙楼一直延伸至会所及两栋板楼，像素的密度随着办公楼的高度逐渐减少。

夜晚，来福士广场用多彩的LED效果来呼应白天的视觉效果，从而变得更具动感。

地下停车场入口
BASEMENT PARKING ENTRANCE
5
ZARA

MANGO

STA COFFEE
ZARA
ZARA

swatch
L'OCCITANE

4楼见
sundian.com
巴厘
泰国
ZA

EVE'S TEMPTATION
X-MOOM
COSTA COFFEE
ZARA

Levi's
ITALIAN TIME

ZARA

G-STAR RAW

Raffles City
PROVOST
L4-05
L3-06
SEPHORA
L1-12a
SEPHORA
SEPHORA
SEPHORA

Tianjin Jin Mao Plaza

天津金茂广场

Project Location: South of Heping District, Tianjin City
Site Area: 24774.9m²
Gross Building Area: 123874.50m², 26211.83m² for Commercial Operation
Real Estate Developer: Tianjin Jin Mao Investment & Development Co., Ltd.
Architecture Design: Liang Huanggu & Yi Heng Design & Consultant Co., Ltd.

地点：天津市和平区南市
用地面积：24,774.9m²
总建筑面积：123,874.50m²，26,211.83m²（商业部分）
发展商：天津金茂投资发展有限公司
建筑设计公司：梁黄顾・艺恒设计顾问有限公司

High-rise: To avoid the clumsy volume high-rise buildings, the plan of this project adopts the pattern of a windmill, and emphasizes a slim volume composed of 4 pieces. 4 independent and separated slim pieces are set well-proportioned, and the lines wrapped with aluminum alloy plates of external walls shows the straight and towering style. In addition, the straight elements symbolizes powerful trunks, which rises from the bottom, and spreads braches at the top as the decorations, makes the mansion become the most characteristic landmark of this region.

Commercial Operation: To associate with commercial activities, and differ from the appearance of Food Street nearby with novel image, external walls of this project adopts geometrical 3D steel frame composed by irregular triangle pieces, aluminum plates embedded with patterns, or characteristic textural glass. Since main commercial activities are to gather characteristic high-grade restaurants in China, corresponding designation of each stall should be applied inside of the building. The internal inclined glass walls of 1st to 3rd floor allow customers

高层：为了打破高层笨重的体量感，高层的平面采取风车型格局，强调四片纤薄的体量。四个独立分离的薄片高度起落有致，再透过外墙铝金属片包裹的直线条，表现出大厦垂直高耸的形态。另外，直立的元素模拟苍劲的树干，从底部往上节节上升，最后在顶层散开为壮茂的分枝作为大厦顶部的装饰，成为该区有特色的地标性建筑。

商业：为了配合商业活动的动态，并务求做到标新立异的新颖形象来对邻近现在的食品街的外观作巨大反差，商业外墙采用了不规则三角碎块的几何立体钢架，中间嵌着有图案的穿孔铝板，或是有特色的质感玻璃。考虑到商业主要是云集各国特色的高级食肆，商业建筑内部应有相应的铺面设计。首先是对内部首层、二、三层区街的斜面玻璃墙体，让顾客可以一面进餐一面俯视商业内街的观景和活动。另外斜面玻璃墙体结合帐蓬式的金属和玻璃框架，让室内的进餐活动可以伸延到半室外的空间，同时遮挡着偶然下雨或下雪的严厉天气，让商业内部的食品街增加不少生气，带动整体商业餐饮活动的蓬勃发展。

24小时自助服务
HOUR BANKING

to overview the scenery and activities inside Commercial Street while they are having diners. In addition, the inclined glass walls and the metal and glass frame shaped like a tent extends the diner rooms to semi-outside space, as well as shelter it from rain and snow. The designation makes the Food Street inside much more vigorous, and leads the entire commercial and dining activities well-developed.

北京同仁堂

总平面图

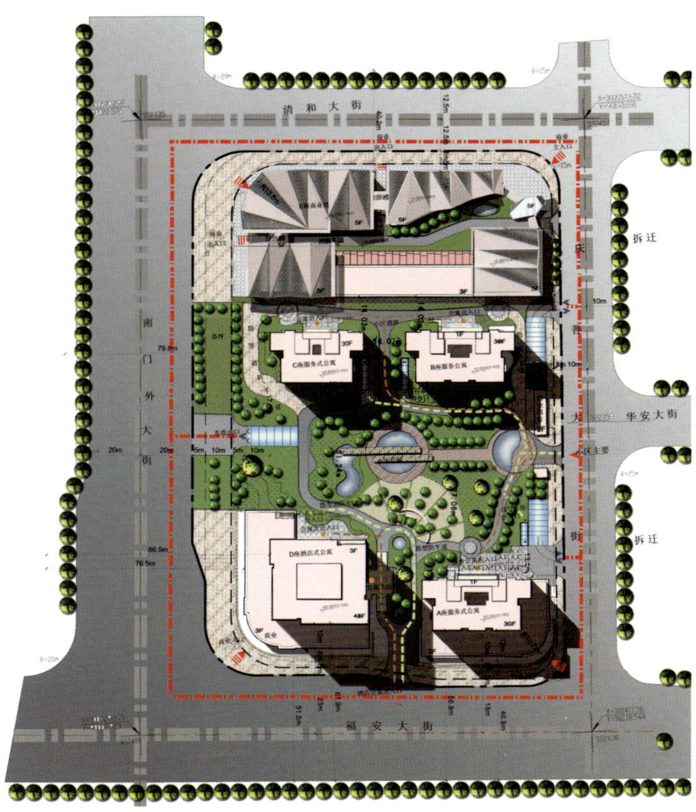

N

图例：

- 新建公寓建筑
- 新建商业建筑
- 原有建筑
- 地下车库
- 规划路中心线
- 规划路边线
- 道路退缩线
- 用地红线
- 小区道路

主要经济指标

序号	指标名称		指标数
1	地块总用地面积		24774.9m²
2	地上总建筑面积(计入容积率面积)		123874.50m²
3	其中	商业部分建筑面积	26211.83m²
		服务式公寓部分建筑面积	49546.05m²
		酒店式公寓部分建筑面积	47916.62m²
		物业管理用房建筑面积	200m²
4	地下总建筑面积(不计入容积率面积)		49550m²
5		地下商业部分建筑面积	1769m²
		地下汽车库建筑面积	44348m²
		地下自行车库建筑面积	1859m²
		地下设备用房	1574m²
6	容积率		5.0
7	建筑密度		35.2%
8	总户数		1339
9	车位数		947
10	绿地率		25.8%
11	建筑高度	A栋服务式公寓	99.95m
		B、C栋服务式公寓	93.85m
		D栋酒店式公寓	155.55m

主要户型指标

序号	类型		户数比
类型	酒店式公寓	带独立厨房	78.95%
		无独立厨房	21.05%
	服务式公寓	小三房	33.33%
		大、中两房	44.83%
		大三房	21.84%

总平面图

比例 0 5 10 15

20060430

发展商：天津市金茂投资发展有限公司

天津市金港·国际项目建筑设计方案

设計顧問：梁黄顧建築師（香港）事務所有限公司

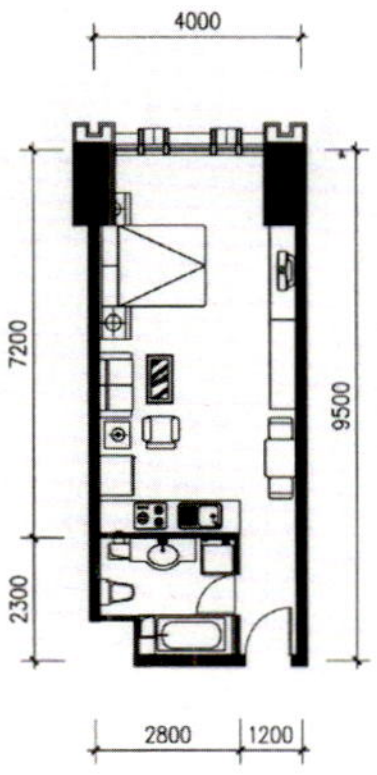

A型

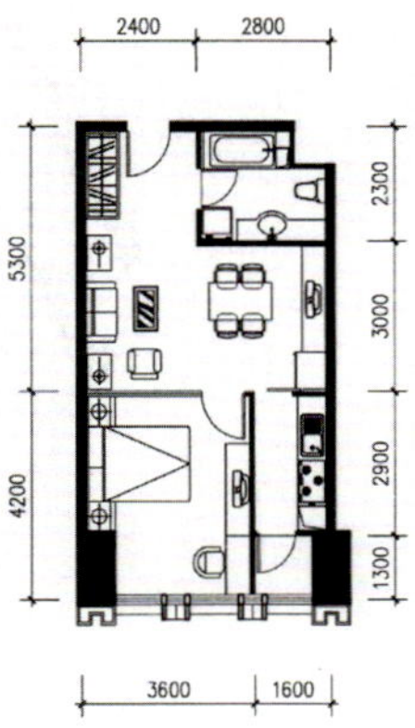

B型

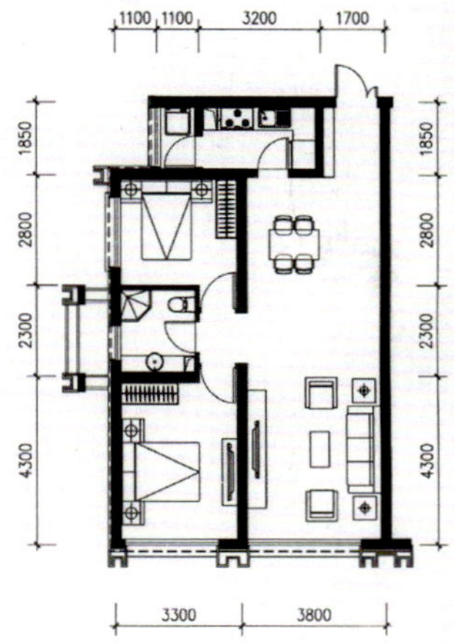

C型

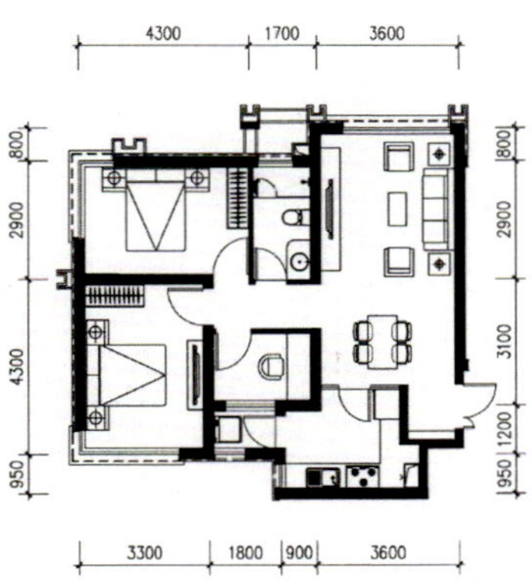

D型

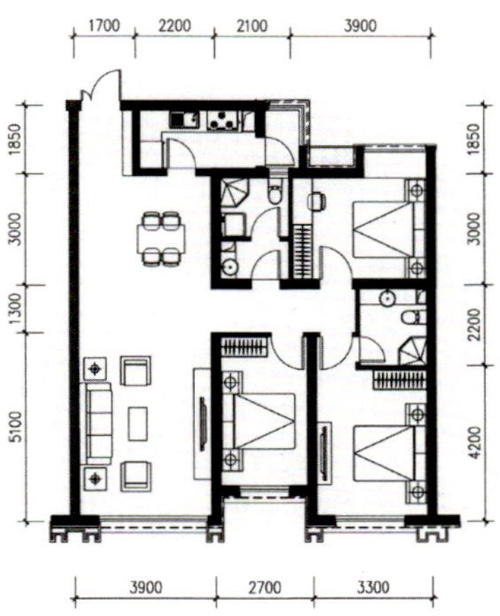

E型

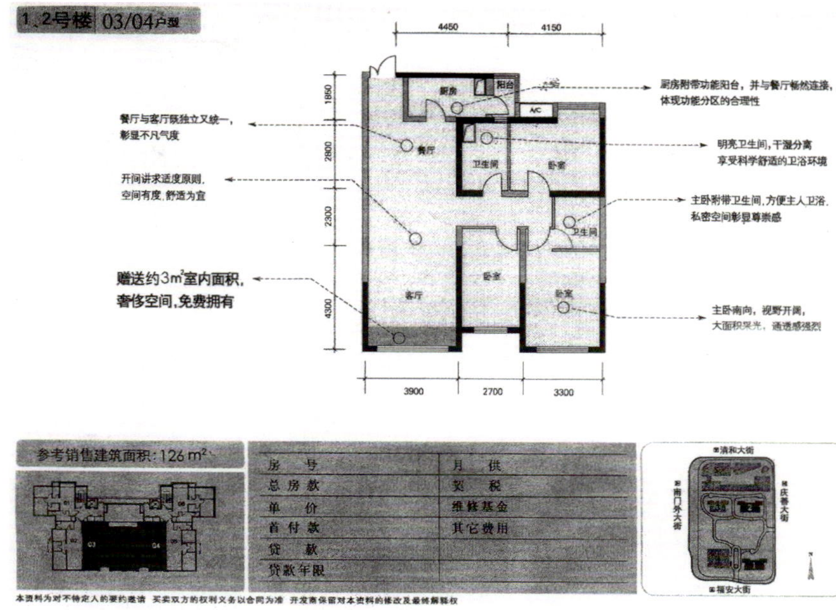
1、2号楼 03/04户型
厨房附带功能阳台，并与餐厅畅然连接，体现功能分区的合理性
餐厅与客厅既独立又统一，彰显不凡气度
明亮卫生间，干湿分离 享受科学舒适的卫浴环境
开间讲求适度原则，空间有度，舒适为宜
主卧附带卫生间，方便主人卫浴，私密空间彰显尊崇感
赠送约3m²室内面积，奢侈空间，免费拥有
主卧南向，视野开阔，大面积采光，通透感强烈
参考销售建筑面积：126 m²
房号
月供
总房款
契税
单价
维修基金
首付款
其它费用
贷款
贷款年限
清和大街
南门外大街
庆春大街
福安大街
本资料为对不特定人的要约邀请 买卖双方的权利义务以合同为准 开发商保留对本资料的修改及最终解释权

EGO Fashion Plaza

EGO潮流广场

Building Area: 16,649m²
Real Estate Developer: Chengdu Pioneer Shangpu Real Estate Co., Ltd.
Storey: 8 floors in total, 3 underground and 5 above the ground

建筑面积：16,649m²
开发商：成都前锋上普实业有限责任公司
楼层：共8层。地下3层，地上5层

EGO Fashion operates 4 categories business, including fashion dresses, characteristic foods, dolls and comic & animation, and sports, provides a fashion industry platform, which combines exhibition, releasing, promotion, experiencing, interaction, entertainment and shopping. EGO Fashion Plaza gets rid of the operations of traditional department store on business layout, provides the whole market a consistent theme of seabed to sky, and names each floor characteristically. The stereoscopic, open and return-around shopping mode contributes in overlapped, deferred and cycled consumption, and optimizes the commercial opportunists. The ramped block is the unique and fantastic commercial body among the city.

EGO潮流广场经营潮流服饰、特色美食、玩偶动漫、运动无极限四大类，全力打造一个集展示、发布、宣传、体验、互动、娱乐、购物于一体的潮流产业平台。在业态分布上，EGO潮流广场抛弃了传统商场的做法，而是将整个卖场统一为一个由海底到天幕的主题，每个楼层的名字也颇有特色，且每个楼层的装修风格不同，带给顾客强烈的视觉感受和欣赏美感。立体开放、动线回返的街区式购物模式，构成叠加消费、递延消费、循环消费，创造最大商机。缓坡街区，全城独潮的超炫超酷商业体。

不潮流
元
攻略
二楼
服饰
皮具
美食
133

壹啊潮流广场
上东大街·主题潮流街区
非EGO，不潮流！
100元
消费攻略
负一负二楼
300家潮流小铺
盛情开放
服饰
皮具
美食
淘淘EGO
EVERYTIME U CAN EGO
JUST FOR NEXT
VERY!
壹啊潮流广场
133

非EGO，不潮流！
100元
消费攻略
负一负二楼
300家潮流小铺
盛情开放
服饰
皮具
美食
EGO
EVERYTIME U CAN EGO
JUST FOR NEXT
潮流广场

「X」JTYS&TYC
全场 6.8 折

L-LARGE
L-LARGE

91

↑WRANGLER
↑LEE
A1-10-A1-15
L1
SWISSUS

时尚 第一

Wii

尚出色
流专卖
A4-12~A4-24
L4

Shenzhen King Glory Plaza

深圳金光华广场

Project Site: Luohu District of Shenzhen City
Gross Site Area: 7,335.8m^2
Gross Building Area: 101,051m^2
Floor Area Ratio: 13.7
Real Estate Developer: Jinguanghua Real Estate (Group) Co., Ltd.
Design Consultant: Hong Kong Liu Rongguang & Wu Zhenmin Architecture Office

建设地点：深圳市罗湖区
总用地面积：7,335.8m^2
总建筑面积：101,051m^2
建筑容积率：13.7
开发商：深圳市金光华实业（集团）有限公司
设计顾问：香港刘荣广伍振民建筑师事务所

King Glory Plaza is the largest modern shopping mall in Shenzhen Special Economic Zone. It is located on the site of originally famous Nanguo Film Studio in Shenzhen, surrounded by landmark buildings in early reform and opening -up times, such as International Trade Mansion and Development Center Mansion, etc. However, the neighborhood was in declining for its backward facilities & traffic situation with the urban extension towards Futian. The completion of project promotes the restoration of commercial & economic vigor in the entire area.

A huge city square and underground pedestrian promenade face to the south of Renmin Road, extend the external space to the internal space. A large courtyard is set opposite to the city square, and 3 sets of courtyards are layout inside to form a movement space in front, rear, top and bottom, as well as guarantees the continuity of commercial streamline. B2 and International Trade Mansion station of Subway No.1 are connected to form the first subway shopping facility in Shenzhen. The shape of the building is simple and exquisite, the internal space is various and prosperous, 100% lease ratio & efficiency in operation confirmes the importance of a successful design for commercial project development.

金光华广场是深圳特区内最大的现代购物中心。地处深圳南国影院原址，周围遍布深圳市改革开放早期的地标性建筑物如国贸大厦、发展中心大厦等，但随着城市向福田延伸，该区域因城市设施的落后及交通条件的恶化正面临城市的衰退，金光华广场的建成带动了整个区域商业经济活力的恢复。面临人民南路一侧形成巨大的城市广场和下沉式步行街，使所有的室外空间要素都被延续到室内，面对城市广场处设置了大型中庭，而向内延伸则布置了三组中庭从而形成上、下、前、后的流动空间以保障商业流线的延续性。地下二层更与地铁一号线国贸站平接，从而形成深圳首个地铁商业空间。建筑造型简洁精细而室内空间丰富热闹，其百分之百的出租率和良好的经营效率均说明成功的设计对于商业建筑项目的开发是何等重要。

光华璀璨
金光华广场盛大开幕

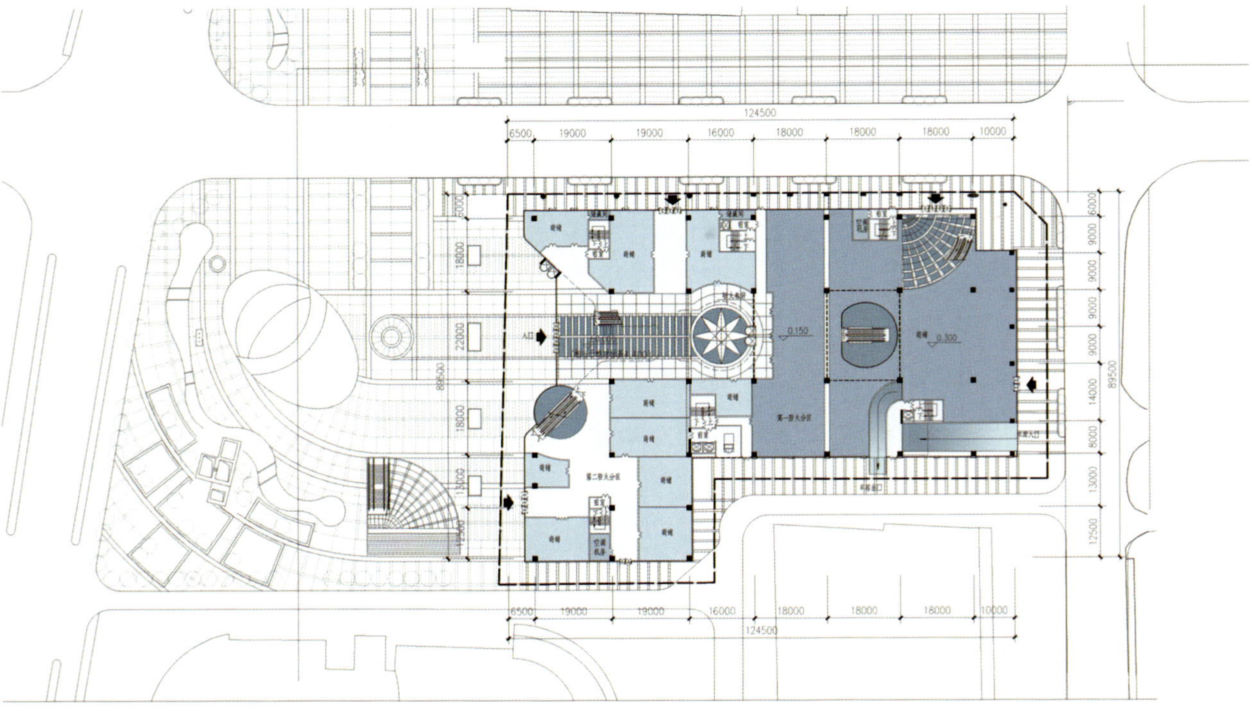

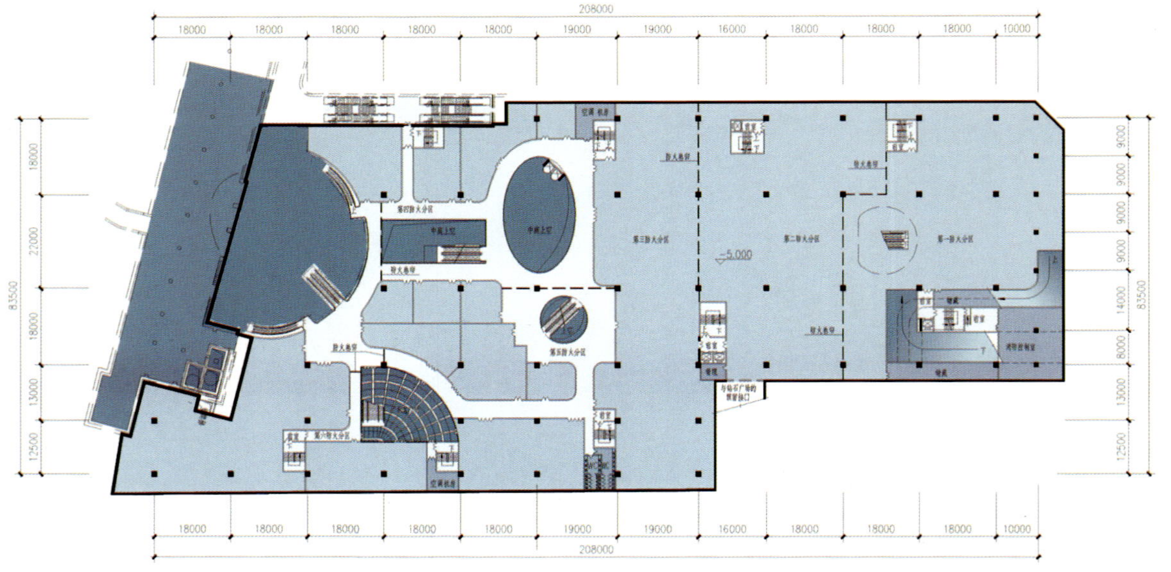

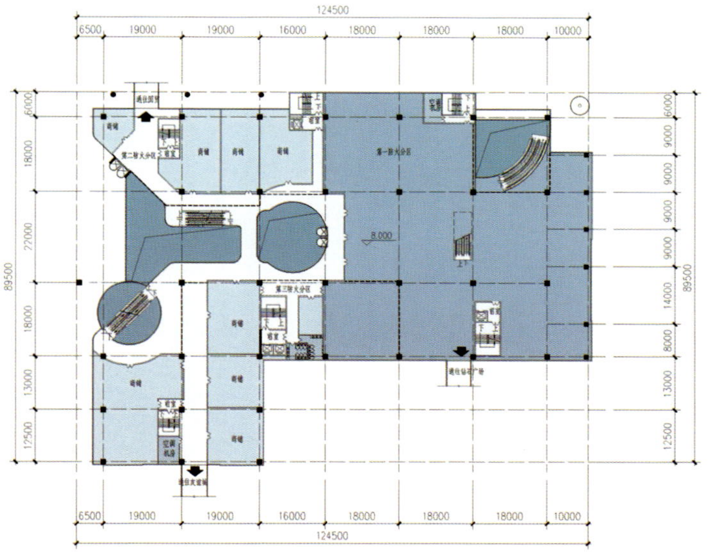

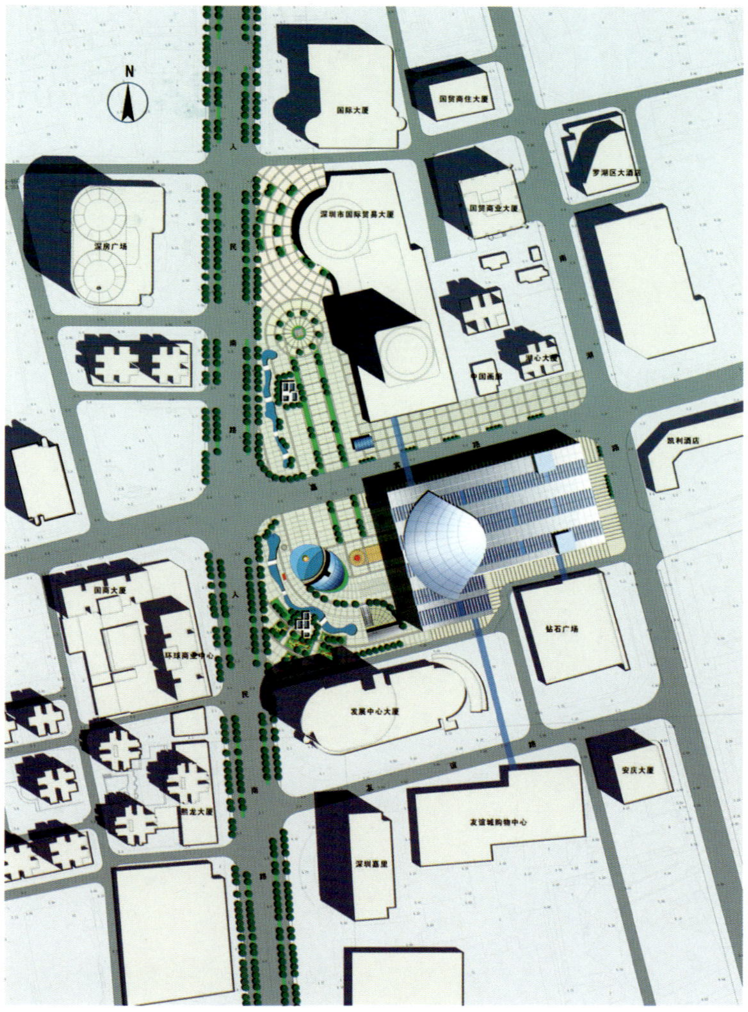

CHOW TAI FOOK 周大福 CHOW TAI FOOK

ESPRIT

MAYBELLINE
MIKIMOTO

Xishuangbanna Shawan International Commercial Plaza

西双版纳莎湾国际商业广场

Location: Jinghong City of Yunnan Xishuangbanna
Client: Xishuangbanna Haicheng Real Estate Developing Company
Site Area: 48,000m²

项目位置：云南省西双版纳州景洪市
发展商：西双版纳海诚房地产开发有限公司
占地面积：4.8万平方米

In Xishuangbanna, Shawan International Commercial Plaza consists of shopping, tourism, lie fallow, entertainment, catering and firstly created Shopping Mall. Created a different commercial shopping liftstyle and space character, it has kept the history culture and development, creating a new era culture in order to make the environmental culture stand out. According to the street situation and original road structure of Shawan International Commercial Plaza, they got a concept of Shawan International space arrangement and construction character. Picking up the Dai nationality construction character; Setting up the entrance fountain square according to the Dai nationality history of nine walls and twelve towers. And also built Chinese (Dai nationality), Burma, Laos and Thailand street in order to establish Shawan International into a commercial plaza of internationalization. The proper integration of commercial and residential space got an effect of neat space and various sight views. And meanwhile the commercial constructions are connected to the residential constructions forming a multi-level space which made up of various formats of Shawan International Commercial Plaza.

在西双版纳，莎湾国际·商业广场集购物、旅游、休闲、娱乐、餐饮为一体，在版纳首创SHOPPING MALL。通过创造不同的商业购物活动方式和空间环境特征，保留其历史文化和发展，创新时代文化，突出其环境的文化特征。根据“莎湾国际”商业广场街区现状空间环境特点及原有道路的组织结构，提出了“莎湾国际”商业广场空间布局和建筑特点。艺术提取传统傣族风格建筑的符号；利用傣族“九井十二塔”的历史设置入口水景广场；设置中（傣）、缅、老、泰四国风情街，塑造“莎湾国际”商业广场成为一个国际化的商业城。商业空间与住宅空间有机组合，得到了空间收放有序，景观变化丰富的效果。同时，将商业建筑与住宅建筑连成一体，形成多层次的空间网络，建立“莎湾国际”商业广场丰富的空间结构。

以“莎湾国际”商业广场为主线，再现中（傣）、缅、老、泰四国的风貌。保留了版纳大厦，在街道格局的基础上，重点对体现特征的屋面、墙面和地面进行设计，留下西双版纳重要的历史信息和耐人寻味的文化内涵，使“莎湾国际”商业广场与景洪市环境形成反差，用对比的手法显现景洪市的发展变化，

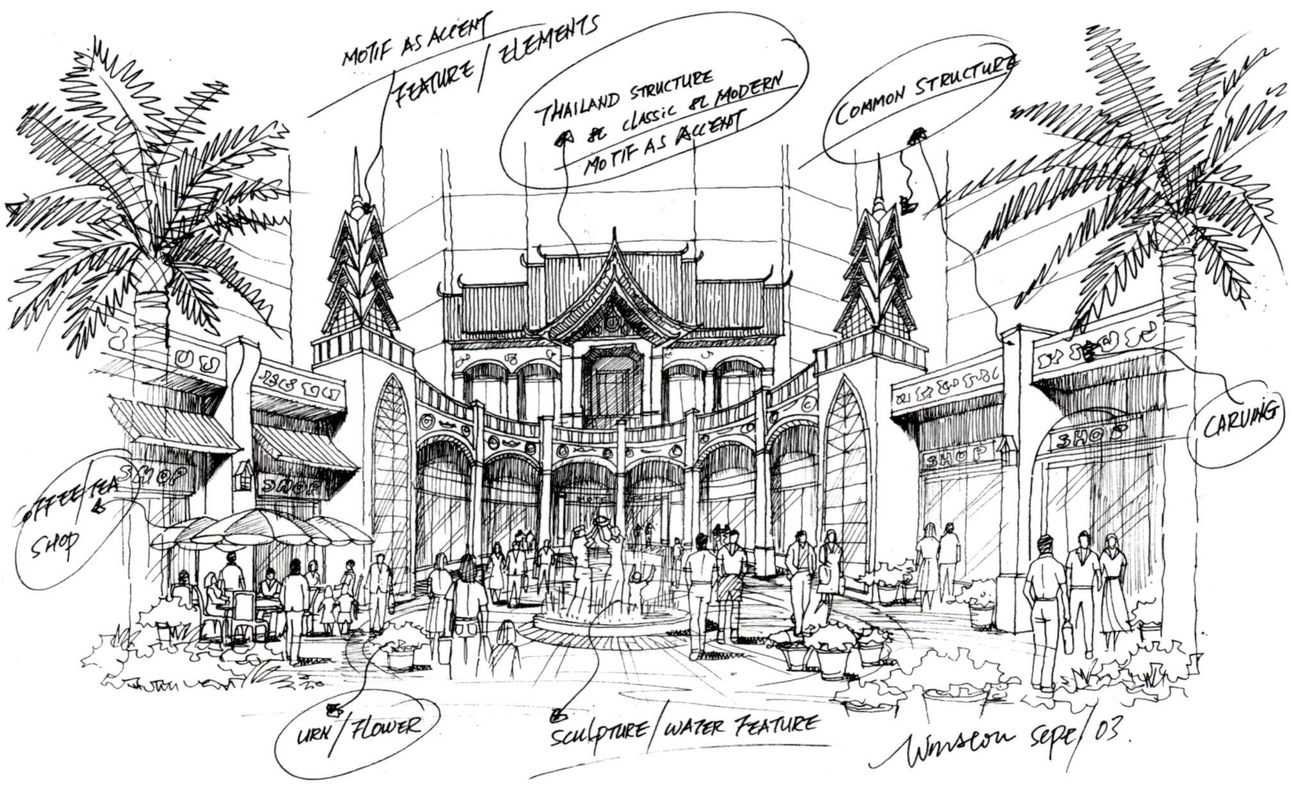
MOTIF AS ACCENT
FEATURE / ELEMENTS
THAILAND STRUCTURE & CLASSIC & MODERN MOTIF AS ACCENT
COMMON STRUCTURE
CARVING
COFFEE/TEA SHOP
URN/FLOWER
SCULPTURE/WATER FEATURE
Winston sepe/03.

LIGHTING
CARVING
PALM TREE
COMMON STRUCTHRE
MOTIF AS ACCENT
"Interior Concept"
Winston sepe/0
FEATURE W/FLOWER

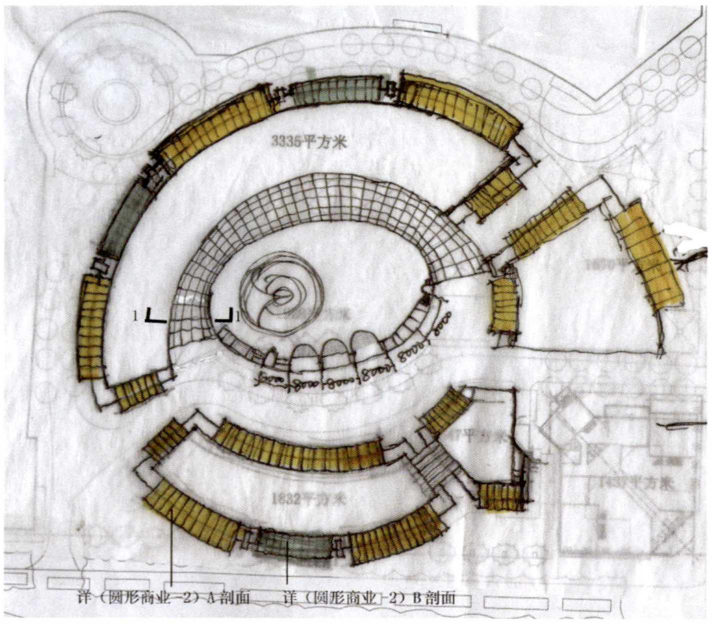

3335平方米
1832平方米
详（圆形商业-2）A剖面
详（圆形商业-2）B剖面

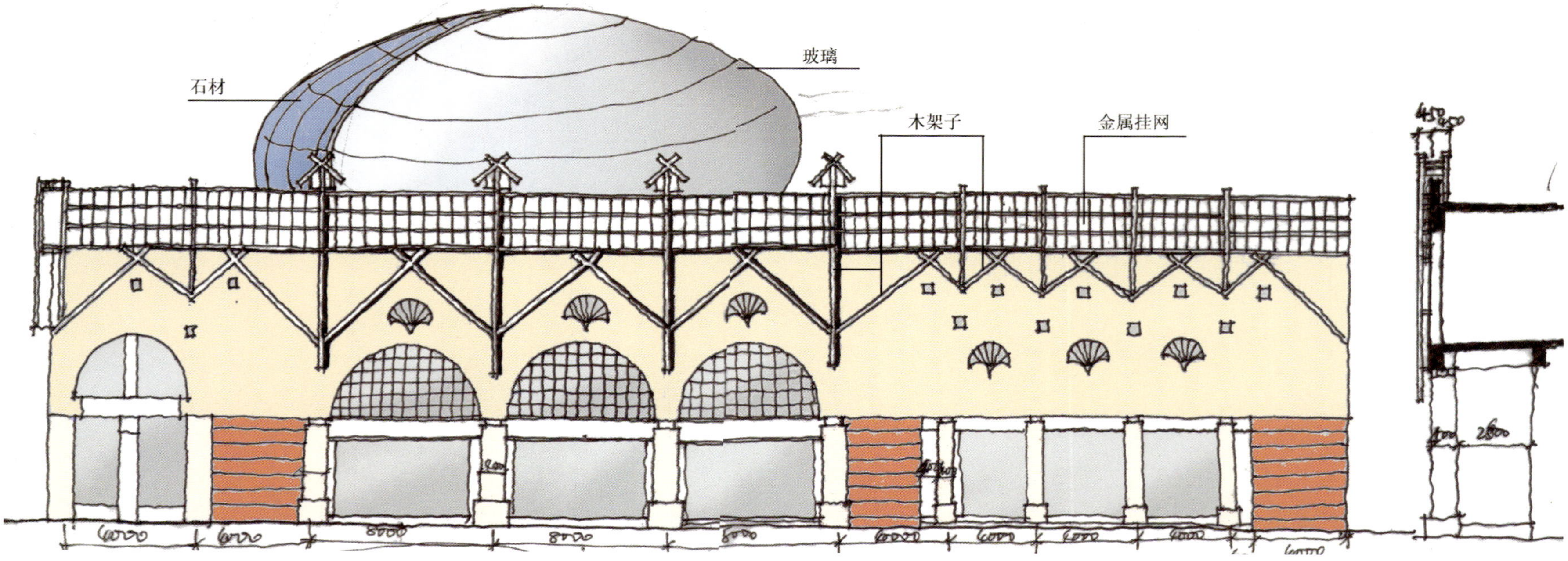

石材
玻璃
木架子
金属挂网

Taking Shawan International Commercial Plaza as the main clue, the master plan represents the various styles of Dai nationality, Burma, Laos and Thailand. On the basis of the street pattern, keeping the existing Banna Tower, the project redesigns the roofing, wall and floor, which show the main features of the building. It forms a stark contrast between Shawan International Commercial Plaza and the context of Jinghong city to restore the significent historcial information and the charming cultural connotations. It is by means of contrast to represent the development of Jinghong city, so as to drive the furthur development of the city, even that of the whole region of Xishuangbanna. Meanwhile, by its theme (with the traditional style of Dai nationality) elements of the architectural space, the project keeps in line with the dimension, style and contents of the Banna Tower, becoming a charateristic and symbolic complex architectural space, which represents the unique style of Xishuangbanna. The design strengthens the core subjective function and the space logics; by building its surrounding environment and guiding the business development, it forms an area which has a powerful tourism function.

从而带动景洪市乃至整个西双版纳的发展。同时，通过建筑项目作为商业街区内主题性（具有傣族传统风格）建筑空间的构成元素，使之与版纳大厦共同形成一组尺度、风格、内容较为协调的、特色突出的、且具有西双版纳形象代表和象征意义的建筑群体空间。方案强化核心空间主体职能和空间逻辑，并通过对其环境的塑造，商业业态的引导，形成极具旅游功能的地区。

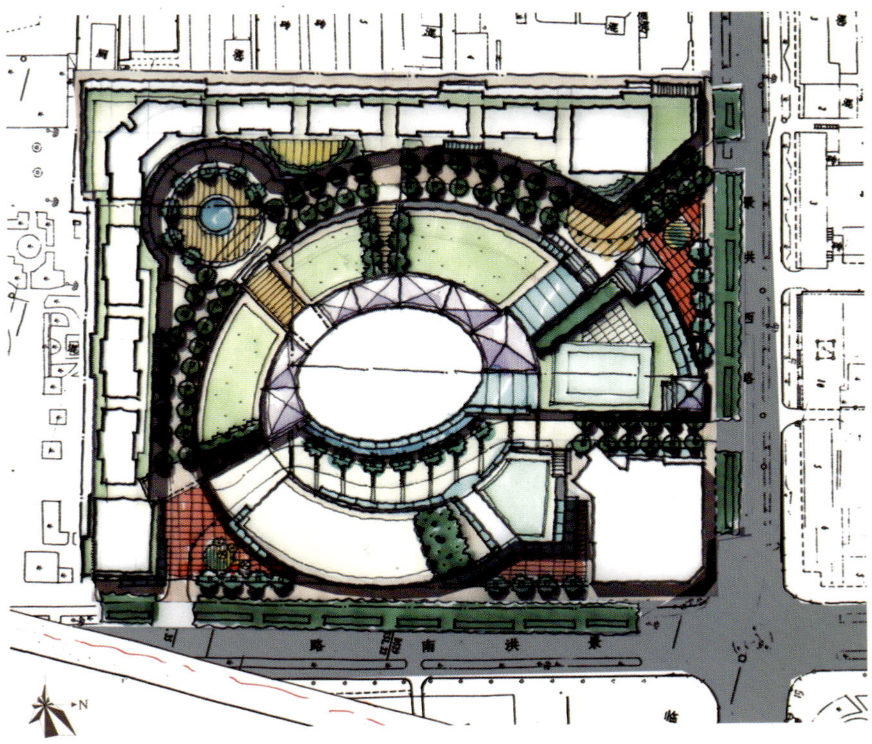

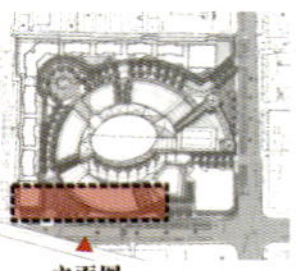

立面图

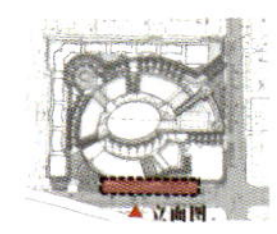

立面图

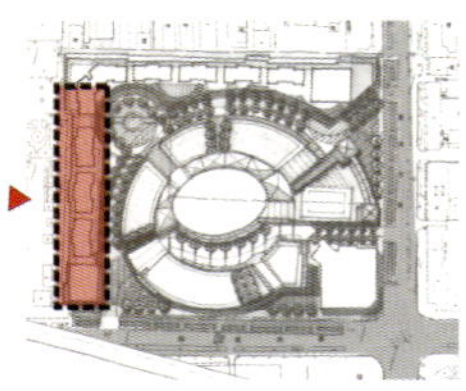

立面图

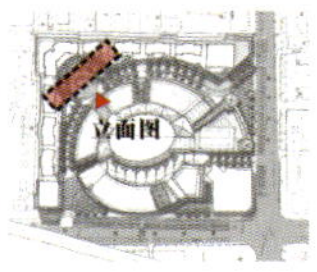

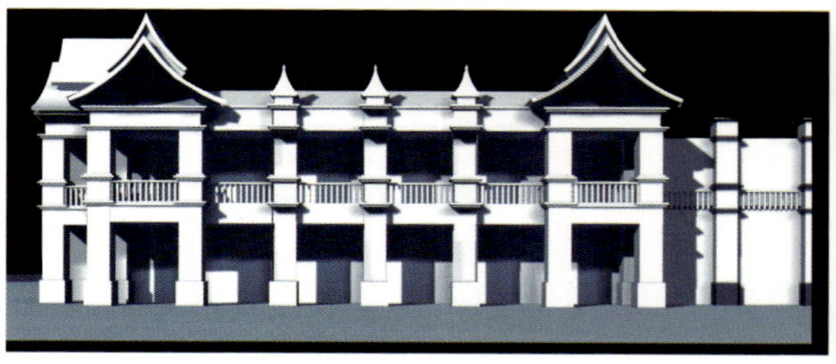

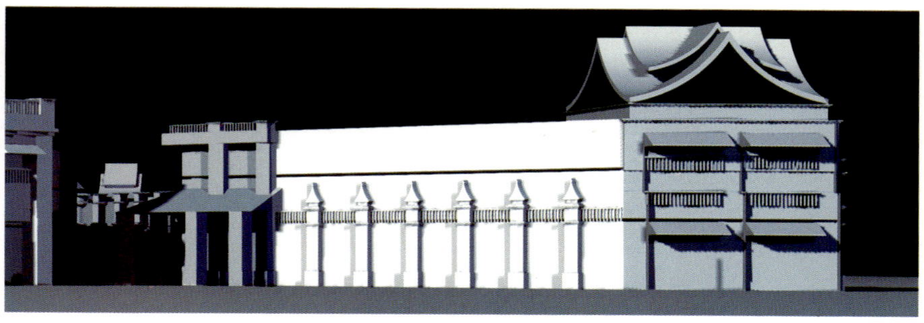

立面效果圖
Elevation

立面效果圖
Elevation

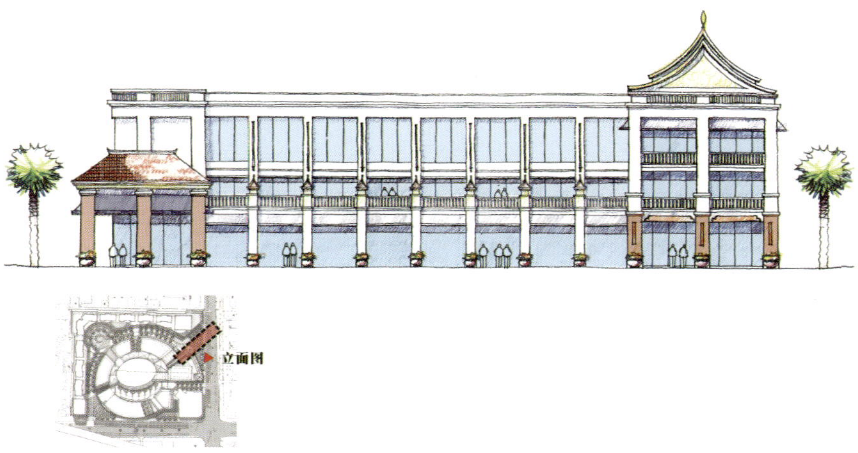

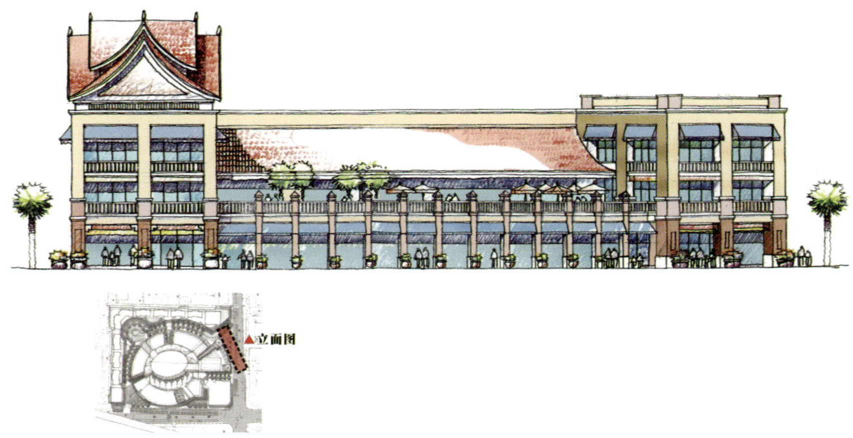

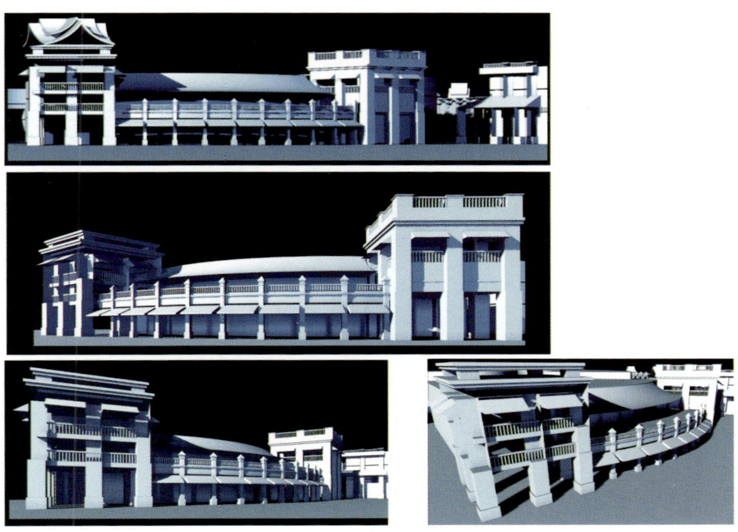

立面模型效果圖
Elevation

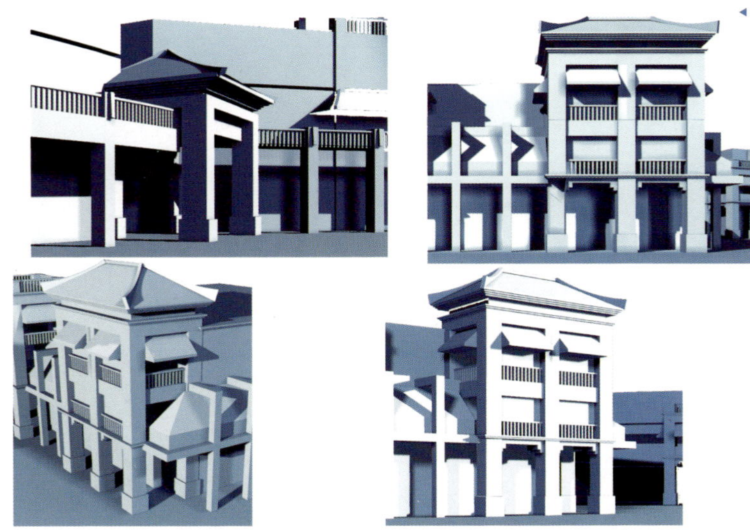

立面模型效果圖
Elevation

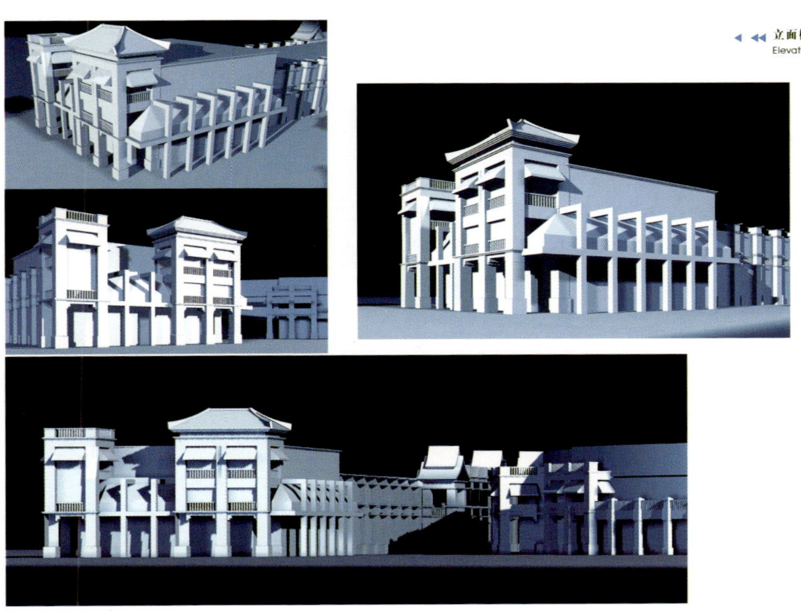

立面模型效果圖
Elevation

Jiaozhou old city commercial center

胶州老城区商业中心

Address: Jiaozhou, China
Project type: commercial center
Size: 205,020m²
Building design: Hanshi (international) Building Design

地址：中国胶州
项目类型：商业中心
规模：205,020m²
建筑设计：翰时（国际）建筑设计

The base is located in old city commercial center in Jiaozhou, with planning size of 66,000m².

This project includes business center, hotel and residence. Business center is located in the south of the planning plot. Designers try their best to strive for more frontage shop in order to improve the commercial value of the shops. People can get into the inside of business center, with more capacity to evacuate crowd. The centralization business principle entrance is located in southwest of the plot, and bigger business circle is shaped with Qingniao Store on the opposite side. Located in northwest, the hotel has a good visual field from the entrance because of the tall spot. The hotel will be the new landmark of the city. The residence is located in quiet northeast, the separate residence zone, which is easy to manage and communicate. The residence in axle wire of Chenghuang Temple should correspond with Chenghuang Temple in the sight and landscape, with the Subtraction of size.

基地位于胶州市原老城区商业中心区，规划总用地面积6.6万m²。

本设计主要功能划分区为商业、酒店、住宅。商业基本布置在地块的南侧，最大限度地争取沿街商业的长度，提高店铺的商业价值，人们可从多个点进入到商业内部，更具有快速疏散人流的能力。其中集中式商业主入口位于地块的西南侧，与对面西侧的青鸟国货形成更大的商圈。酒店布置在西北角，为用地内的制高点，入口具有良好的视野范围，将作为新的城市地标出现。住宅布置在相对安静的东北侧,易于形成比较独立的住宅区，易于管理和交流。位于城隍庙中轴线上的住宅，在体量上做减法，最好的在视线和景观上与城隍庙相呼应。

SHOWCASE

A&S
胶州老城区商业中心改造项目总体规划设计
JIAO ZHOU OLD TOWN COMMERCIAL CENTER
RENOUATION PROJECT COMPREHENSIVE PLANNING DESIGN
总平面图
MAIN PLAN

首层平面图

酒店
酒店式公寓
超级市场
商场
商铺
住宅

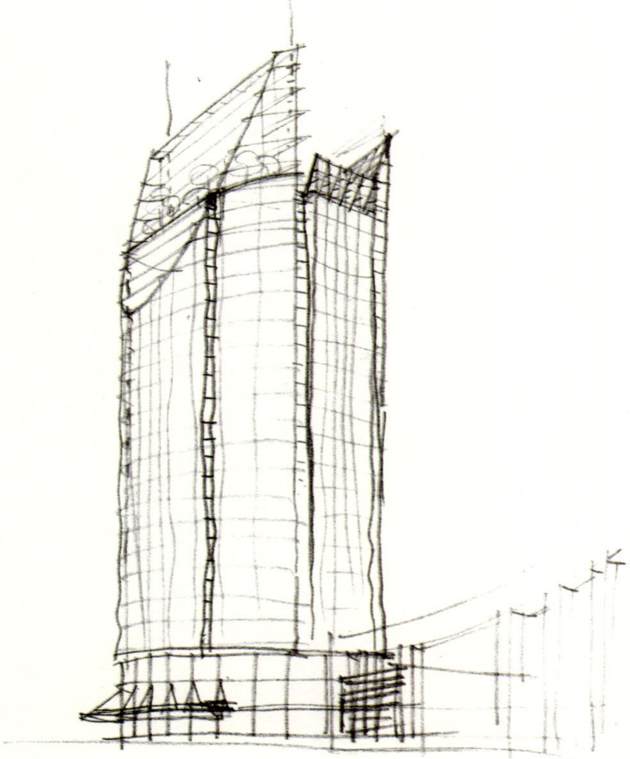

海岸城（西座

招商银行
CHINA MERCHANTS BANK

骏马
海岸物业

海岸会所
海岸城（東座）
BBFLY
KTV

华尔街英语
ICBC 中国工商银行

ABN·AMRO
荷兰银行

圣罗兰

BBFLY
量贩式
KTV

海岸城(西座)

KTV

TILEY
天利名城

Poly Cultural Plaza

保利文化广场

Real Estate Developer: Poly Group
Architecture Design: Guangzhou Zhujiang Overseas-funded Architectural Institute & Guangzhou Urban Planning & Surveying Designation Institute
Site Area: 54,000 m^2
Building Area: 150,000m^2

开发商：保利集团
建筑设计：广州珠江外资建筑设计院、广州市城市规划勘测设计研究院
项目占地：5.4m^2
建筑面积：15万m^2

The project is located in the Nanshan Business and Cultural Center Zone of Shenzhen City, which is a landmark building, combines theaters, cinemas, commercial operations, restaurants, culture and arts. It features with culture, and embodies with commercial activities, promotes the concept of cultural commerce creatively. The fresh commercial model interests famous merchants in worldwide. The planning focuses on cultural functions, and emphasizes the integrity of cultural and commercial equipments. In addition, the planning of central area focuses on ecological environment construction, and forms a complete greening system through connection of several greenbelts. Due to relative gathering of super high-rise buildings in the architectural layout, it highlights the landmark building, forms a well-proportioned space, and creates a perfect image of central area. Modern and three-dimensional transportation system of the transportation plan emphasizes the construction of subway and public traffic equipments, and creates a convenient transportation environment in central area.

Nanshan Business and Cultural Center Zone functions as Service Center of

该项目位于深圳南山商业文化中心区，是集剧院、影院、商业、餐饮、文化、艺术等多项功能于一体的标志性建筑。主要以文化为特色、商业为主体，创造性地提出文化商业概念，开创新的商业形态，吸引了全球知名商家入驻。规划重视文化功能，强调文化设施与商业设施的有机融合；另外中心区的规划特别重视生态环境的建设，通过绿化带的连接，形成了完整的绿化体系。建筑布局规划中超高层建筑相对集中，突出设置了标志性建筑，形成重点突出，错落有致的空间形态，塑造了完美的中心区城市形象。在交通规划中设计了现代化、立体化的交通体系，强调地铁与公交设施的建设，创建了中心区便捷的交通环境。

在深圳带状城市结构中，南山商业文化中心区处于城市副中心位置，是南山区区级服务中心。在南山区政府“三点一线一中心”发展战略中，中心区承担了重要的角色。作为深圳西部未来的商业、商务和文化中心，南山中心区的目标是建设成为具有国际一流水平的现代化海滨城区，成为南山区建设的精品，中心区的长远规划让人们对其未来发展潜力充满信心。

透视图

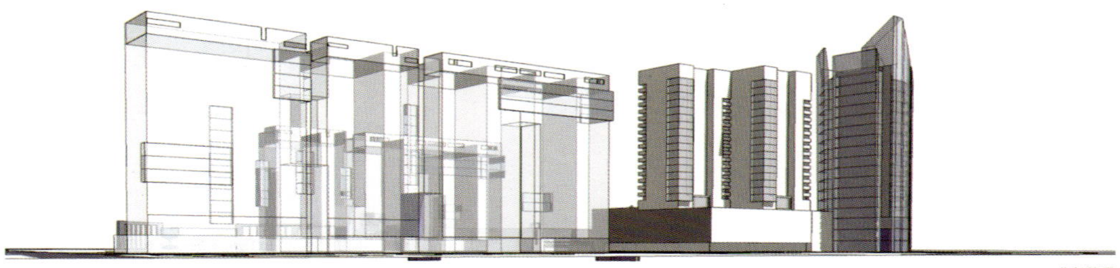

住宅透视图

商业中心及酒店透视图

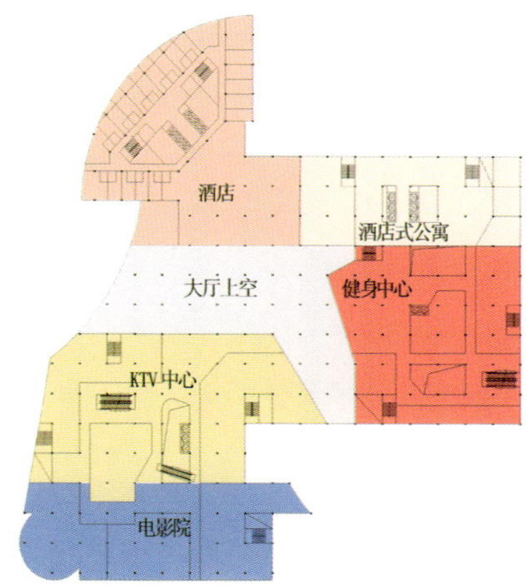

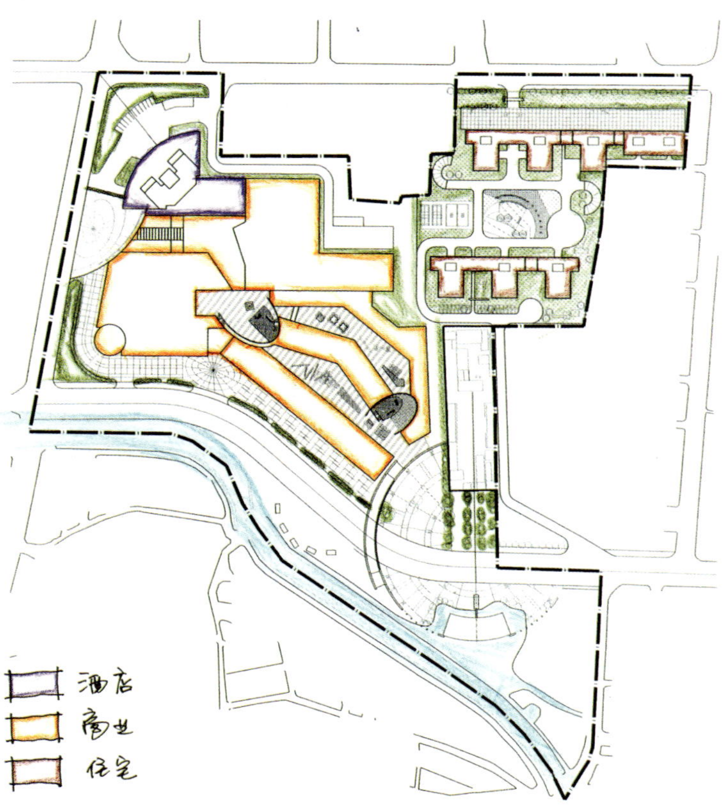

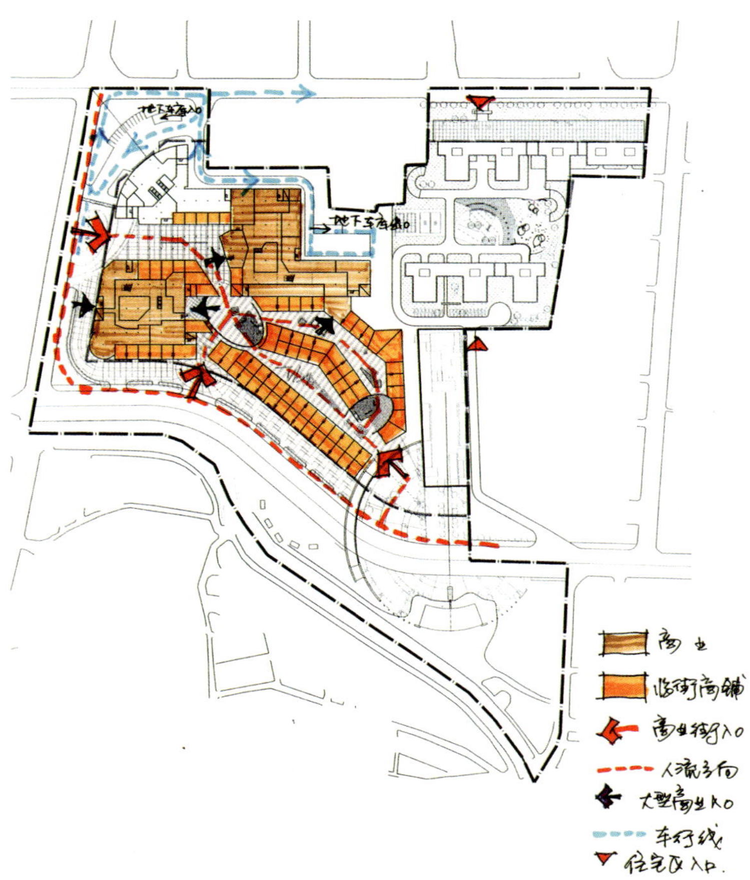

Life Enjoying Villa's Living Room

慢城岛屿生活的起居厅

Developer: Zhuozhou Nanhaiyucun Real Estate Ltd.co, First Branch Office
Project location: North Yingshi Road, Zhuozhou
Project size: 10,194m²
Building size: 15,000m²
Designer company: Beijing Jingaokaisi Design Ltd.co,
Chief designer: Hu Dawei, Ren Ling
Design team: Xu Shaomei, Jiang Wei, Yang Wenxuan, Meng Yueling, Chen Hejin, Hou Chunlei, Wang Fen

开发商：涿州市南海鱼村房地产开发有限公司第一分公司
项目位置：涿州市影视路路北
用地面积：10,194m²
建筑面积：15,000m²
设计单位：北京京澳凯芬斯设计有限公司
主设计师：胡大伟、任凌
设计团队：徐绍梅、姜衡、杨文选、孟月玲、陈合金、侯春雷、王芬

Life Enjoying Villa, located in Zhuozhou Yingshi Road, is a xanadu of "Slow down" which has been a luxury word for the modern people. People in here can peruse the style outside the body, and also can enjoy the village nutriment inside the body. People enjoy the holiday and weekend waked up by birds and the happiness with parents and kids. People feel the friendship from neighborhood; have comfortable private room for their life. Coming back rural life is opening up slowly. Flirtatious Business Road, as an important part of totally new ideal, becomes a landmark of land life.

Speak of road, people will think of the picture of romantic old city and the popular roads. There would be electric car with stories, warm street lamp, and all kinds of walking people. We will not move all the appearance of the old city streets, because they belong to another time and culture background. We will learn the feature of filling with stories, plenty and humanism of the old cities. We will provide a road room where people can slow down the footstep, stop to watch and do shopping. It not only provides people a basic place to go shopping and

一个位于涿州影视城，将“慢”这已经成为现代人一个奢侈的字眼，正在以拥有一个可以荡涤心灵的世外桃源，从身体之外的形式追求，到身体之内的原乡滋养；享受被鸟儿叫醒的假日与周末，与父母、孩子共享天伦之乐；感受久远的邻里情感；体现私密舒适的生活空间；回归田园质朴的生活本质正在慢城岛屿徐徐展开。作为一个全新的生活理念的重要组成部分之一——风情商业街，更加成为了岛屿生活当中一个地标。

提起街道，很容易想到老城市浪漫、充满人气的街道画面，有充满故事的电车、温暖的街灯、行走的各色人。我们不是要把老城市的街道面貌照搬过来，因为它属于另一种时代背景、文化背景。我们需要学习老街道的故事性、丰富性、人性。提供一个可以放慢脚步、停下来休息观赏、购物的街道空间，不仅满足基本的购物、餐饮，更多提供一处休闲、观赏，可以引起遐想的生活化的街道，同时通过设计语言的整合和变化，形成自己的特色风情。于是，我们摈弃华丽的设计元素，选择充满历史感的电车轨道作为整个街道的主题元素，它连接整个街道，形成一个个休息停留空间，最后自然消失在绿地中或者融入特色“电车酒吧”中，仿佛它早存在于此。

catering, but also is a life road where people can have leisure, watch, and think. Meanwhile, according to integrate and change the design language, the road has formed its own flirtation. So we drop the luxury design element, and choose the electric car orbit which is full of history sense to the principle element of the whole road. The orbit links the whole road, and some rest room will be formed. At the end of road, the orbit will be disappear naturally, or integrate into "electric bar", even if it has been here for a long time.

In order to build a picture of reminiscence atmosphere, memory and warmness, as if the picture is growing up from the earth, the design chooses "electronic orbit" as the element, through the whole road, integrating with shops, plants, logos and advertisement brands. In the side of every building, the design will be done by the life style, life rhythm, and motional world, and the buildings will be international galleries, fashion shops, restaurants, coffee bars……In the outside, it is flirting allays, but inside is totally modern life style. Just a step, it is totally different, and it really makes people feel like crossing space time. The splendid street show and creative fashion show bring motive modern fashion feeling. It will be the leader of modern fashion.

In order to give expression of leisure and independence, the whole structure and centralized cubage of the city business center is totally different: the size is separated in to several strips and blocks. Every part is relevant separate and local is connected. The whole building expands coming along the street and contacts with the environment.

Because the all the units is shaped from topological transformation, there is no tall comparison, and volume subordination. It means the same species growing up with balance and harmony in the nature. According to the aesthetics of building, the splendid outline is shaped--revolution, rising, crossing, and gets integrated with landform appearance of the field. With the peaceful and far-reaching character, talking with the nature, it will be integrate ordinary environment, no matter the size or space organization.

提取"电车轨道"为设计要素贯穿街道，结合铺装、小品、植物、灯光、标识、广告牌等元素，意在构建一个充满怀旧气氛、充满回忆、温暖的画面，仿佛画面从这块土地生长出来……每座建筑内部，则按照21世纪现代都市人的生活方式、生活节奏、情感世界度身定做，成为国际画廊、时装店、主题餐馆、咖啡酒吧……门外是风情万种的弄堂，门里是完全的现代化生活方式，一步之遥，恍若隔世，真有穿越时空之感！而精彩纷呈的街头表演、新意迭出的时尚活动，带来富有动感的现代时尚风采，成为现代潮流的领导者。

为了体现闲逸独立的特点，建筑整体格局与城市商业的集中型容积迥然不同：体量被剥分成若干个条、块，各部分相对分离却又相互顾盼呼应，局部交接。整个建筑沿着街角展开，与环境充分接触。

由于体块是通过拓扑变形而生成，没有高度上的对比，也没有体积上的从属关系。象征着自然世界里同类物种间平衡而和谐的生长。在建筑美学层面，又形成了丰富的外部轮廓——回转、升起、贯通、与平原地带的地貌现象达到默契的统一。具有平静、深远的性格，与自然对话，无论是外在体量还是空间组织，都充分地融入原有环境。

每个体块看似不经意的摆放，打破通常城市里所见的平行罗列，紧贴建筑边沿的水院、星罗棋布的树木，都暗示着自然法则下，从泥土里孕育的生命，这种立意作为设计师对设计的一种反思出现在本项目中。人工环境需要相对连续、可识别性的界面，院落的错动依靠中间连结的两个体块造成空间的变换。

影视城的一街之隔给予街道天然的文化氛围和气质；区域环境赋予场地独特地理优势和商业潜力；利用、挖掘区位优势和环境特质，营造富含自身特色和韵味的文化气质至关重要。

提炼电影的视觉符号作为丰富街道空间的关键点；演变为雕塑、标识、广告牌、地雕、门牌，甚或是你手中的菜单，来一次视觉、味觉的大餐吧……

VGY FSD

Every unit looks like setting casually, breaking the rule which is familiar in the city. It is no longer to paralleling and spread out. The pools beside building and the sufficient trees suggest the life was born from the earth with the natural rules. This conception turn to be this project as a kind of designer's self-examination to the design. The man-make environment should be connected, and recognizable interface.

The Show city crossing the road brings this street natural cultural atmosphere and quality. The region environment brings this place unique geography and business potency. It is important to use and dig the region advantage and environment quality, and build the culture character splendid of its own feature and quality.

Abstract the vision symbol of the movies is the important spot for enriching the street: evolve into sculptures, loges, advertisement brands, door brands, even the menus in your hands. Enjoy yourself with vision and taste……

The "plaza" between two buildings in the entrance of the building group, is not only the square for people group concentrating and spreading out, but also the open field to have the nature view. The prospect beside the bank is connected into man-make city, and it is the surprise for people who come first time. This is the designer's new understand which the city buildings probably take parts. The plaza is relatively private, without the sense of big square no matter the real size or visual impression. It is more like a courtyard for small group of people concentrating, communication, running across, and experience quietly. It brings people safety and make people want to stay with the sense of belonging. In the summer, because of the small size of building, the plaza is quiet and cool. In this efficient business place, the outdoor is the expanding of indoor.

Do a heavy native project, and tell a reminiscent, romantic and traditional story.

Outside street space is built with simple, bright and romantic quality. Separated by green land, the bend shopping galley will be shaped. The street will be friendly, beautiful. Meanwhile the outside street will provide the showing room for the whole business region, with itself feature. The logo of entrance will be emphasized, and the leading to inside the street will be enhanced. The design

在建筑主入口的两个建筑体之间是“活动场”（Plaza），这个相对开放的场地既是人群集散的广场，又是观测天然景观最好的开阔地，岸边的细腻景致渗透到人造的城市中，它是人们来到时初见的欣喜，是建筑师对城市建筑中可能扮演角色新的理解。另外院落相对私密，无论是真实的尺度还是制造的视觉效果上都不追求广场的感觉，相对前者，更像是小规模人群聚集、交流、偶遇、静静体验的小院（Courtyard），它给人能够停留的稳定感，是能产生归属感的小型场所。夏日里，由于建筑之间的尺度较小，院子里静谧阴凉，在这样一个高效的商业场所，户外场所是室内空间的延伸。

做一个厚重、质朴、沉着的项目，诉说一个怀旧的、浪漫的、能传承的街道故事……

营造简洁、明朗、充满浪漫气质的外街空间，通过绿地的分隔，形成曲折趣味的购物通道，提供宜人、亲切，活动的街道场所，同时为整个商业区提供展示空间，突出该区域自身特色。强化各入口标识感，加强对内街的引导。细节上休闲设施、商业广告的景观化处理；通过材质的细节对比、原生大树的点缀，让景观空间自然融合，相互渗透。

入口留声音乐喷泉广场拉开了表演的序幕，流水、灯光、音乐、表演构成商业街最动人的一幅画面；铸铁门架爬满了玫瑰，牵引我们进入街道，高大浓荫的法桐树、野草让你的心灵一下放松，流水门架好似一个相框，框住远处的美景。节日时，时尚走秀，水幕电影，抽空来坐坐吧……

电车轨道贯穿整个街道，连接各个休息空间，述说一个个街道故事。电车轨道起点在电车酒吧，终点在“暗香”水池，轨道串起电影雕塑、灯柱、花池、花钵、花架、休息座椅、地面雕塑、枕木、人等丰富的景观元素，构建一幅怀旧、浪漫的街道画面。漫步其中，感受街道怀旧气息，享受来自水面的微风；同时，你很可能邂逅一场电影对白，置身其中，遐想可能发生的人与人的故事，或许这就是你此次怀旧街道之旅的最大收获。质朴的自然锈板悄悄融入室内、绿地中、墙体上，好像它长在这里……啤酒桶形的花钵溢出的花墁，还有温暖的灯光，热情的店主，让你不知不觉放慢脚步，坐下来喝一杯，慢慢品味窗外的风景……

plans to deal with the detail of material and lighten the view with ordinary trees. All these make the building integrating with nature.

The music fountain begin the show. The running water, light, music and the show make the most attractive picture of this business street. The iron door are full of roses, leading us into the street. Tall French parasol shade and the green grass will give you a kind of relax for the heart. The door frame of fluent water just like a picture frame, form which, you can see the view far away. In the holidays, there will be fashion shows, movies. You can come to have a rest here.

The electric car orbit are through the whole street, connecting every rest place, and telling every street story. The start point is at the electric car bar, and the end is flavor pool. The orbit connected all the abundant view item like movie sculptures, light pillars, flower pools, flower shelves, rest seats, crossties and people. A historic street picture is painted. Walking in the street, you will feel the historic atmosphere in the street, and enjoy the breeze from the water. Meanwhile, maybe you drown into a step of story in the movie, and think if it probably happen, and this is the biggest harvest for the historic street. Native rust board infiltrate slowly and quietly into the indoor, grass, wall, and like it is just born here……flowers come from the bowels with the frame of beer cask, with the warm light and the passionate barkeeper, which makes you cannot help to slowing your step, having seat and having a cup of drink, enjoy the view of outside slowly.

Natural gentle slope and the buildings hold themselves slightly aloof. By means of native trees, the buildings sink into the nature seamlessly; the rest platform beyond the water provides the relative private room for leisure and enjoying the nature. The other side of water is thick forest, which makes the villas more implicative.

With the time going, the building will be old. They will be spirit and culture place, and will be in people's heart for long. In the remote future, people think about it, they also will smile from the heart.

自然缓坡驳岸和建筑若即若离，借助原生大树、自然缓坡的处理，建筑悄然融入自然中；滨水休息平台为享受自然。提供休闲活动的相对私密；对岸密密的树林，使得别墅建筑更加含蓄，两岸互为对景、借景……

随着时间的流逝……建筑会陈旧，藉人之活动形成的精神与场所文化，将长存于曾在此停留的人的心中，在遥远的将来想起，还会露出会心的微笑……

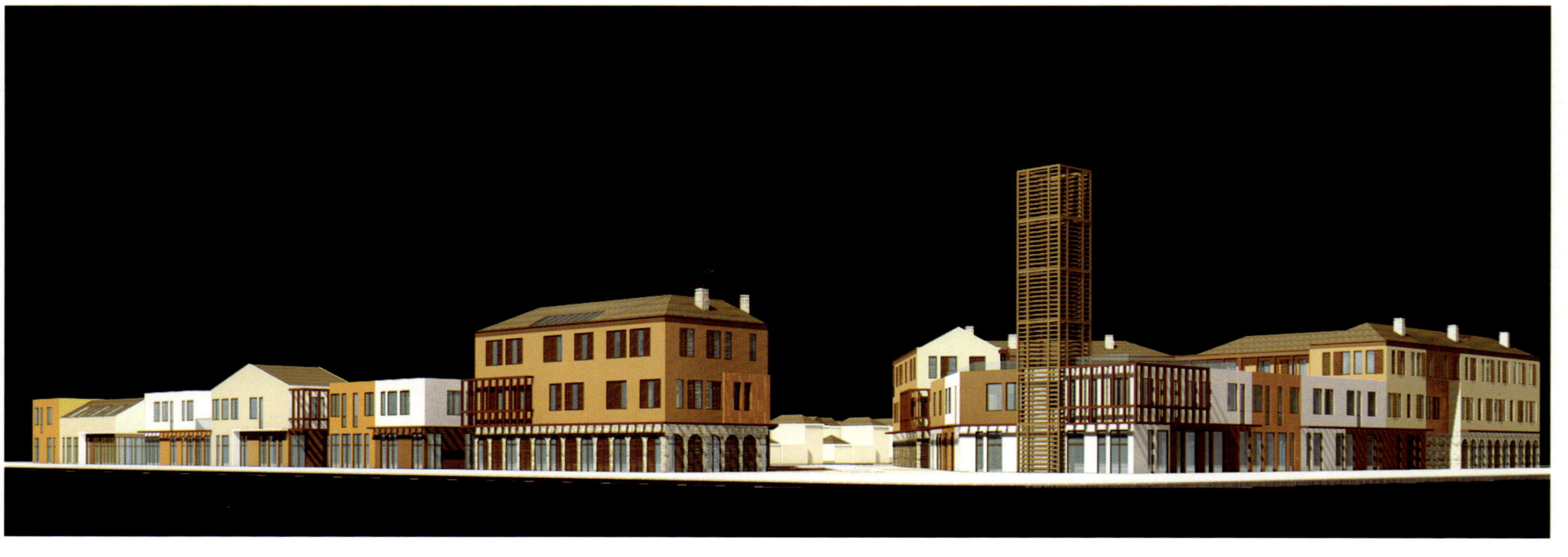

SUBWAY
Dior

SUBWAY
coffee

coffee

SUBWAY
Coca-Cola

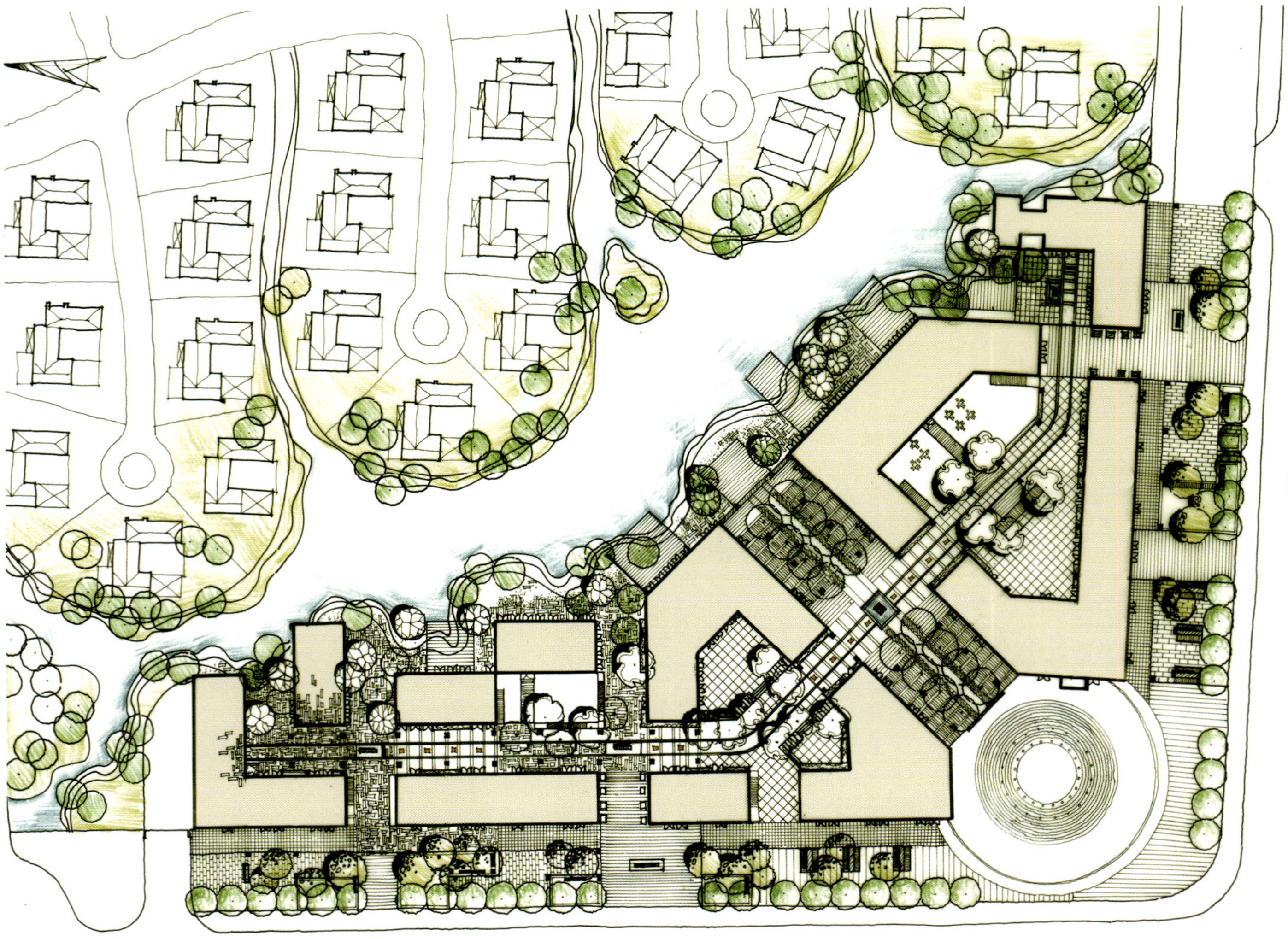

加拿大杨
高草
置石

暗香.

都市夜归人

铸铁花架.
休闲坐凳
高脚
木地板
干杯

架
花钵.
地灯.
邂逅
DE 地方

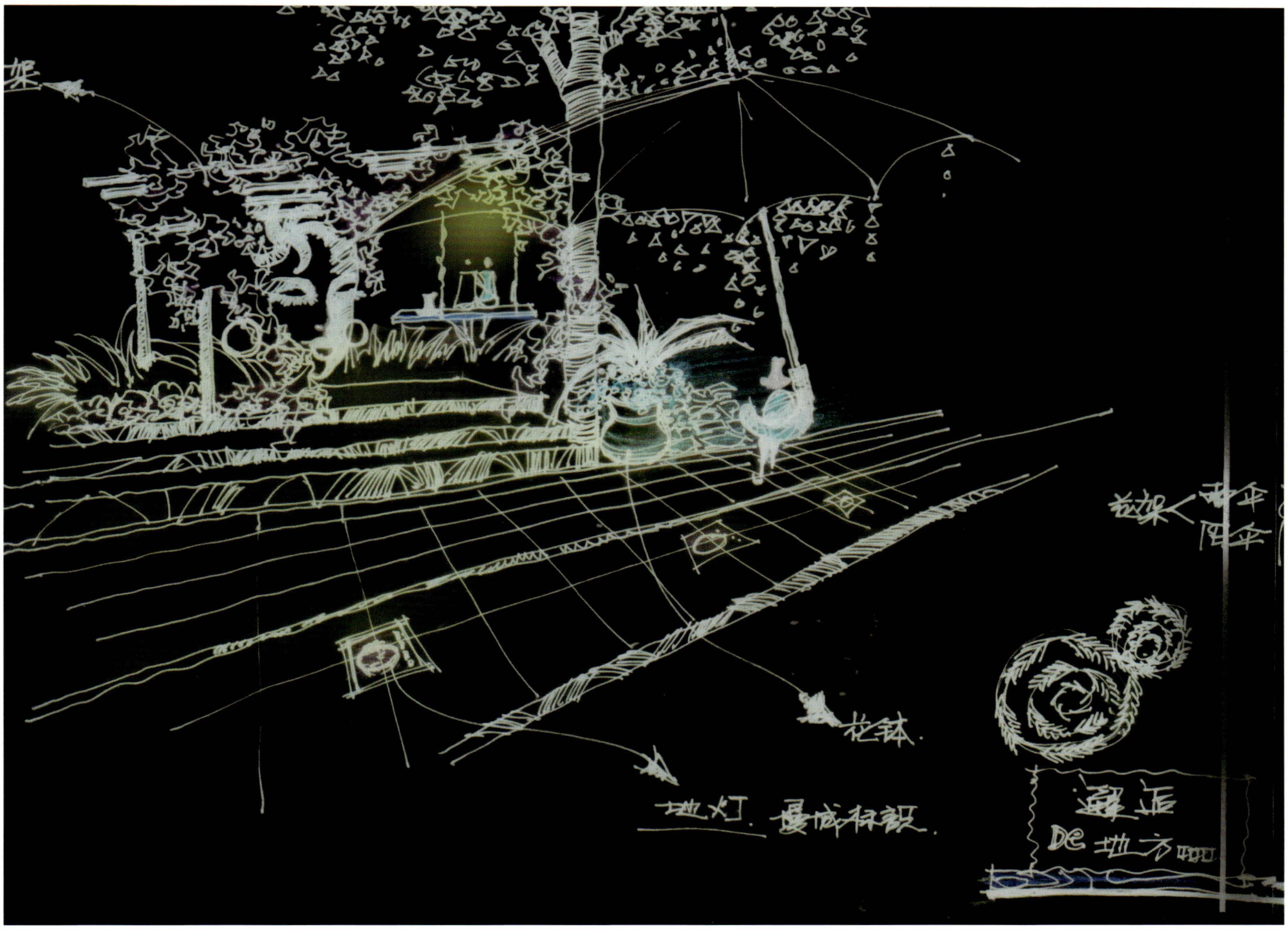
架
花钵.
地灯.
邂逅
DE 地方

Shanghai pentagon world trade centre

上海五角世贸商城

Design Company: Architecture & City plan college, Tongji University
Project address: north Nanhui Huinan New City, Shanghai
Project size: 273,795.9m²
Building size: 486,607.8m²

设计单位：同济大学建筑与城市规划学院
项目地址：上海市南汇惠南新城北部
用地面积：273,795.9m²
建筑面积：483,607.8m²

The project of Shanghai pentagon world trade centre is located in north Nanhui Huinan New City of Shanghai. It is the extra large business centre with the business, office, leisure, and shop. It mix variety functions as one body.

Nanhui and Shanghai district are separated by a river. It is connected with Pudong New District in the north, and East Sea in the east. In the south is Hangzhou Valley. In the north is city outer ring which connected the two air ports. In the north east is the modern Pudong Air Port, and the southeast is the important door Luchao Port. The geographic location determines its traffic advancement. The main roads inside this district are Shanghai city outer ring, Shanghai suburb ring, Jinan Road, and South Six Road, etc. The Hulu High speed road is in the plan. These roads form the traffic network system in here.

Shanghai is becoming the place full of transnational seller and supplier. After Chinese coming into WTO, Shanghai takes bigger footstep to build the international big metropolis, and do its best to be international centre city of economy, finance, business and transport. The domestic companies move

五角世贸商城工程项目位于南汇惠南新城北部，为汇集商业、办公、休闲、商务及研发等多种功能于一体的特大商业城。

南汇与上海市区隔江相望，北接浦东新区，东濒东海，南临杭州湾。北翼，有连接上海两大航空港的城市外环线；东北端，是现代化的浦东国际航空港；东南端，是上海的重要出海门户芦潮港。地理位置决定了它的交通优势。境内主要公路有上海城市外环线、上海郊区环线、沪南公路、南六公路等，规划建设的有沪芦高速公路等，形成了四通八达的交通网络体系。

上海正在成为跨国采购商、供应商云集之地，入世后，上海加大了建设国际化大都市的步伐，力争建成国际经济、金融、贸易和航运的中心城市。国内企业纷纷把营销总部迁往上海，以便进入国际市场；国外客商也争相进驻上海，把产品和服务推向全国。不久上海将成为跨国采购链中的重要一环。南汇是上海新一轮对外对内开放的重点区域之一，是以航运市场、航运信息及物流为中心，集运输仓储、临港加工、商业居住、金融服务、娱乐旅游等功能为一体，逐步发展成为具有综合功能的港口城市。五

their marketing head office into Shanghai one after another, to coming into international market. The guest customers are coming into Shanghai as well in order in push their products and service into China. It is very soon that Shanghai will be the important point of transnational business chain. Nanhui is one of the important area which open up inside and outside. It is comprehensive function area with the transport market and transport information as the principle business. It concentrates the transportation, processing beyond the port, business residence, financing service, and travel. The pentagon world trade centre building is an important part of new type logistics business. It must push the location economy.

The principle design of pentagon world trade centre is multiwall building with the shape of pentagon circle. The circle around central sightseeing tower, separated into five zones, with the names Asia zone, Australian zone, African zone, European zone and American zone. The zones become five angles of the business center. Every zone are separated from inside to outside. There are zone A, zone B, and zone C. The bottom floor, second floor and third floor of zone A and zone B are merchandise wholesale; the fourth floors of zone A and B are business offices. In the middle area of business offices roof is the roof garden with the width of 18m. Zone C is the mating service zone of zone A and B. The main functions are restaurants, leisure, sports, banks, communication service and meeting.

角大楼商城的建设，是新型物流业态建设的一个组成部分，必然给当地的经济带来促进作用。

五角世贸商城总体设计为五角环形的多层建筑，围绕中心观光塔，分设5个区，命名为亚洲区、澳洲区、非洲区、欧洲区、美洲区，成为商城的5个角。每个区自外向内又分为A区、B区、C区。其中A、B两区的底层、二层、三层均为商品批发展示；A区四层、B区四层、五层为商务办公，三层屋顶两排商务办公之间为宽约18m的屋顶花园。C区为A、B两区的配套服务区，主要由餐饮、娱乐、休闲、体育、银行、通讯服务、会议、精品展厅等功能构成。

该商城以五角中心位置的“绿核”和B、C区之间的“绿环”为主规划景观，在A、B区之间设置“绿带”，加上屋顶绿化，全方位美化购物环境，创造出一组完整，亮丽的商业景观，在为整个商城提供均匀等值的购物空间的同时，也为广大顾客与员工提供极佳的休闲场所。大型中心绿化与屋顶花园、带状路边绿化的结合使绿化真正为人所享。

五角世贸商城为五角、五圈环状、辐射型的建筑形态，中央高塔引领全城。五角的形态使建筑本身对城市道路及道路交叉口做出了不同程度的退让，在留出广场和停车位的同时，形成丰富的沿街界面效果。同时，通过局部退让和虚实对比，化解整个商城的巨大体量感。

The principle landscape is the "green core" in the centre of pentagon and the "green belt" between zone A and B. With the green roof, the completive and bright business view is formed. At the same time of providing uniform equivalent shopping room for the whole business centre, it provides excellent leisure place for customers and employees. The large size of central green and roof garden, mixed with green beside the road are really enjoyed by people.

Pentagon business center is the shape of pentagon, five cirle and radial pattern with the central tall tower leading the whole city. The shape of pentagon gets the building itself make room to the road of the city and road crossing. It not only gives room for square and parking place, but also makes affluent road view.

The building modeling design focuses on time feature and business feature which business architecture should be. The design pursues natural and elegant modern architecture style according to unique art processing. The business elevations along the street are connected by three floor arcade, forming the large scale. The elevation design of business buildings are concerned adequately about the modern composition elements. The architecture material reflecting blocks and opposition of false or true, are designed very carefully. According to elegant detail handling and coordinated property relation, the excellent view is made out. It is full of rhythm, express the new modern style, and conclude Shanghai regional culture.

The entrance and exit for cars and people handling make the elevation modeling more plentiful and the entrance road more outstanding. The whole elevation structure modeling is not only changeful, but also organized. The combination relationship of the room shape is more coordinately, and the form and content of the architecture are centralized. The style is elegant and scholar, novelty and outstanding. It is the completed building form and unique quality.

Visit tower is standing in the middle of the business center. Five pieces of frameworks are surrounding the elevator and evacuation stairs. It faces to the five zones, as the symbol of customers from the world, coming to the Nanhui treasure place to make wonderful career. The frameworks contracted upside, with the strong and tough shape, support the visit hall and diamond hall above. It is the center and tallest point of the business center with the imposing manner of flying up to sky.

The integral environment of business center adopts the mention of mixing the points and lines. The green point and green belts create vision room with style conformity and various forms. Large piece of green grass is inset four season flowers, boulder strips, crushed stone walking trails. It enrich the elevation of the building in the angle of environment. So as to it weaken the cold and tough sense which the building takes to people. With organization formed by the pools and valleys, it makes the former dull center green place more active.

Different materials are interlinked as the terrace, which bring the visual beauty. The floor design of can road makes the use of leading, and this is a part of marking view in both side of the road. The garden lamps and grass lamps make the night unique atmosphere get more completely show. The light mention of visit tower in the center is adopted by the combination of spot light and interior light, with the twinkle diamond, becoming the logo and bright spot of the business center in the evening.

The double functions of view and health is paid attention to for the plant position. The four season effect is made with the flowers and trees as the subject of four seasons. Some tall arbors are in the belt region along the road to embody the nature air of the business center.

The whole vision design is to embody the nature, with the modern life atmosphere. It is abundant and elegant to build a warm and healthy nature business room.

建筑造型设计注重商业建筑所应有的时代气息、商业气息，经过独特的艺术处理，追求纯朴、典雅清新的现代建筑格调。沿街商业立面由三层的过街楼连为一体，形成了较大的规模与气势。商业用房部分的立面设计充分考虑现代构图要素，采用反映块面和虚实对比的建筑材料，设计精心入微，注重细致的刻划。通过运用细腻、精致的细部处理和协调的比例关系，创造出良好的视觉效果，富有节奏韵律，既表现新颖现代的风格，又不失海派文化底蕴。

车行出入口和人行出入口的处理使立面造型更加丰富、入口大道更加突出。整个立面的结构形态既清晰有序，又富有变化，空间形态组合关系有机协调，建筑形式与建筑内容相统一，风格端庄大方，高雅脱俗又清朗细致、新颖夺目，创造了完整的建筑形式和独特气势。

观光高塔傲立商城中央，五片构架包围住中间的电梯和疏散楼梯，面向五区，象征来自五湖四海的商家，汇聚南汇金石宝地，共创美好事业。同时构架向上收分，以强有力度的造型，托起上部的观光厅和钻石厅，以一冲飞天的气势，形成商城中心点和制高点。

世贸商城的整体环境采用点线结合的方式，点状的中心绿化结合带形的环状绿化，创造出风格统一，形式多样的景观空间。大片的草地加以四季花草、条石、碎石步道作为镶嵌，从环境的角度丰富了建筑的立面，从而削弱了建筑带给人们冷、硬的感觉。结合水池和道路形成有规律的布置，使得原本单调的中心绿地更添生气。

地坪材质的铺设采用不同材料的相互衔接，达到视觉上的美感。车行道的铺地主要起到导向作用，如出入口；园林步道特色铺地可以形成良好的地面景观，丰富和美化环境空间。

采用造型简洁明快的步道灯作为世贸商城主要的光源，形成道路两侧具有强烈标识性的景观组成部分。庭园灯以及草坪灯使夜间的特有氛围也能够得到充分的展现。中心观光塔采用投光灯和内透光相结合的照明方式，加以钻石闪烁，在夜空中形成商城的一个标志和亮点。

植物配置注重观赏和健康的双重功能，并配以春夏秋冬为主题的花卉或树木，营造出四季缤纷的效果。在沿路的带状区域中使用一定数量的高大乔木以体现商业城的自然气息，为商业城的对外形象增添亮丽的一笔。

整个景观设计以自然为主，又不失现代生活气息，丰富而又简练，营造出一个温馨祥和、自然健康的生态型的商业空间。

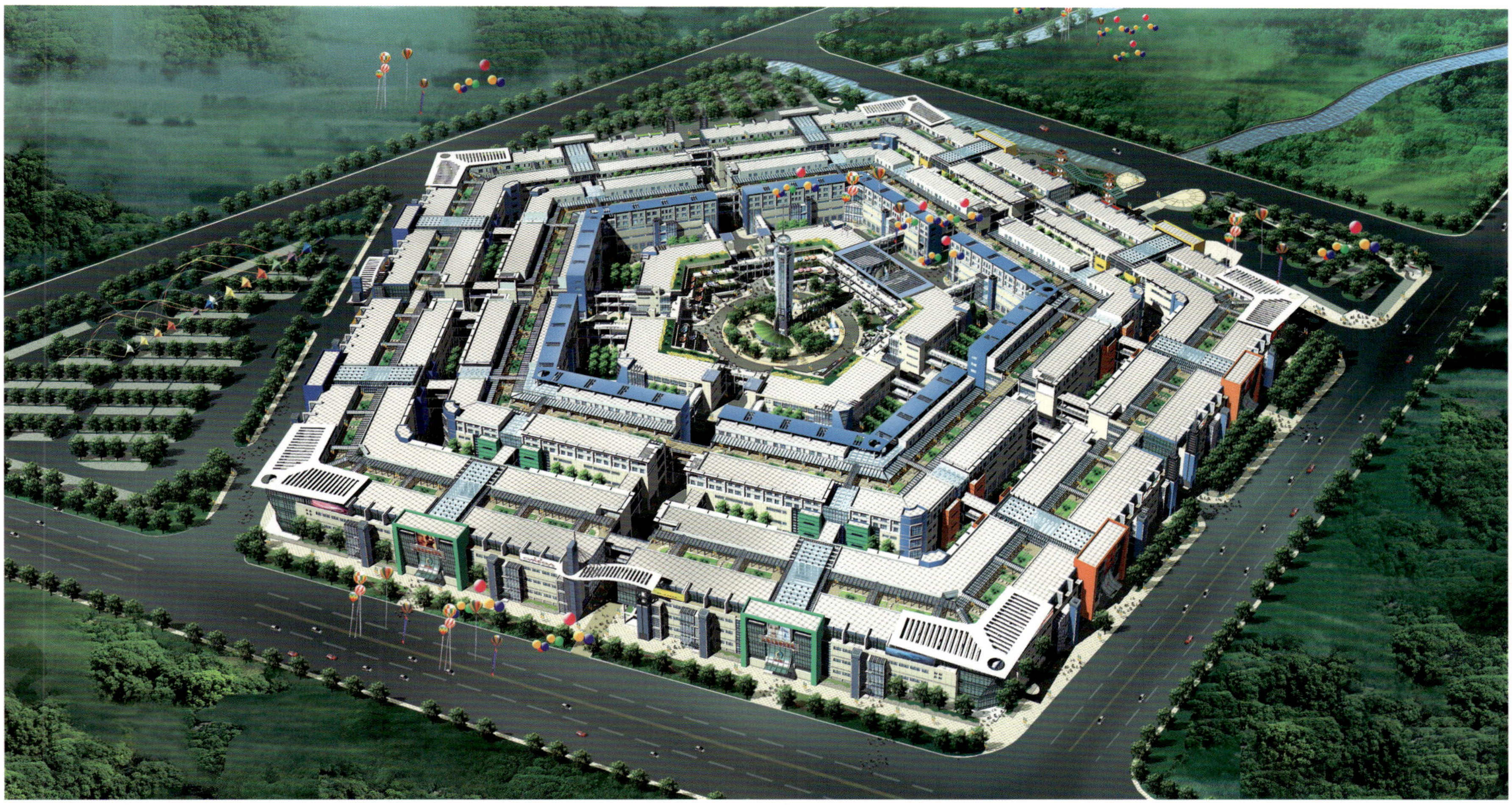

PATATIN
PATATAA
OMEGA
Fine

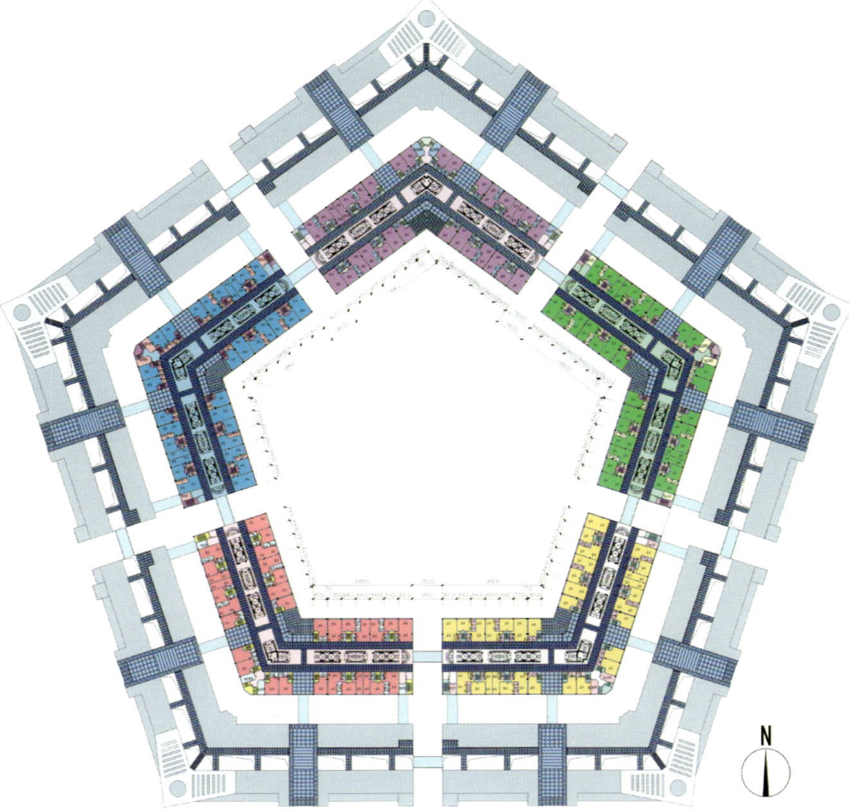

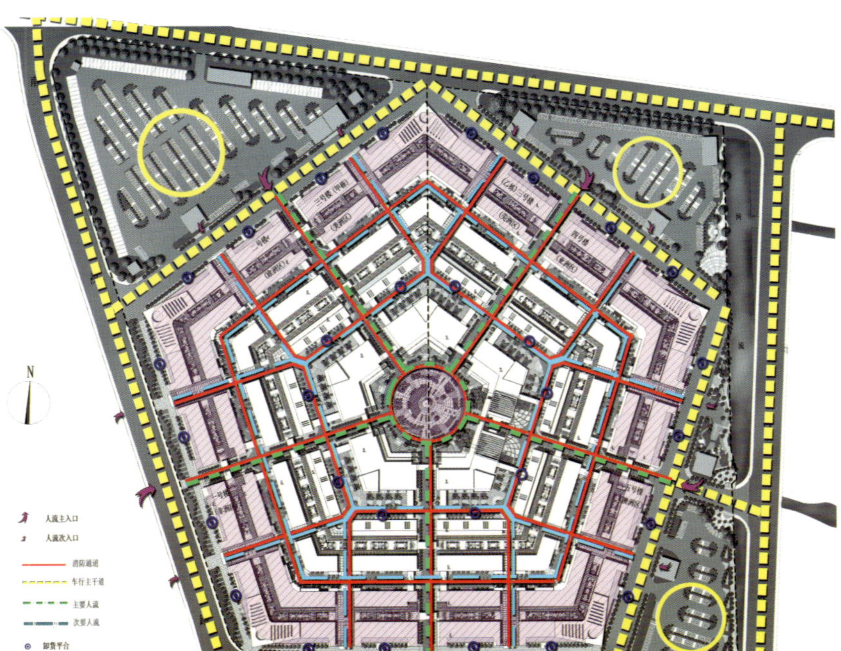

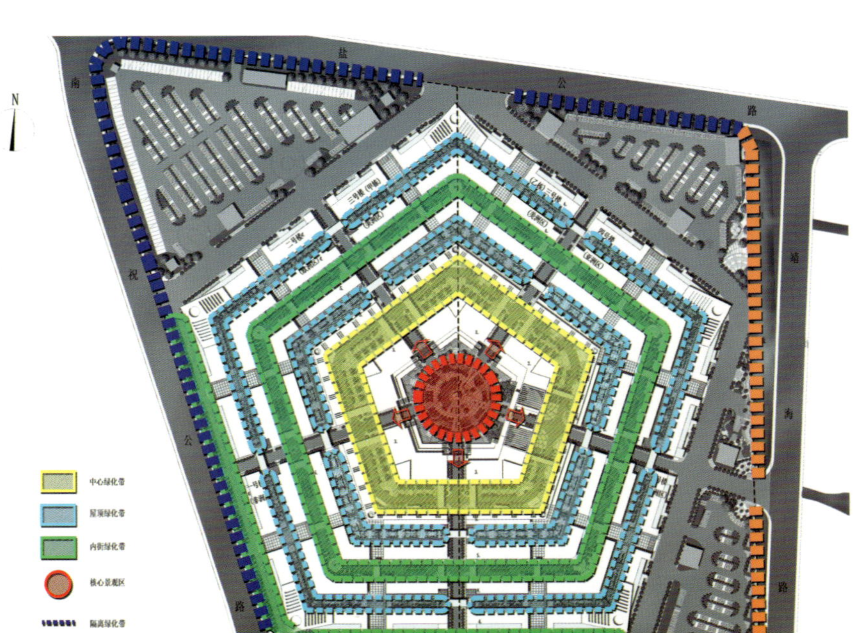

City Crossing

华润中心

Location: Shenzhen, China
Client: China Resources (Shenzhen) Company, Ltd.
Services: Master Planning, Architecture, Environmental Graphic Design

地点：中国・深圳
业主：华润（深圳）有限公司
服务：总体规划、建筑设计、环境标识设计

In heart of a former fishing village, RTKL was commissioned to design City Crossing, a world-class mixed-use development. As a flagship model for future projects of its kind, the three-part, 77,000-SM plan is a widely appealing mix of retail, residential, and hospitality options.

The business-oriented northern section of the development links to the middle section by a six-level retail and entertainment center that houses international retailers, a multiplex cinema, banking, a supermarket, and an Olympic-size skating rink. The middle section contains additional retail and entertainment options. Conceived as an urban neighborhood, the southern sector of the project showcases a 300-room hotel and terraced, lower-height residential buildings offering scenic views of the surrounding landscape. Bold, expressive design elements are consistent throughout the project, establishing the development as a focal point for the city of Shenzhen.

业主委托RTKL在原有渔村的旧址上设计华润中心，这是一个世界级的综合功能开发项目。它将成为今后相关项目的设计典范。项目分为3个部分，共77,000m²，包括商业、住宅和酒店等。

以商务为主的北部地块通过一栋6层的商业和娱乐中心同中部地块连接。中心包括国际零售商业、多银幕电影院、银行，超市和溜冰场。中部地块包括更多的商业和娱乐。作为城市边缘地块，基地南部建立了一个300间套房的酒店和带露台的低层住宅，为周围景观提供了极佳的视野。大胆而生动的设计元素贯穿于项目始终，为深圳建立了城市亮点。

the mixc

the mixc

BreadTalk

KENZO

East Media

东方梅地亚

Developer: Beijing East Media Real Estate Ltd.co,
Building design: Earth(international) building design office
Project size: 119,775m²
Building size: 86,650m²

开发商：北京东方梅地亚置业有限公司
建筑设计：大地（国际）建筑设计事务所
占地面积：119,775m²
建筑面积：86,650m²

This project is located in the heart of CBD in Beijing. In the north is Guanghua Road, in the east is guanghui Road, in the west is Langjiayuan Road, and in the south is the Tieyin South Road. The business in here is very prosperous, and the traffic is so convenience.

The project is extended along the Guanghui Road, and separated into four building as A, B, C, D. Two towers, building A of 25 floors and building B of 27 floors, with one hundred meters will be set in the road crossing of Guanghui Road and Guanghua Road. Between the two towers, there is a mating annex (buiding B) which will connect the building A and building B. A business apartment (building D) of 8 floors will be set along the planning road to the west. There is a quiet and comfortable yard in the west of this project, which is the humanity room for the people rest who work here. The exit of this project is on the Guanghua Road, and the entrance is in the Langjiayuan Road, which connected the traffic inside and the city outside. The first floor of this whole building group is designed with pillars and aisles, in order to provide a comfortable and convenience walking space for people.

本项目位于北京CBD中心地带，北临光华路，东临光辉里二号路，西临郎家园路，南临铁印南路，商业繁盛，交通十分便捷。

项目沿光辉里二号路展开，共分为A、B、C、D 四座。在光辉里二号路与光华路及规划路交口处布置了两栋标志性百米高塔：25层的A座和27层的C 座。高塔之间由一栋 5 层的配套裙房（B座）所相连接，并沿规划路向西沿街布置了一栋8层独栋商务公寓（D座）。项目西侧构造一了个安静舒适的院落，为商务办公人员的放松提供了人性化的空间。出入口分别设置在光华路上和郎家园路上，将内部交通与城市交通有机地联系在一起。整个建筑群首层均为柱廊式设计，以便为人们提供一个宜人的、便利的城市步行空间。

A、C座为高层超5A级豪华写字楼，均为敞开式自由组合式办公区，两栋楼的1~2层为商务、金融等高档配套服务区域，并配有员工餐厅；B 座地下1~4为车库，地上1~5层为商业娱乐服务区，主要作为项目的整体配套用房，有多功能会议厅、视听室、新闻及广告信息发布中心、传媒专业的培

Design sketch in the southwest. 西南方向效果图

The two building standing at night are like shining peals. 屹立在夜色中的大厦像两颗璀璨明珠，耀眼夺目

Night scene sketch of the buildings 大厦夜景效果图

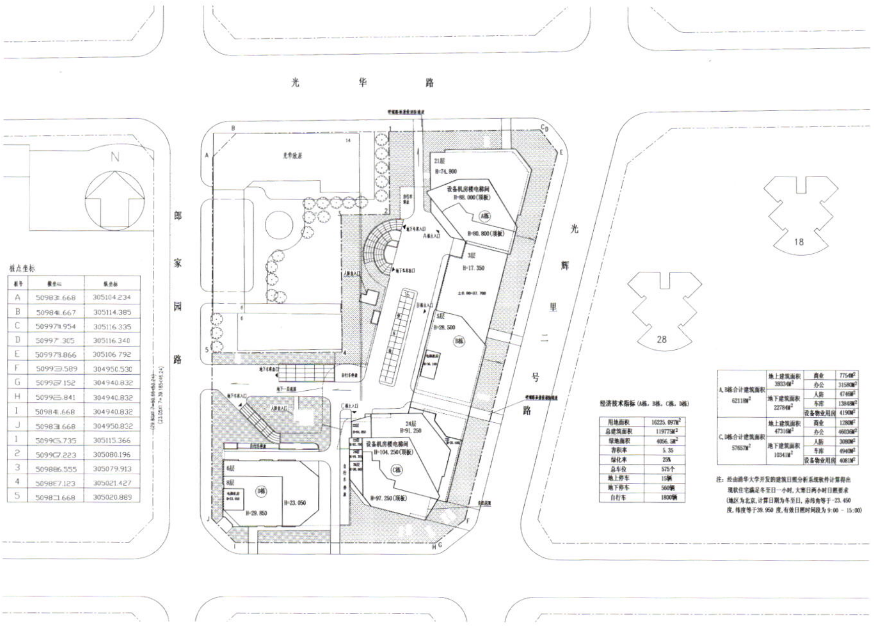

Site plan　总平面图

Buicing A and C are super 5A office building, which are open and combination free. The 1-2 floors of these two buildings are top mating service region for buisness and finance, with the employee dinning romm. The undergroud 1-4 floors of building B are car port, and overground 1-5 floors are buisiness entertainment service region, with multi-fuctional meeting room, audio visual room, new and advertisment broadcast center, trainning center for media, high rage Chinese and west resterants, various bars, audiovisual product selling hall, and etc. The first, second and third floors of building A, B, C are connected with each other by a open hall designed inside along the road with hundred length and three floor height. No matter for the location, size, but also the traffic, it is the deal logo office place for big enterprises.

The main body of tower building A and C are built with two elegant triangle shapes. The form of these two main building have its own feature. The design concept of modeling and aesthetics quality is coming from "model diamond body", which means make every block unit are elegant like diamond. One of the modeling makes stress to the exact angles and lines when the diamond cut, and it will embody the huge size and sculpture effect; the other modeling embodies the smooth and bright of diamond. The two shining each other, express the strong visual shocking. The two tall point combination of tower A are facing the Guanghui Road in the east, the tall point combination of tower C is facing the planning road in the south. According to the opposition effect of true or false of various folding pieces in the detail of building structure, the cutting of every angle and colored glaze between the building main bodies using will express the importance of entrance and the hall. It will embody the inherence visual contact between the center of East Media and CCTV which has been built. Power of nature is used as hard as possible to shape beauty and general character its own.

To show the visual shocking of the building, the business offices at the bottom of building B are facing to the main street (Guanghui Road) totally. According to the huge visual attracting and aesthetic quality which composited by life atmosphere, environment design, media and motive staircases inside the bottom room, the building group will be the center of the whole street.

The main body of building outside finishing is composited by three kinds of high rage glass walls showing different appearance and several metal extra wall broads. The structure form is framework to be the main body and some places are shear wall. The room indoor can be cut as possible as you want.

Solid face of the building design enhances the concept of energy conservation: aluminum alloy adopting the structure of heat insulation. The coefficient of heat transfer is small, and preserves the heat. The glass of the building is insulating glass whose radiation is very low. The argon gas inside can improve the quality of keeping warm and preserving heat.

训中心、高档中西餐厅、各类咖啡厅、酒吧、形象设计中心、健身中心、影视书店、音像制品销售大厅等。其顶部两层设置为全方位、多功能、高科技会议商务中心。A、B、C三座大楼首层、二层及三层彼此连通，其沿街一侧内部设计有长达百米、三层高的开敞中庭，以彰显其商业氛围。D座则为独栋临街办公大厦，不管是位置、面积及交通均可成为大型企业总部理想的标志性办公场所。

A、C栋塔楼主体由两个优美的三角锥体组合而成，两个主体建筑形式各具特点，造型和美学品质的设计概念来自“塑造钻石般的身体”，使每个体块都如同钻石一样优美。其中一个形态强调钻石切割的精确角度和线条，中间开洞更好地体现了这一斜向切割的原理，显现出巨大体量的雕塑效果；另一个形体则体现了钻石的光洁和明快。二者相互映衬，显示了强大的视觉震撼力。A塔楼的两个形体高点组合面向东侧光辉里二号路方向，C塔楼高点组合面向南侧规划路方向，两栋形成了完美的建筑构图。在建筑构成的细节上通过各折面的虚实对比，各角度的切割以及建筑主体之间的彩釉玻璃的运用来表达入口和广场的重要性以及体现东方梅地亚中心与将建成的中央电视台之间内在的视觉联系，最大限度运用环境的力量来塑造自身的实美和共性。

为创造建筑的视觉震撼力，将B座建筑底商裙房部分向主街（光辉里二号路）完全敞开，通过裙房内部的生活气势、环境设计、媒体以及动感的扶梯形成对周围环境巨大的视觉吸引和美学品质，使建筑群成为将来整个街区的中心。

建筑外饰面主体由三种呈现不同外观的高级玻璃幕墙以及少量的金属外墙板组成。结构形式为主体框架局部剪力墙结构，室内可自由分割。

项目立面设计增加了节能内涵：铝合金型材采用断桥隔热结构，传热系数小，保温隔热性能好；玻璃采用低辐射中空玻璃，中间充氩气，保温隔热性能有很大提高；玻璃幕墙层间位置外侧采用中空玻璃，背面加100mm厚防火保温岩棉；外墙上分布不规则的铝合金构架除了丰富建筑立面，还可以提高建筑遮阳效果。

Design sketch in the northwest　西北方向效果图

Design sketch in the southeast　东南方向效果图

Design sketch in the northeast　东北方向效果图

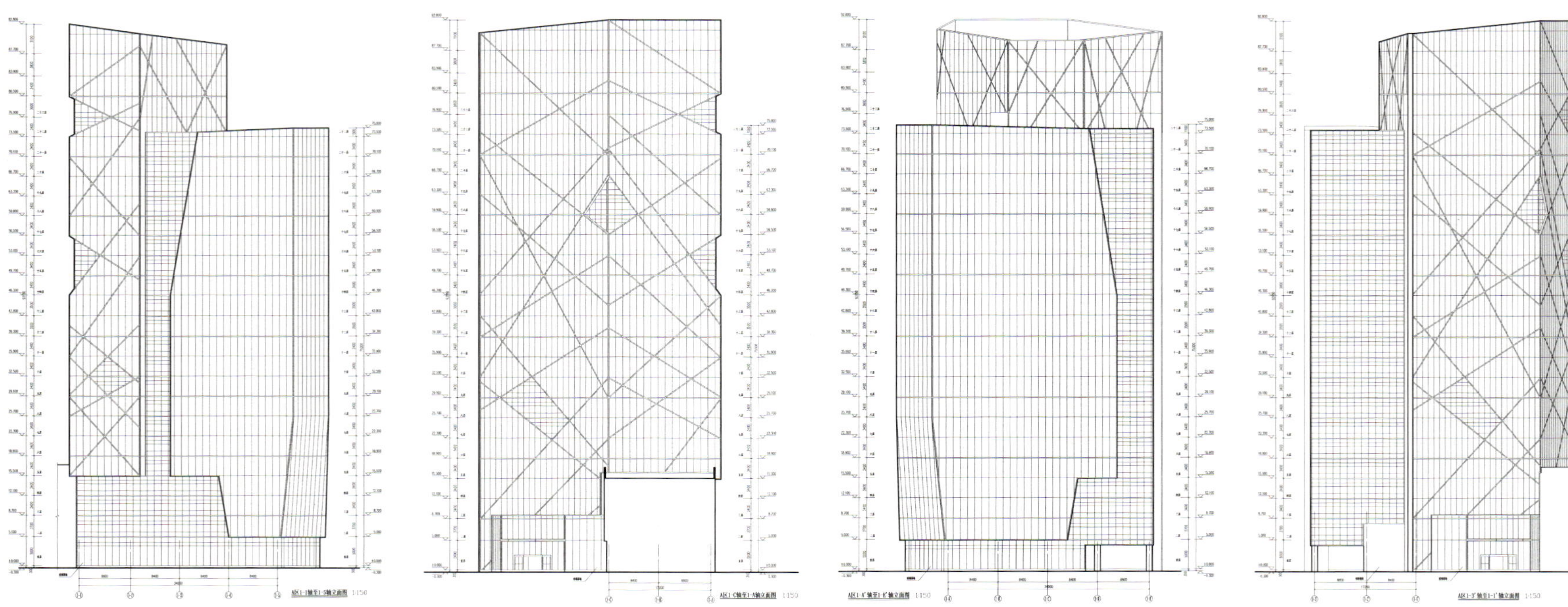

Building A elevation sketch　A立面图

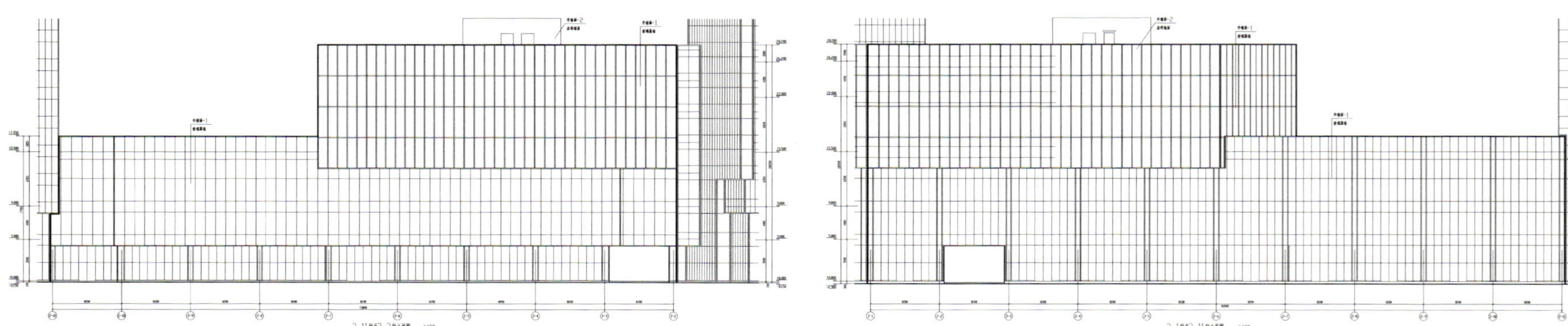

Building B elevation sketch　B立面图

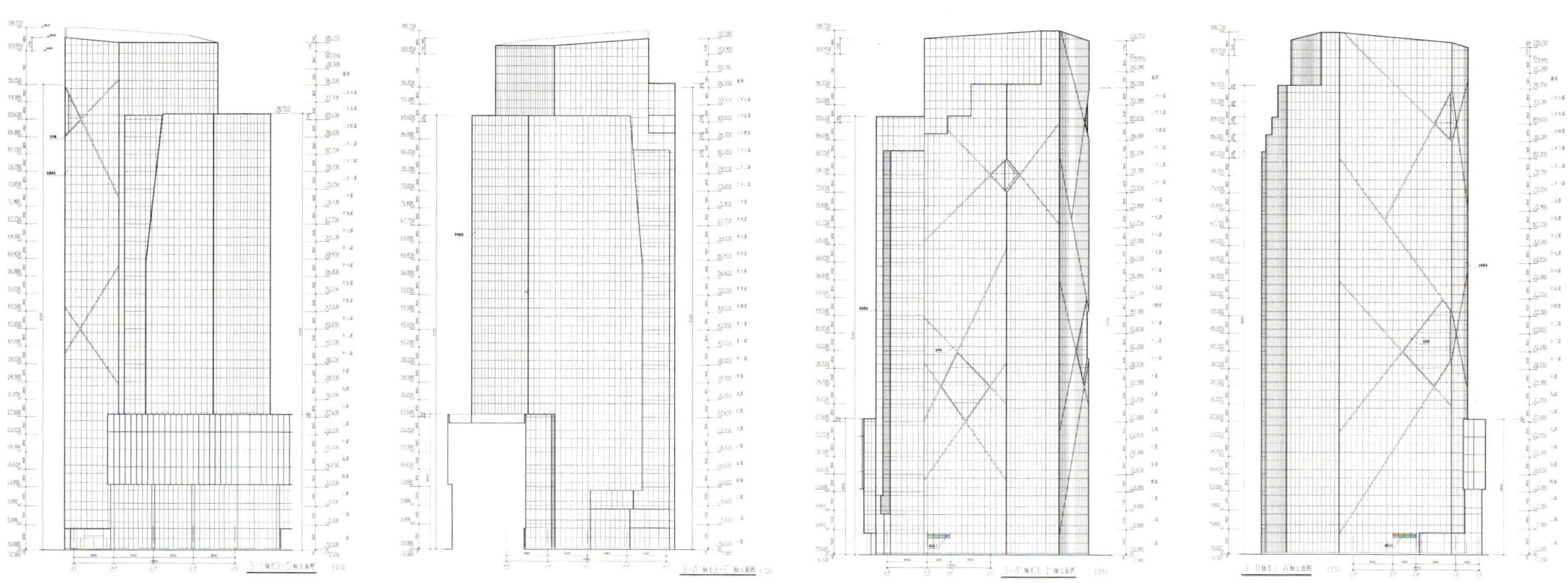

Building C elevation sketch　C立面图

The design concept of modeling and aesthetics quality is coming from "model diamond body", which means make every block unit are elegant like diamond.

建筑造型和美学品质的设计概念来自"塑造钻石般的身体"，使每个体块都如同钻石一样优美

The perfect visual sketch is formed with building A and C.

A、C两栋建筑在视觉上形成了完美构图

One of the modeling makes stress to the exact angles and lines when the diamond cut, and it will embody the huge size and sculpture effect; the other modeling embodies the smooth and bright of diamond.

一个形态强调钻石切割的精确角度和线条，中间开洞更好地体现了这一斜向切割的原理，显现出巨大体量的雕塑效果；另一个形体则体现了钻石的光洁和明快

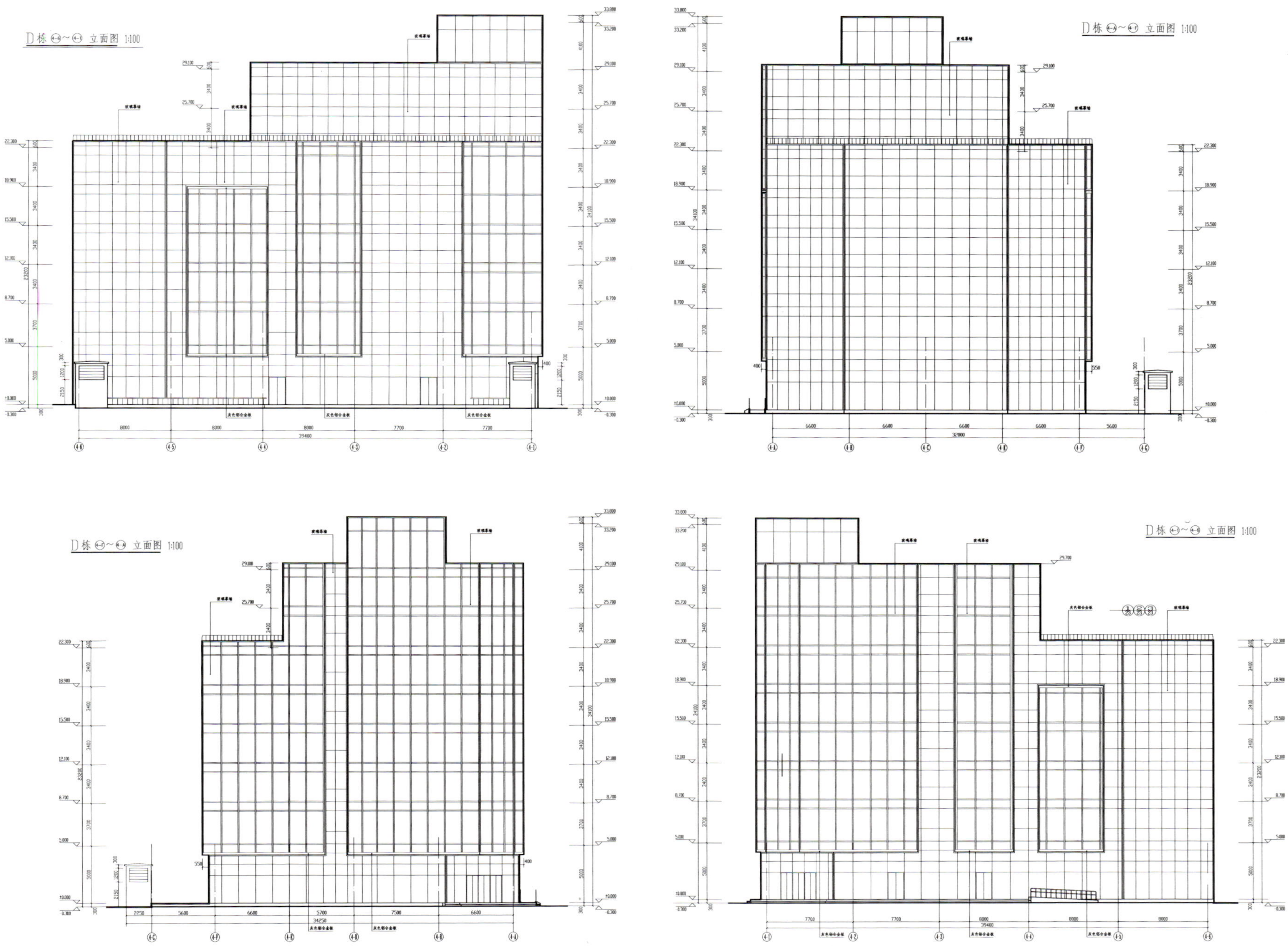

Building D elevation sketch　　D立面图

Beijing Yintai Center

北京银泰中心

Designer: China Electronics Engineering Design Institute
Main designers : Bo Chen, Yu Lou, Junqi Zhou, Ying Zhang, Guangpeng Zhao, Hejun Han
Building address: Jianguo Gate Street, Beijing, China
Building size: 350,000m^2
Building floor: 63 above the floor, 4 under the floor
Building height: 250m

设计单位：中国电子工程设计院
主要设计人：陈柏、娄宇、周军旗、张颖、赵广鹏、韩合军
建设地点：中国北京建国门外大街
建筑面积：350,000m^2
建筑层数：地上63层、地下4层
建筑高度：250m

Yintai Center is a comprehensive building group composited with HYATT, international 5A range office building, and international high quality commercial center. It is 350,000m^2. It stands on the east extending line of Chang'an Street, as the simple founder geometry group. Its imagine subject is Chinese lantern shining Beijing, and it is the core region of central business region in Beijing.

The building group contains three tower buildings and annexes. Two office tower building of 186m and 42 floors in each side of the principle tower building, and the skyline of here is outlined. The principle tower building of 250m is composited by hotel public area, 237 sets of customer rooms, 267 sets of luxury apartments, and special member club.

The business centers, international meeting centers, the business facilities of catering and clubs are in the annex building, which connect the three tower buildings as a completed group. Entire the principle entrances and exits of various building function, with a tall plaza hall. The water view and greening extend from the building inside to outside and roof, just like the building is in

银泰中心是一座由世界顶级超豪华酒店柏悦级酒店（HYATT集团）、国际5A级写字楼、世界高品质商业中心组成的35万m^{2}的综合性建筑体，以极其简洁方正的几何体组群，纯净清晰地矗立在长安街的东延长线上，以照耀北京的中国灯笼做为形象主题，成为北京中央商务区的核心地带。

建筑包括三座塔楼和裙房，二座186m高的42层的办公塔楼衬托呼应着主塔，构勒出该区域的天际线。250m的主塔由酒店公共区、237套客房、276套豪华公寓和特殊会员俱乐部组成。

裙房布置商业中心、国际会议中心和餐饮、俱乐部等多种商务设施，将三座塔楼联系成为一个完整的群体，通过一个架高的广场大堂，将各种建筑功能的主出入口统一起来。建筑的水景及绿化从建筑内延伸至建筑外和屋顶，如同水和绿化中的建筑，提供清新的环境氛围，将建筑与繁忙的城市区别开来。

water and greening. The fresh environment atmosphere is provided, and the building and busy city are separated by the water and greening.

The stones and glass combine into unit wall of the building, which is founder square, light yellow color, smooth and coarse soil. In the atmosphere of air and sunlight, the stones have lots of changes with the neat and simple. The elegant quality is embodied by the implication gesture. The huge lantern and glass tower in the top floor of the building, and light shining in the evening and the huge framework in the daytime, they all try to wake the potential idea of traditional East media.

Double deck elevator is set in the office building, which improves the economy of the office building plane. The lobby in the air show the concept of second hall, bring the building more flexibility and variability.

It is different with the original hotels that the lobby of this hotel is set in the lantern of the top floor. Two station elevators get the lobby. With the huge building framework with the height of 50m and bright glass room, the subject water and greening are integrated into the building again. Excellent and shocking panorama view shows the vigor and perfect of the city.

建筑使用石材与玻璃的组合单元式幕墙，方正的分格，淡黄色的石材，光滑与粗质的肌理，在大气环境的映托下，于简洁纯净中蕴含丰富的变化，以含蓄的姿态体现优雅的品质，建筑顶层的巨大灯笼和玻璃塔，夜间明亮的灯光以及白天巨型的构架，试图唤起对传统东方情调的潜在意念。

办公建筑内设有双层轿厢电梯，使得办公楼层平面的经济性被改善，空中大堂提供了二次门厅的概念，为建筑带来了更多的灵活性和变化性。

与一般酒店不同的是，酒店的大堂被设置在顶层的灯笼内，通过双站式高速电梯抵达，50m高的巨型建筑构架和明亮的玻璃空间，水和绿化的主题再次被溶合于建筑内，提供了优秀的、震憾性的全景视野，展示城市的勃勃生机和完美形象。

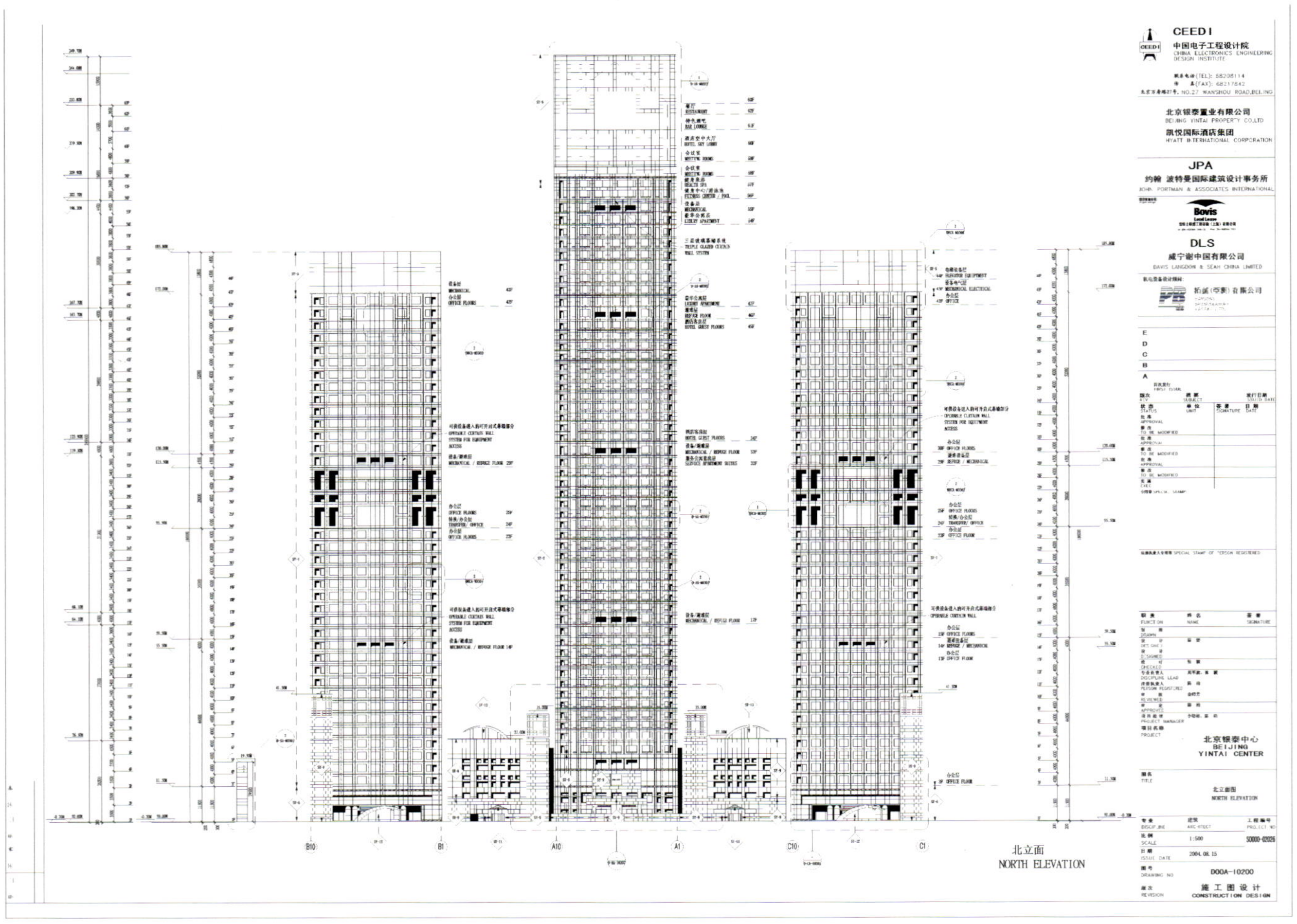
CEEDI
中国电子工程设计院
JPA
DLS
北京银泰中心
BEIJING
YINTAI CENTER
北立面
NORTH ELEVATION

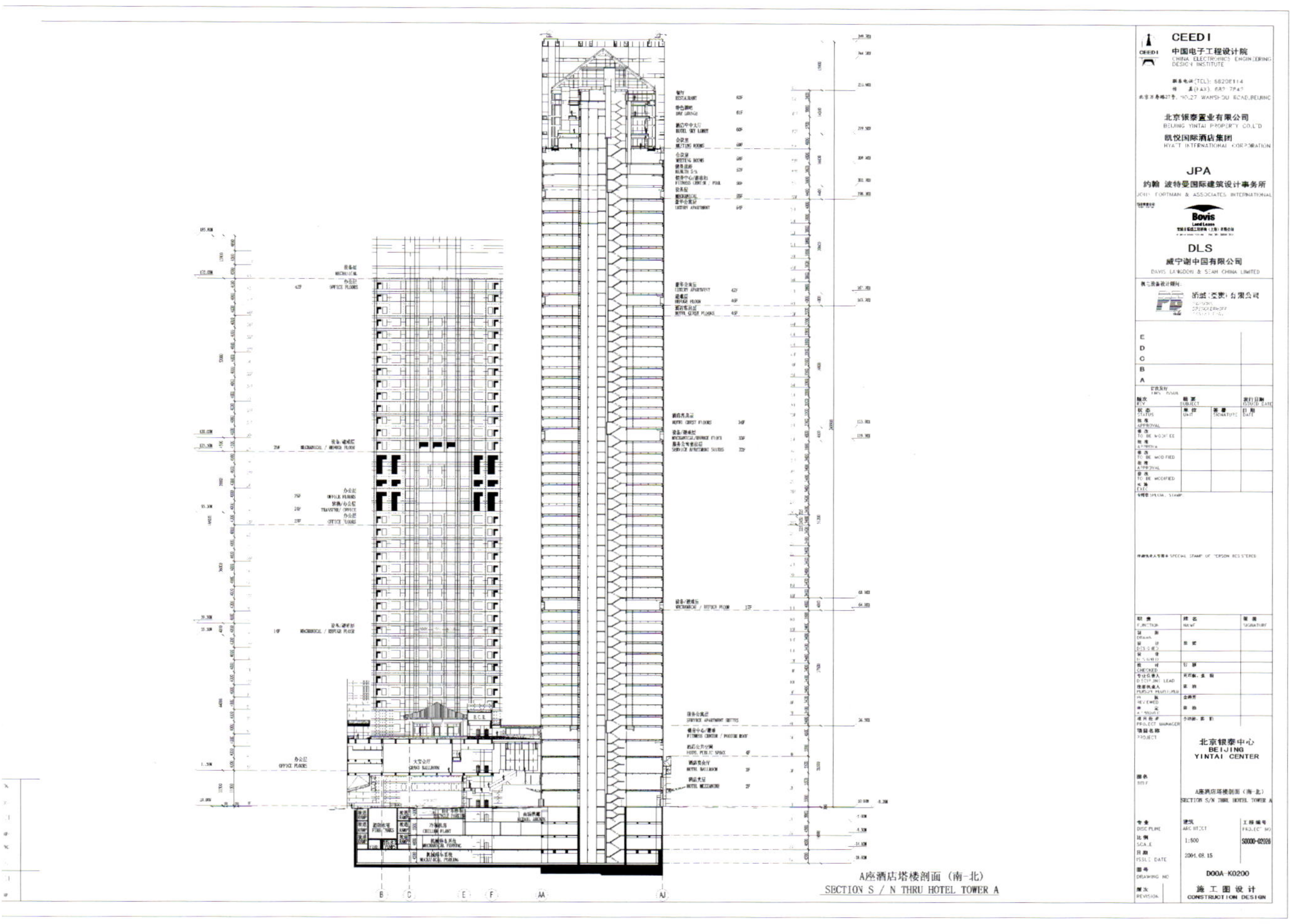

A座酒店塔楼剖面（南-北）
SECTION S / N THRU HOTEL TOWER A

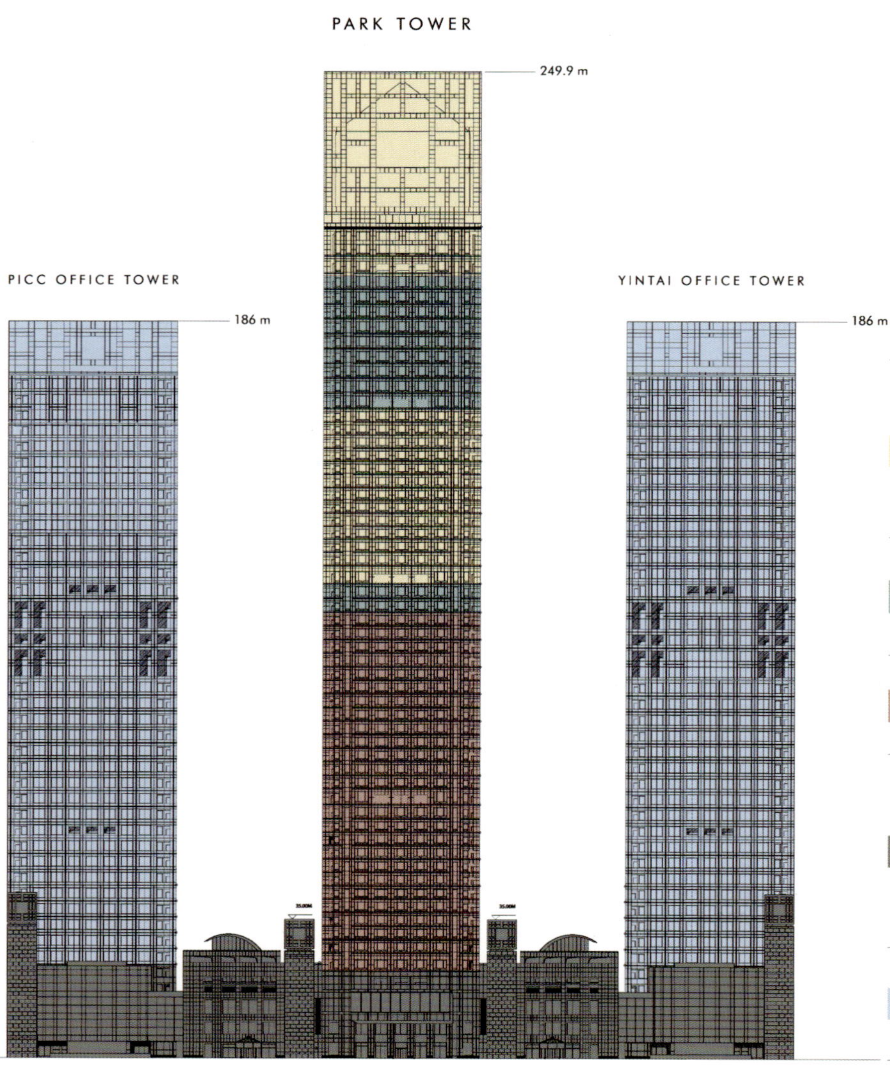

	Area	Levels
PARK HYATT BEIJING	37,018m²	**Level 66:** *Restaurant* **Level 65:** *Theme Bar Lounge* **Level 63:** *Sky Lobby* **Level 61/62:** *Function Rooms* **Level 59/60:** *Fitness Centre/Health Spa* **Level 37-49:** *Hotel Guestroom Floors*
PARK HYATT PENTHOUSES	17,410m²	**Level 50-58:** *Park Hyatt Penthouses* **Level 35/36:** *Park Hyatt Penthouses*
PARK HYATT RESIDENCES	39,212m²	**Level 7-33:** *Park Hyatt Residences*
PARK LIFE	52,199m²	**Level 6:** *Fitness Centre/Roof Garden Bar* **Level 5:** *Chinese Private Dining Restaurant* **Level 3:** *Lifestyle Retail/Ballrooms & Meeting Rooms* **Level 2:** *Lifestyle Retail* **Ground Level:** *Lifestyle Retail* **Lower Ground Level:** *Lifestyle Retail*
OFFICE TOWERS	151,669m²	**West Tower:** *Yintai Office Tower* **East Tower:** *PICC Office Tower*

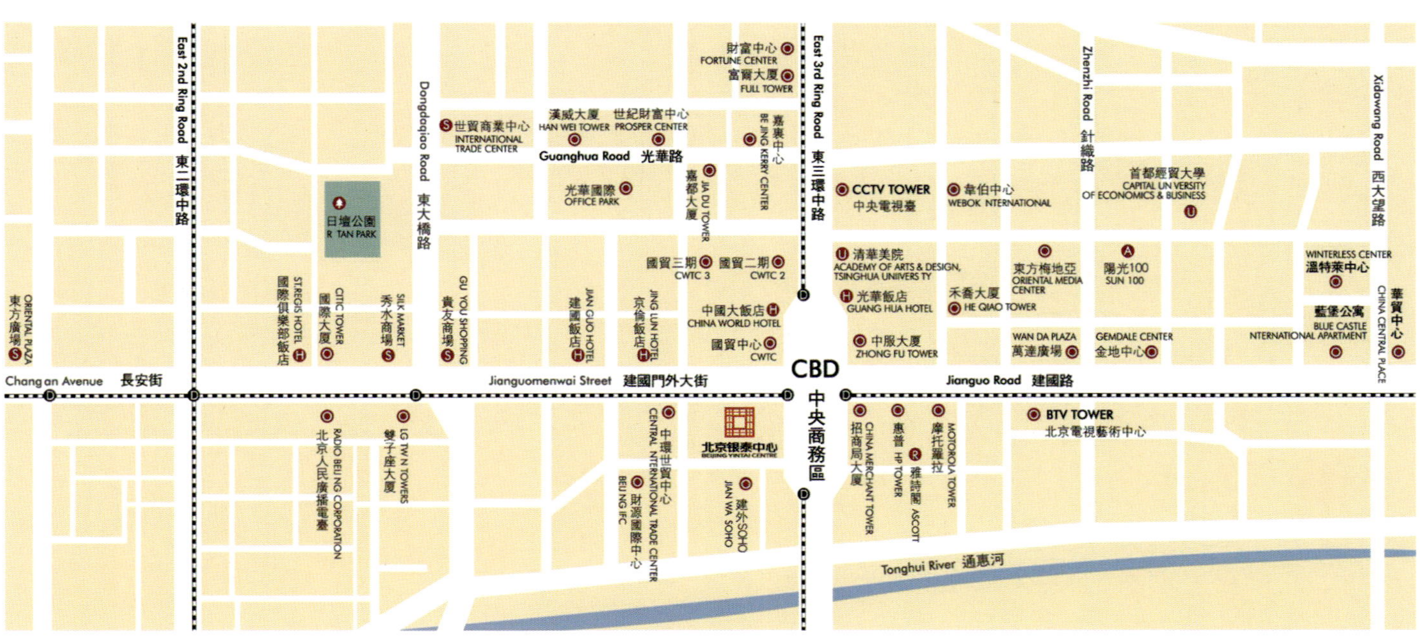

East 2nd Ring Road 東二環中路
Dongdaqiao Road 東大橋路
East 3rd Ring Road 東三環中路
Zhenzhi Road 針織路
Xidawang Road 西大望路
Guanghua Road 光華路
Chang an Avenue 長安街
Jianguomenwai Street 建國門外大街
Jianguo Road 建國路
Tonghui River 通惠河
CBD 中央商務區
財富中心 FORTUNE CENTER
富爾大厦 FULL TOWER
世貿商業中心 INTERNATIONAL TRADE CENTER
漢威大厦 HAN WEI TOWER
世紀財富中心 PROSPER CENTER
嘉裏中心 BE JING KERRY CENTER
光華國際 OFFICE PARK
嘉都大厦 JIA DU TOWER
日壇公園 R TAN PARK
國貿三期 CWTC 3
國貿二期 CWTC 2
中國大飯店 CHINA WORLD HOTEL
國貿中心 CWTC
東方廣場 ORIENTAL PLAZA
國際俱樂部飯店 ST.REGIS HOTEL
國際大厦 CITIC TOWER
秀水商場 SILK MARKET
貴友商場 GU YOU SHOPPING
建國飯店 JIAN GUO HOTEL
京倫飯店 JING LUN HOTEL
北京人民廣播電臺 RADIO BEIJ NG CORPORATION
雙子座大厦 LG TW N TOWERS
中環世貿中心 CENTRAL INTERNATIONAL TRADE CENTER
財源國際中心 BEIJ NG IFC
北京银泰中心 BEIJING YINTAI CENTRE
建外SOHO JIAN WA SOHO
CCTV TOWER 中央電視臺
韋伯中心 WEBOK NTERNATIONAL
首都經貿大學 CAPITAL UN VERSITY OF ECONOMICS & BUSINESS
清華美院 ACADEMY OF ARTS & DESIGN, TSINGHUA UNIVERS TY
光華飯店 GUANG HUA HOTEL
禾喬大厦 HE QIAO TOWER
東方梅地亞 ORIENTAL MEDIA CENTER
陽光100 SUN 100
WINTERLESS CENTER 溫特萊中心
藍堡公寓 BLUE CASTLE NTERNATIONAL APARTMENT
華貿中心 CHINA CENTRAL PLACE
中服大厦 ZHONG FU TOWER
WAN DA PLAZA 萬達廣場
GEMDALE CENTER 金地中心
招商局大厦 CHINA MERCHANT TOWER
惠普 HP TOWER
雅詩閣 ASCOTT
摩托羅拉 MOTOROLA TOWER
BTV TOWER 北京電視藝術中心

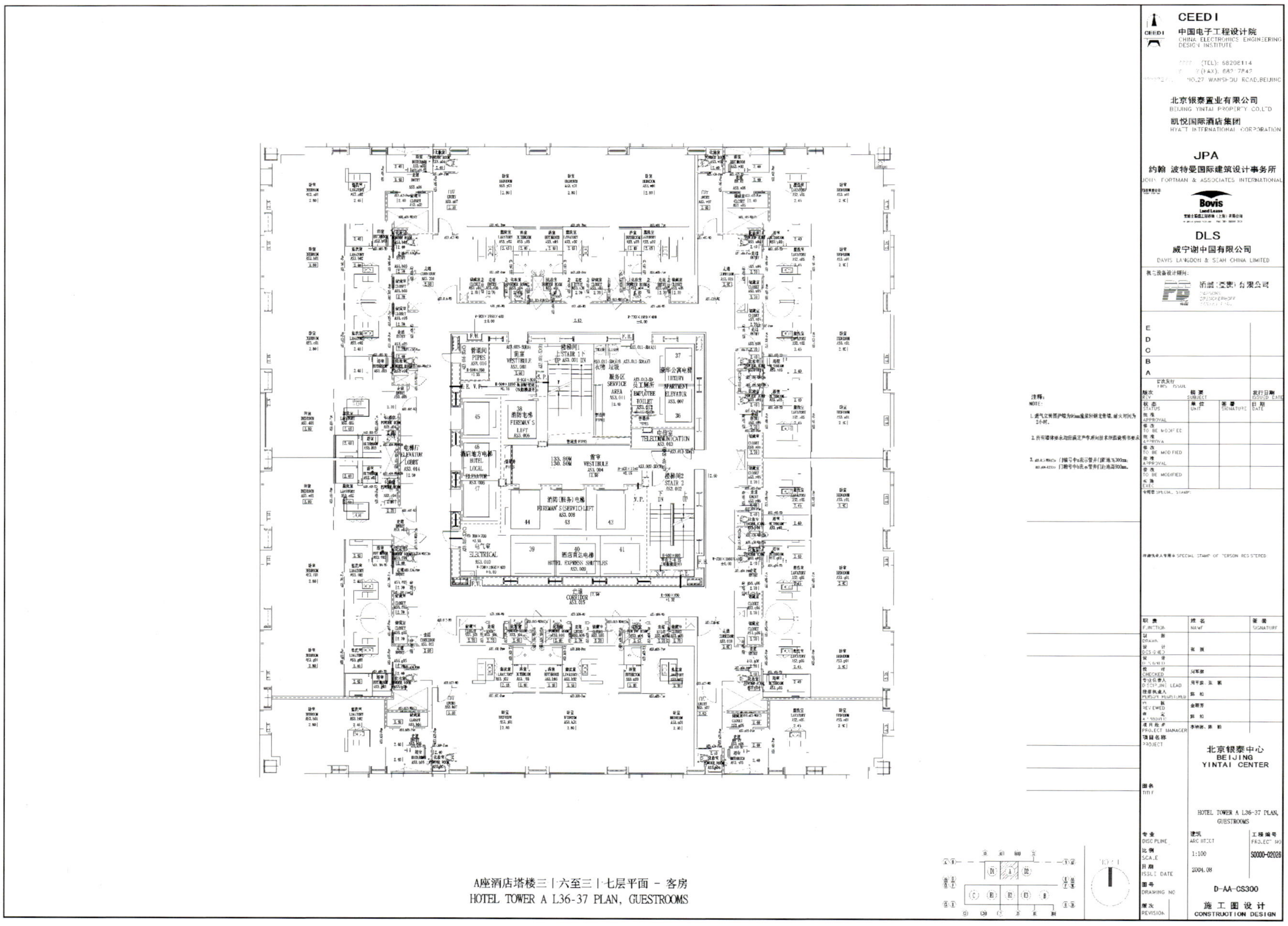

A座酒店塔楼三十六至三十七层平面 - 客房
HOTEL TOWER A L36-37 PLAN, GUESTROOMS

CCTV

中央电视台

Project: New Headquarter of CCTV
Client: CCTV
Location: CBD, Beijing, China
Managing Partner: Ole Scheeren, Rem Koolhass

项目：CCTV中央电视台新址
业主：中国中央电视台
地点：北京CBD区
主管合伙人：奥雷・舍人， 瑞姆・库哈斯

CCTV building is becoming one of the CBD regions. We are all striving to be unique - all different expressions of verticality.

The tragedy of the skyscraper is that it marks a place as significant, which it then occupies and exhausts with banality that is twofold: in spite of their potential to be incubators of new cultures, programs, and ways of life, most towers accommodate merely routine activity, arranged according to predictable patterns.

Instead of competing in the hopeless race for ultimate height - the project proposes an iconographic constellation of two high-rise structures that actively engage the city space: CCTV and TVCC.

CCTV combines administration and offices, news and broadcasting, program production, and services the entire process of TV making - in a loop of interconnected activities. Two structures rise from a common production platform that is partly underground. Each has a different character: one is dedicated to broadcasting, the other to services, research and education; they join at the top

CCTV 大楼将成为北京CBD区众多高楼大厦中的一员，我们力求独特，给予竖向不同的诠释。

摩天楼的悲剧在于它使一个地方变得意义重大，但很快被平庸淹没。这种平庸由两部分组成，在功能上，大多数摩天楼只按照其用途提供常规使用空间，而没有体现建筑作为城市新文化、新内容和新生活孵化器的可能性；而在形式上，摩天楼执着于对垂直的表达，已经被证实阻碍了想象力，摩天楼拔地而起，创造力轰然倒地。

摩天楼在绝对高度上的比拼全无希望，高度只能取胜一时，但很快将会被新大楼的高度所代替。由此，我们设计了一个更具标志性的建筑群，两座高楼渗透入城市的空间之内：CCTV大楼和电视文化中心。

CCTV大楼是一个具有行政办公，新闻广播制作，和全程电视节目制作的综合大楼。在互相联通的功能流线中，两座塔楼从普通制作平台拔地而起，在空中相连接，形成了一个悬臂形式的管理层，至此，一个新的标志

to create a cantilevered penthouse for the management. A new icon is formed; a truly 3-dimensional experience, a canopy that symbolically embraces the entire population. The building itself contributes to the coherence of the organization. Two function of the two tower building is significant. One is for broadcasting, and the other is for comprehensive service, research and education function. The two tower buildings connect each other. The employees cooperate the TV shows progressing in the same building, and everyone knows the concept of cooperation is the chain of relation. The chain is uniting, not the isolate; it is teamwork, but not the opposition. So the building plays an important role for the employees to enhance the cohesiveness. CCTV is the building with lots of confidential facilities, but in the design progress, it is open to public. The visit line will be through all the function region, with which the visitors not only can know the production progress, but also feel the CBD and overlook the Forbidden City through the wall.

The Television Cultural Center (TVCC) is an open, inviting structure. It accommodates visitors and guests, and will be freely accessible to the public. On the ground floor, a continuous lobby provides access to the 1500-seat theater, a large ballroom, digital cinemas, recording studios and exhibition facilities. The tower accommodates a five-star hotel; guests enter at a dedicated drop-off from the east of the building and ascend to the fifth floor housing the check-in as well as restaurants, lounges, and conference rooms. The rooms of hotel are located in both sides of the building, rising vertical, and the grand atrium is formed.

In the southeast of the new location, Media Park is designed to be the extending of green backbone of the CBD. It provides place of having parties and ceremonies, even the take shot outdoor.

形成了，它不是可预见的二维的，涨满天空的大楼，而是真正意义上的三维建筑，它的悬臂部如同华盖，象征着拥抱所有的人。两座塔楼，分工明显，一座塔楼是主供播出功能，而另一座综合服务，研究和教育功能为一体。两座塔楼相互贯通，员工在一座大楼内共同完成电视节目的制作，每个人都了解到合作的本质是相互联系的链，是团结而非孤立，是协作而非对立，因此大楼本身对于加强员工凝聚力起到了十分重要的作用。CCTV是一座保密措施极严的大楼，但在设计中，它将对公众开放。参观流线将贯穿所有的功能区，除可了解电视制作过程外，观众更可通过幕墙感受CBD和眺望紫禁城的壮观景象。

CCTV电视文化中心是一个开放的，引人注目的结构，它主要用于接待访客，同时对公众开放。在首层，人们可穿过大堂直达1500座的剧场，大型宴会厅和数码影院，录音棚，展览设施。大楼包含为2008奥运会特别建立的国际演播中心。还设有5星级酒店，客人从大楼东侧的专属停靠站进入可直达位于五层的登记处及餐厅，休息室和会议厅。酒店房间位于大楼两侧，垂直而上，形成一个壮观的中庭。

在新址的东南部，媒体公园被设计成CBD区的绿色中枢的延续，可为公众提供聚会和庆典的场所，甚至用于室外拍摄。

国贸大厦2座

LEADER
LOUNGE
MEETING
NEWS RELEASE
RESEARCH & TRAINING
新闻区
NEWS
CANTEEN
入口大堂

领导
LEADER
NEWS RELEASE
RESEARCH & TRAINING
CANTEEN
制作
CCTV
STUDIO
入口大堂

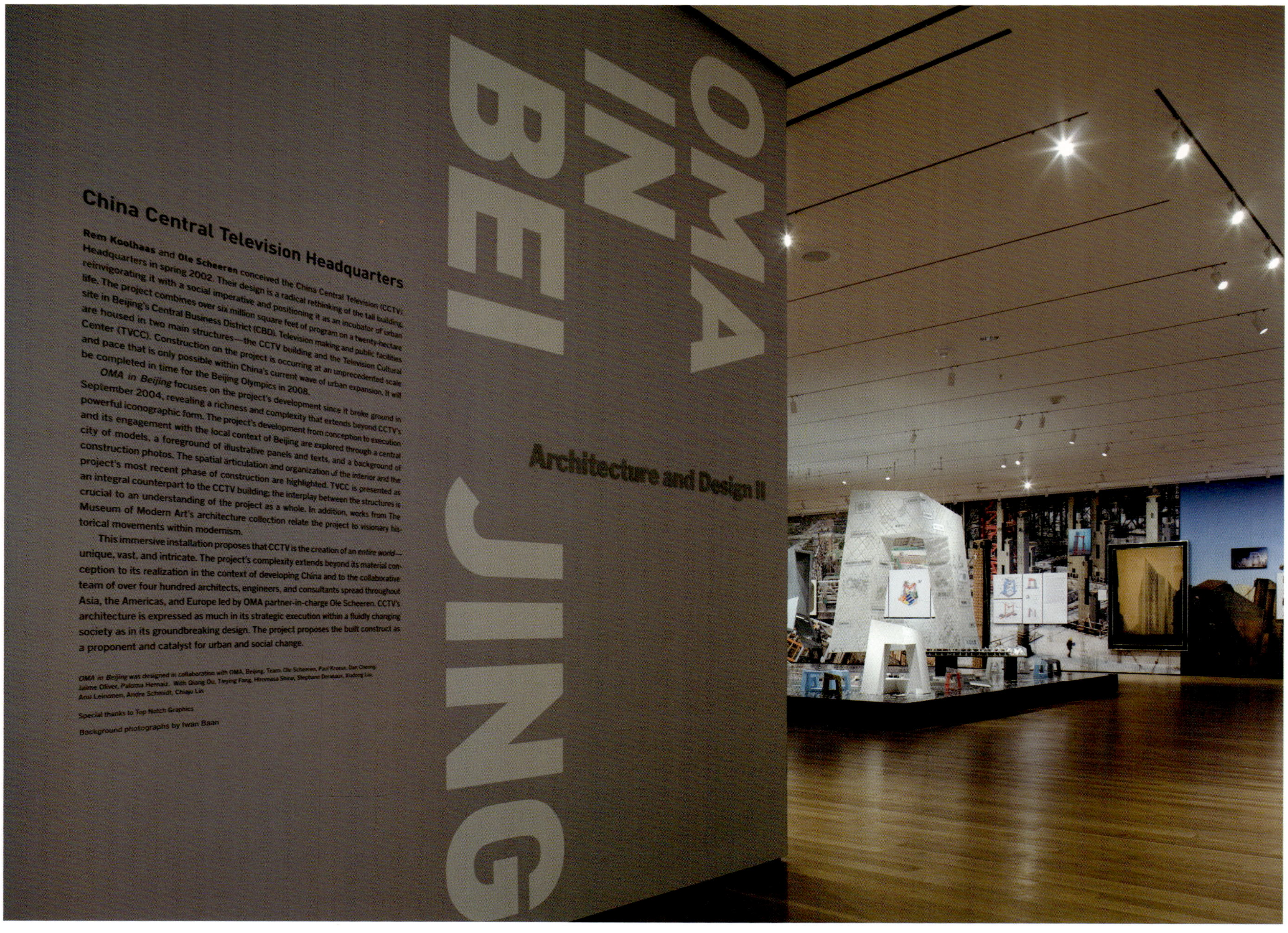
OMA
IN
BEI
JING
Architecture and Design II
China Central Television Headquarters
Rem Koolhaas and Ole Scheeren conceived the China Central Television (CCTV) Headquarters in spring 2002. Their design is a radical rethinking of the tall building, reinvigorating it with a social imperative and positioning it as an incubator of urban life. The project combines over six million square feet of program on a twenty-hectare site in Beijing's Central Business District (CBD). Television making and public facilities are housed in two main structures—the CCTV building and the Television Cultural Center (TVCC). Construction on the project is occurring at an unprecedented scale and pace that is only possible within China's current wave of urban expansion. It will be completed in time for the Beijing Olympics in 2008.
OMA in Beijing focuses on the project's development since it broke ground in September 2004, revealing a richness and complexity that extends beyond CCTV's powerful iconographic form. The project's development from conception to execution and its engagement with the local context of Beijing are explored through a central city of models, a foreground of illustrative panels and texts, and a background of construction photos. The spatial articulation and organization of the interior and the project's most recent phase of construction are highlighted. TVCC is presented as an integral counterpart to the CCTV building; the interplay between the structures is crucial to an understanding of the project as a whole. In addition, works from The Museum of Modern Art's architecture collection relate the project to visionary historical movements within modernism.
This immersive installation proposes that CCTV is the creation of an entire world—unique, vast, and intricate. The project's complexity extends beyond its material conception to its realization in the context of developing China and to the collaborative team of over four hundred architects, engineers, and consultants spread throughout Asia, the Americas, and Europe led by OMA partner-in-charge Ole Scheeren. CCTV's architecture is expressed as much in its strategic execution within a fluidly changing society as in its groundbreaking design. The project proposes the built construct as a proponent and catalyst for urban and social change.
OMA in Beijing was designed in collaboration with OMA, Beijing. Team: Ole Scheeren, Paul Kroese, Dan Cheong, Jaime Oliver, Paloma Hernaiz. With Qiang Ou, Tieying Fang, Hiromasa Shirai, Stephane Derveaux, Xiadong Liu, Anu Leinonen, Andre Schmidt, Chiaju Lin
Special thanks to Top Notch Graphics
Background photographs by Iwan Baan

COMPETITION MODEL
31st MARCH '03
02 APR '03
14 MAR 03
03 Mar Arup

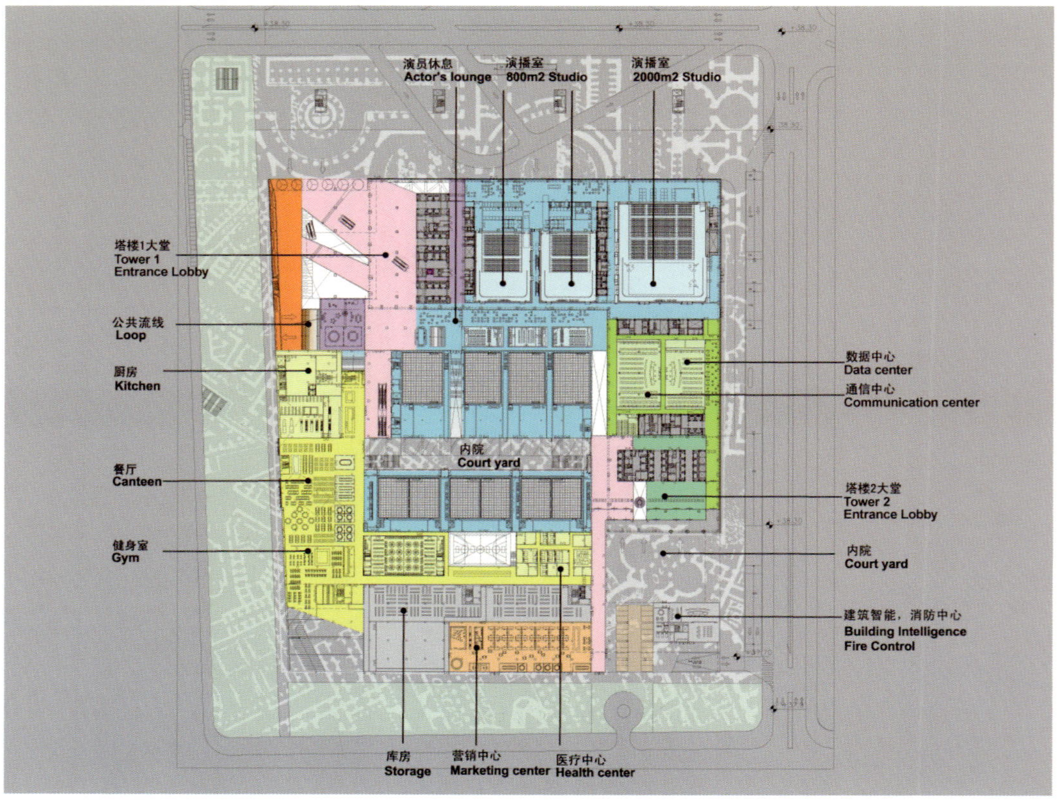
演员休息
Actor's lounge
演播室
800m2 Studio
演播室
2000m2 Studio
塔楼1大堂
Tower 1
Entrance Lobby
公共流线
Loop
厨房
Kitchen
餐厅
Canteen
健身室
Gym
内院
Court yard
数据中心
Data center
通信中心
Communication center
塔楼2大堂
Tower 2
Entrance Lobby
内院
Court yard
建筑智能，消防中心
Building Intelligence
Fire Control
库房
Storage
营销中心
Marketing center
医疗中心
Health center

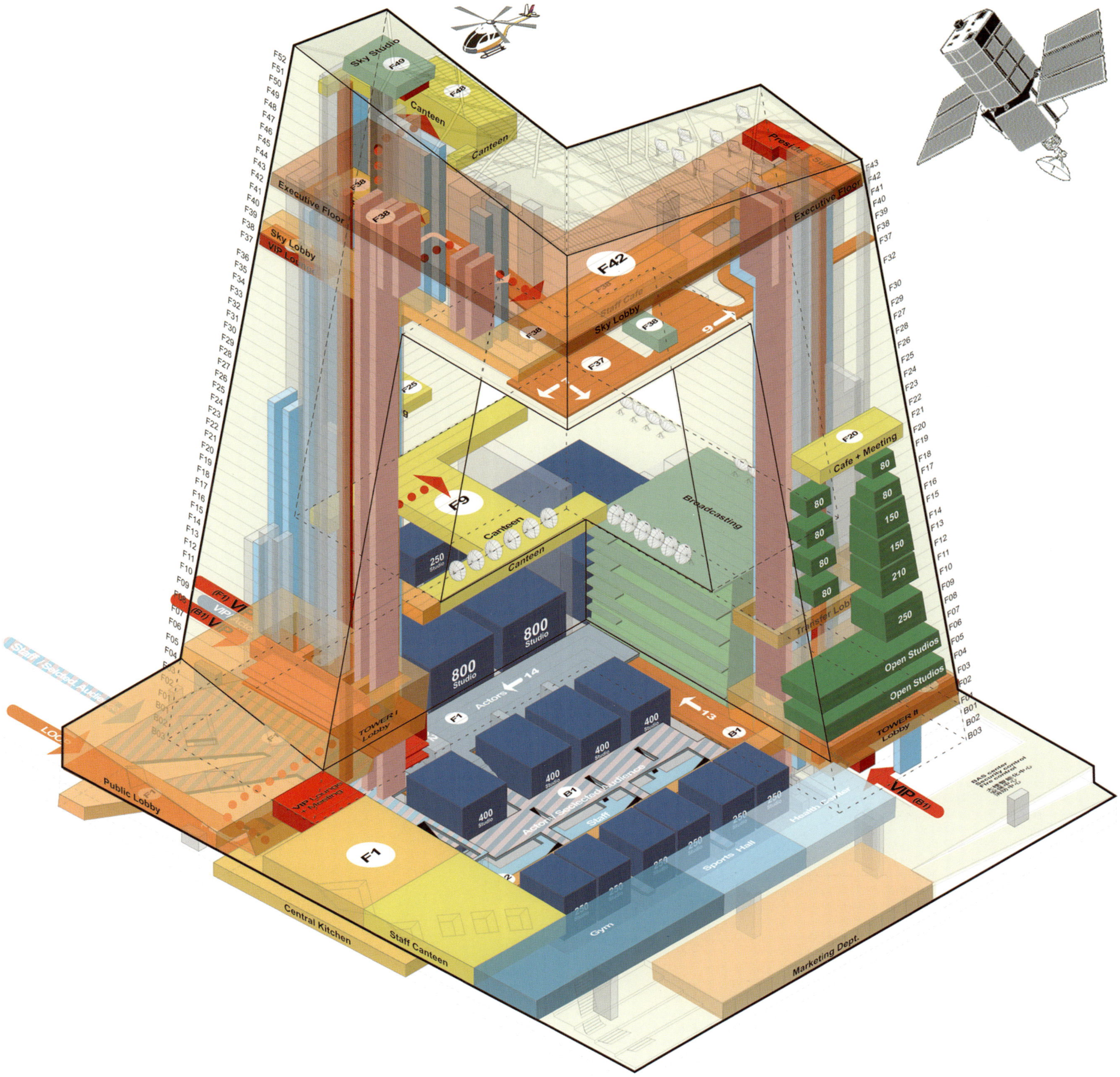

Sky Studio
F49
F48
Canteen
Canteen
Executive Floor
Executive Floor
Sky Lobby
F42
Sky Lobby
F37
Broadcasting
Cafe + Meeting
F20
F9
Canteen
Canteen
Open Studios
Open Studios
800 Studio
800 Studio
400 Studio
250 Studio
Actors
TOWER I Lobby
TOWER II Lobby
Public Lobby
VIP Lounge + Meeting
F1
Central Kitchen
Staff Canteen
Gym
Sports Hall
Marketing Dept.
Staff
VIP (B1)

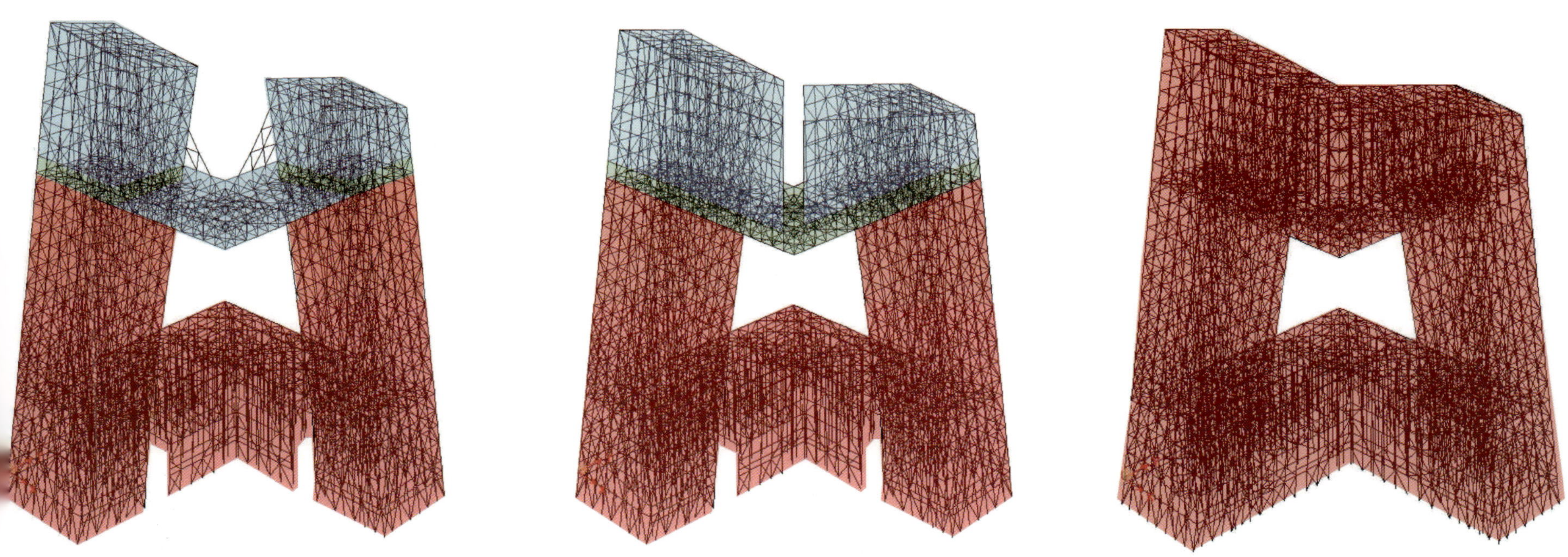

TVCC
行政办公区
Void
Mechanical Area 演播区
29F(+130.00m)
Restaurant Area 演播区
27F(+122.00m) to
28F (+126.00m)
行政办公区
Atrium
行政办公区
5★ Hotel
Hotel Rooms North 演播区
F15(+58.00m) to
F25(+98.00m)
Mechanical Area 演播区
F13(+50.00m) &
F14(+54.00m)
行政办公区
Theater
Hotel Rooms North 演播区
F6(+22.00m) to
F12(+46.00m)
Fly Tower 技术用房
F1(+0.00m)
Atrium Lobby 演播区
Hotel Lobby 演播区
F5 (+18.00m)
Spa & Gym 演播区
F4(+22.00m)
行政办公区
Cultural Center
News Release 演播区
F2(+7.10m)
Recording Studio 技术用房
F1(+0.00m)
Ballroom Area 演播区
Hotel Drop Off 演播区
F5 (+18.00m)
Understage 技术用房
F1(+0.00m)

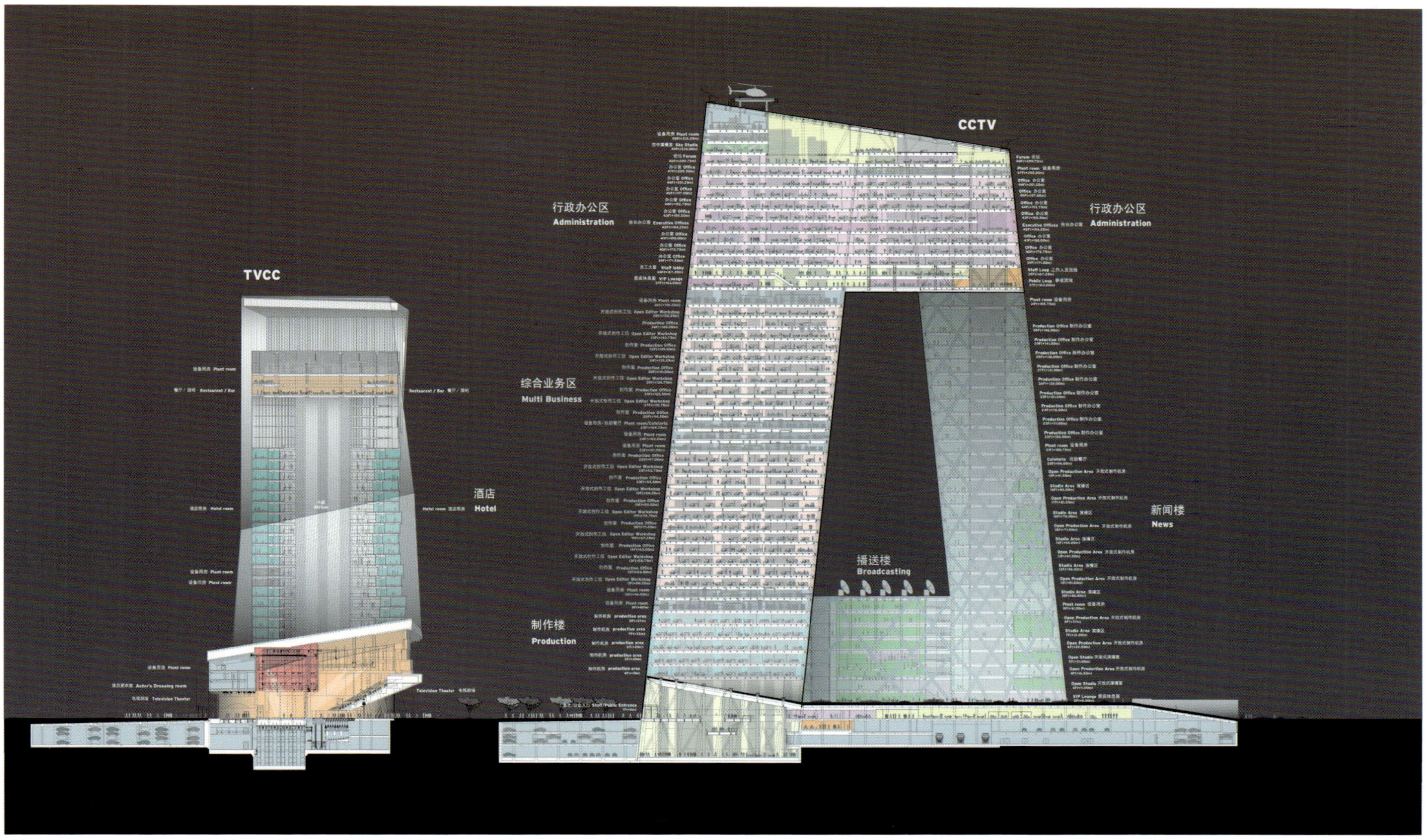
CCTV
TVCC
行政办公区
Administration
行政办公区
Administration
综合业务区
Multi Business
酒店
Hotel
新闻楼
News
播送楼
Broadcasting
制作楼
Production

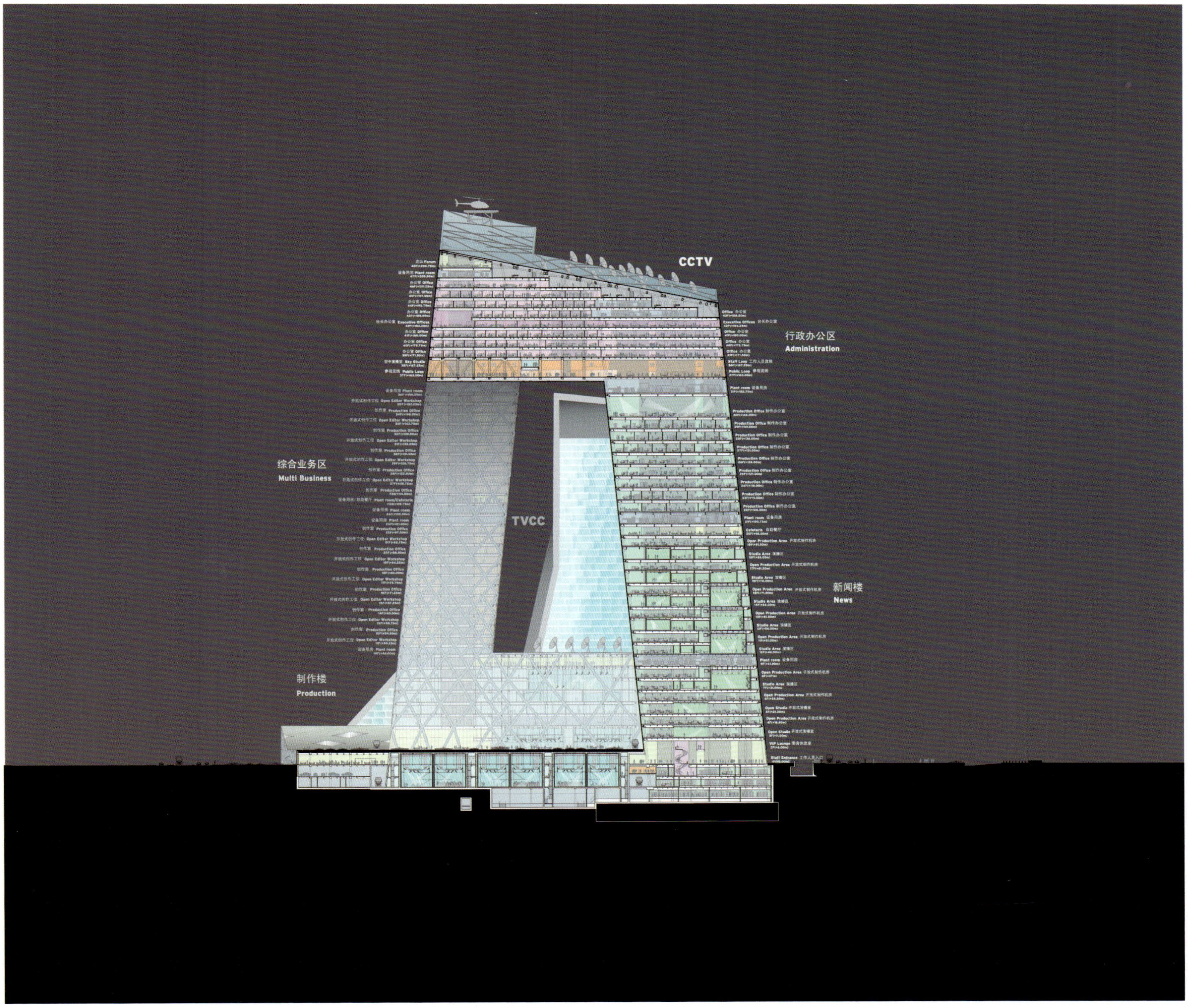
CCTV
行政办公区
Administration
综合业务区
Multi Business
TVCC
新闻楼
News
制作楼
Production

Tianjin Inner Mongolia international center

天津内蒙古国际大厦

Developer: Inner Mongolia Nailun (Group) Beijing Nailun Real Estate Development Company
Project address: Binhai New Park Xiangluowan business zone
Floor size: 2,960m²
Building size: 155,812m²
Landscaping ratio: 30.6%

开发商：内蒙古奈伦（集团）北京奈伦房地产开发公司
建筑设计：汉森国际·伯盛设计集团
项目地址：天津市滨海新区响螺湾商务区内
建筑占地面积：2,960m²
总建筑面积：155,812m²
绿化率：30.6%

Inner Mongolia international center is located in Binhai New Park Xiangluowan Business zone, "other provinces business zone". The west and south are the principle streets that the municipal government planed and the east and north are the subordination roads. It seats in the southeast of business zone center. The southeast of this place is the direction of river into sea; the northwest is the center of Xiangluowan Business zone, surrounded by the similar office building and business mating building in this zone. The project is requested that the building should match the image of the entire business zone, brilliance and sense of times. Specific landmark is also requested to be. It should stand for the image of Inner Mongolia autonomous business region staying in Tianjin.

The large-scale office and small-scale office, as well as the functions of mating business, leisure and entertainment will be set. The 1-3rd floor are business or large–scale office. The 4-14th floors is large-room office, 15-18th floors are the service area of meeting, catering, fitness and so on. 19-44th floors are the small-scale office for the companies staying Tianjin. The 15-30th floor are for refuge,

内蒙古国际大厦位于天津市滨海新区"外省市商务区"响螺湾商务区内21号地块，西、南为市政规划干道，东、北为市政规划次路。地处商务区中心偏东南部，地块东南为海河入海口方向，西北为响螺湾商务区中心，周边在商务区内为其它省市同类办公写字楼建筑及商务区配套建筑，场地及周边地势平整。项目要求建筑符合商务区的形象要求，鲜明而具现代感，同时亦要求其有明确的标志性，能够代表内蒙古自治区驻天津国际港的商务形象。

设置大型办公和小型高级办公及一定面积的配套商业、休闲娱乐功能。1~3层为商业或大型办公、4~14层为大空间办公、15~18层为会议、餐饮、健身等服务设施，19~44层为驻津企业小型高级办公，15层和30层为避难层，部分设置设备间、游泳池、健身房、高尔夫练习室等。

按照《天津塘沽响螺湾地区城市设计控制导则》规定，建筑向南退让道路红线20m，东、西、北面退让道路红线15米，北面与地块19-01在四、五、六层设通廊相连。地下一层设人行通道与地块20相连，地下二层设车行通道与地块19-01、19-03相连。

some parts of where are set to facility room, swim pool, fitness room, and golf exercise room.

According to the rule of < The Tianjin Tangku conch shell bay area city layout controls the guidance >, the building should back off 20m from the road red line south, 15m to the east, west and north. The building connects the fourth, fifth, and sixth floors of 19-01st block in the north. The first floor underground is connected with block 20th, and the car road in second floor underground is connected with block 19-01st, 19-03rd.

With the active and aggressive image, imbricate walls of the building symbolized the metal knitting armors of the ancient Inner Mongolia warriors. The prolongation sprite of preserving and braveness is reflected.

With the rive relief, the large-scale open glass walls are set in the southeast, and the large-scale real walls and window holes are set in the northwest. Every three to five floor own the air garden together, which is gotten through the steel stairs outdoor. The desks and chairs, coffee and drinking zone should be set here. It will get soft nature breeze in south yard. The user of north room can enjoy the sunshine in the rest time, which will fix the shortage of room facing direction, and a communication room with vigor will be formed.

The floor surrounding the area is opening greening and some entire hard floor alternates in. in the south is greening view car parking area and a large-scale opening green. The west and east can see green everywhere and there is all view, according to the green area planed by municipal government. The plants, grass, and leisure facilities are considered adequately in the greening area, plaza and the outdoor car park.

The principle entrance is in the north, the secondary road of municipal government. The car entrance for underground carport is in the west, which having the distance above 7m with the road. Surrounding the building is fire controlling road, with the width of 4m, and there is no obstruct 4m above.

The portrait traffic design is another significant feature of this building. 16 high-

本建筑运用简洁而积极开拓进取的形象，鳞状的幕墙象征了古代蒙古武士的金属编织铠甲，反映了一种勇往直前的开拓精神和百折不挠的民族精神。

结合江河地势，东南设置开敞通透的大面积玻璃幕墙，西北设置大面积实墙和窗洞。在满足采光遮阳的功能要求下形成强烈的虚实对比。建筑每三到五层设置一座共用空中庭院，相邻层通过设置在庭院中的室外钢楼梯到达和利用这个庭院。庭院中可设置桌椅、自助咖啡、饮料食品区等，设置在南面的庭院获得了良好的自然通风，使北向房间的使用者可在休息时间享受到阳光，弥补居住朝向的不足，形成一个有活力的交流空间。

建筑四周设开放式绿地穿插整体硬质铺地，南面设置绿化景观停车场及大片开放式绿地，西面东面结合市政规划中心绿地和沿街绿带，做到处处见绿、步步有景。绿化同硬质广场、室外停车场结合，并充分考虑树木、草地、休闲设施的设置。

建筑北临市政次路，设置建筑主入口，西面临市政次路，设置地下车库车辆出入口，地下车库出入口距离道路7m以上，交通便捷。建筑物四周设置环形消防通道，通道宽4m以上，上部4m内无阻碍。

竖向交通设计将成为内蒙古国际大厦的又一亮点，16部1.6t、4m/s的高速电梯可满足人员运输的要求。核心筒创新设计使电梯轿箱在前后两个方向通透，再结合庭院设置，将自然光引入电梯，使乘用者在徐徐上升的同时，能看到各楼层的活动和因庭院位置跳动而引起的光线变化。此举可消除人在封闭电梯中的焦虑感，将高层建筑的电梯由冰冷的集装箱式的运输工具变成人性化的交通系统。新的电梯系统也不同于传统的视野单调的景观电梯，新的电梯系统在不同楼层面运动，为乘用者提供不同场景的动态切换和视线交流，从而丰富人们对大厦人性化的体验，而透明的电梯系统也同时成为休息庭院的动态景观。

标准层面积3100m²,每层分两个防火分区，分别设置两座带前室的防火剪刀梯，两座消防电梯。标准层上下开口部分周边设防火卷帘或防火玻璃隔断。室外广场、公共入口通道及参观路线按无障碍设计要求设置。建筑入口、入口平台及门、水平垂直交通、接待用房、公共用房、公共卫生间作无障碍设计。

speed elevators of 1.6t and 4m/s can satisfied the traffic demand of people. The design of core tube supported structure makes the elevators clarity behind and back. With the setting of yard, introduce the nature view into the elevators. The users can see every floor and feel the light changing with the yard situation, when the elevator rises slowly. This design weaken people anxiety when they are in the elevators, changing the elevators of tall buildings from transport tools of cold contain to traffic system with humanity. The new elevator system is different from the traditional dull visual elevator. It is moving in different floors, providing the motive changing through various floors and visual contract for the users. So the experience of people to the building will be abundance.

Standard area is 3100m2, and each level is divided into two fire compartments, establishing two riggers bands separately fire protection scissors ladder, two fire lifts. The above and under opening parts of standard area are set with fire resistant shutter and flame resistant glass. Plaza outdoor, public entrance and visit line are set with the barrier free design. There is also the barrier free design in entrance of building, entrance of platform, the doors, the receiving rooms, public rooms, and public washing rooms.

Surrounding the building, there are office buildings, hotels, and top-range residences. The whole base is without any pollution. All the offices and public corridors have the window which can open out or the machine air exhausting without any noise. all the material of the building are environmental protection.

The main facing direction is southeast and northwest. The body of tower is simple and elegant, with the structure of concrete. Roof and external wall insulation and efficient extruded paste board, curtain wall can be opened using energy-efficient low-E coating off hot box thermal insulated glass curtain wall with high permeability, through the use of low-E insulating glass with high thermal physical properties, the maximum block the sun radiant heat, while the best solar lighting. Unique design with the use of walls within the courtyard and roof gardens for best air circulation to minimize the central air conditioning and mechanical ventilation dependency.

大厦周边为办公、酒店、高级住宅，基地内无有害污染物。所有办公室及公共走道和卫生间设可开启窗或机械排风而且不产生噪音干扰。建筑材料均采用环保型产品。

项目主要朝向为东南、西北。塔楼形体简洁，以混凝土为主体结构。屋顶及外墙面贴高效保温材料挤塑板，幕墙采用可开启的高效节能断热框镀膜低辐射热高透中空玻璃幕墙，利用低辐射热高透中空玻璃的物理特性最大限度地阻挡了阳光辐射热，同时获得最佳的日光照明。利用幕墙独特的设计配合内天井及空中花园可获得最佳的空气循环，最大限度减少对中央空调和机械通风的依赖。

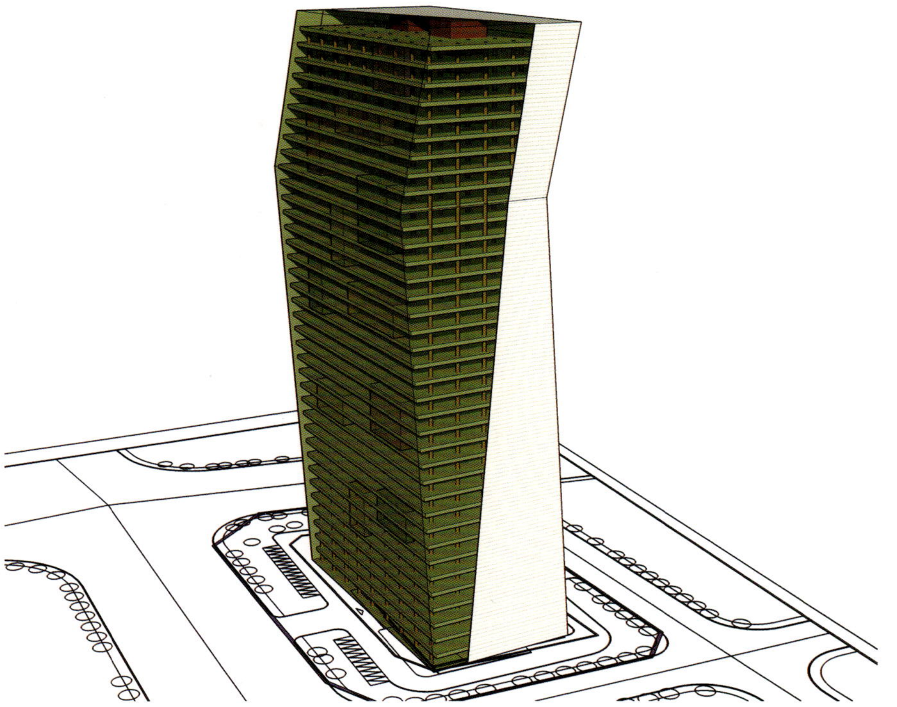

Mac OS

THE 24-HOUR STREET

Beijing Jianwai SOHO

北京建外SOHO

Developer: SOHO Chinese Ltd. Co,
Architect: Shanbenlixian (Japanese)
Domestic cooperation designer: Beijing New Era Design company
Project address: Chaoyang district third ring road 39th, Beijing
Floor size: 122,800m^2
Building size: About 700,00m^2
Landscaping ratio: 33%

发 展 商：SOHO中国有限公司
建 筑 师：山本理显（日本）
国内配合设计单位：北京新纪元建筑设计公
项目地址：北京市朝阳区东三环中路39号
总占地面积：122,800m^2
总建筑面积：约700,000m^2
绿 化 率：33%

Beijing Jianwai SOHO is located on the Chang'an Street, the opposite side of International Commercial Center. The project is composited by 18 apartment buildings, 2 office buildings, 4 SOHO small-scale business houses and lots of annexes. The mating facilities are kinder garden, clubs, residences, offices and shops. There is no outer wall for this project. 16 narrow streets are through and connect each other in the building group. The street culture is full of the project. It is called "the most fashion life show window" in Beijing.

All the shops of SOHO are standing facing the street, coming and going independently. It is a unit for the shops in first and second floors with the compound structure. The third floor gets the public yard with escalator directly, which is very convenience. There are tea houses, coffee houses, restaurants, video shops, beauty salons, and fashion dress shops in this moment. So many business companies have operated in SOHO, and in the future, SOHO will be the new capital of fashion in Beijing.

建外SOHO位于北京的长安街上，国贸中心正对面。项目由18栋公寓、2栋写字楼、4栋SOHO小型办公房及大量裙房组成，配套设施包括幼儿园和会所，集住宅、办公、商铺于一体。建外SOHO没有围墙，16条小街在17hm^{2}的建筑群中穿梭串连，制造出充满人情味的小街文化，被称为北京“最时尚的生活橱窗”。

SOHO商铺全部沿街设立，独立进出，面积从100~600m^{2}不等，设施齐全。一、二层铺面复式结构自成一体，三层铺面由滚梯直达公共院落，出入便利。目前已有茶馆、咖啡馆、火锅店、料理店、音像店、美容院、时装店等多家商业单位进驻SOHO街，未来的SOHO街将是北京新的时尚集中地。

为了保持和谐统一，商店入口层的店铺门面设计了一种模式体系，但为符合当地做法也设置许多展窗和招牌预留位。“我们不得不在秩序和自由之间寻求一种平衡，”Mikan的合作者之一Manuel Tardits解释说。Mikan的系统与山本的简约白色柱梁立面相呼应。“我集中精力创造一个完全抽象的表现形式，这样，建筑就可呈现其中的人及其活动的特征了”，山本说。

SOHO

For keeping the harmony and unify, a kind of mode system has been designed on the entrance floor, but many showing window and signboard reserve also set to coincide local custom. "We have to find a kind of balance between the rule and free." One of the cooperation of Mikan, "Manuel Tardits" explained. The system of Mikan matches the white beam column and reflected with each other. "I focus on building a total abstract expressing style, and then, the building can show the people and what they do will be showed." Shanben said.

In the south and north are two SOHO office buildings. The office size is 98,000m2, with the height of 28 floors. The body of building is so bright and clarity, simple and modern. The entrance of every unit is independent, with cutting mention of flexible indoor room. It is suitable for individuality needs of every company to office environment.

SOHO is mixed. In here, people can live or work. The house in here may be a tea house, a flower shop, a art studio, or maybe a kinder garden. There are countless possibility of life in here. It is clarity in daytime, and shining in the evening. The first to third floors are shops facing the street. The fourth and above are SOHO units. Eighteen building are disorderly and especially vivid.

Many independent units size is big to 280m2 in the apartment of higher floors. They connect the room of living and working, almost without the walls. So the renter can cut the room as they want.

The biggest garden in the CBD area is above 30,000m2, like two man-made stadiums.

280,000 m2, the second and third floors underground conclude the swim pool, racket ball hall, tennis courts, gym, sauna, massage, coffee shop, and bars.

The spirit of Jianwai SOHO is without the center, multi-level, and mixed is the energy. Architectural design approaches a maze and small neighborhood town. The city is filled with small squares. two gardens can random up and down, meandering bridges, streets. Looking up, there is the tall and transparent glass building.

Over the past decade, Shanben is very active in construction stage of Japan, paying attention to the building and the environment mixing. When Jian Wai SOHO was in the designing, he was visiting Beijing's hutongs and courtyard houses. he is very much appreciate the alley can be string together on each side street yard, so that the street will be built into the building in SOHO, with more than a dozen street construction each individual series. Street is irregular, with small squares and small garden, so that the people living in the community can communicate. He said, "Straight business streets are not attractive. The real charm should be twists and turns, always giving a surprise. In this way, it will let customers looking forward to. So the streets are curved, and people do not see the end, which is adding to discover the fun, and the desire to enhance the shopping. In addition, it requires a sophisticated space. It will narrow the distance between the buildings and is very tolerant of the three-dimensional structure." so a lot of 3m × 3m × 3m of the cubes was born into a whole program. One of the cubes is like a cell, with which keeps dividing, the whole unit will expand. Therefore, this program is named "cell city." a sunken green is designed in Underground garage, so that the whole area is full of three-dimensional. SOHO, built in the center of Beijing CBD, these humane neighborhood streets, just busy make up the CBD deserted defect which busy during the day and quiet at night.

Minimalism is the greatest feature of Jian Wai SOHO. From the main road to the ground, to the floors of the building itself, and even underground facilities are reflected. Each area is a combination of housing, shops, offices, with strong individuality. However, they also contact with each other to form a whole.

Opening should be another big design features. Shanben said: the spirit of Jian Wai SOHO inspired by the Moroccan city of Ceuta. "My approach is to build to be open, interlude the construction into the street, breaking the closed concept of the hutongs and courtyard." Jian Wai SOHO included residential, retail, office and garden and many other functions, and all of space but also integration, so they created an open community, rather than the monotony of the traditional closed residential area, which is kind of in Beijing new concept.

In here, high-rise apartments and small shopping malls combine to form a loose checkerboard pattern. The entrance of Service area is set in commercial facilities and layout the entrance of car dealers in the underground parking of each apartment. The entire region of floor space turns to be pure space for walking. There are a number of sunken garden on the ground floor layout. The gardens connect the two levels together, so the sun can be transmitted to the lowest recesses.

Jian Wai SOHO is the inclined large building that Beijing never has. In the historic city of Beijing, almost all of the buildings are south and north; the inclined large building is nowhere to be found. But the direction of Jian Wai SOHO Building is south and east as 30 °, conforming to the surrounding streets and Tonghui River flowing through the sidelines. This level of rotation and scattered apartment towers (the farther away from the water, the higher the building) make full use of the base, and each apartment of the building can enjoy plenty of sunshine, eliminating the shadows affected between buildings, and maintenance the tenants privacy further step.

南北两座SOHO办公楼，办公面积98,000m²，楼高28层。楼体通透明亮、简约现代。各单元均有独立入口，配合灵活的室内空间分割方式，适合不同公司对于办公环境的个性化需求。

SOHO是混合的，在这里可以办公、可以居住，这里的房子可以是一个茶室，一个花店，一个艺术工作室，也可以是一个幼儿园，这里的生活有无数种可能性。体型细长，白天透明，夜晚发亮。公寓楼1～3层为临街商铺，4层以上为SOHO公寓单元，18栋楼高低错落，异常生动。

高层公寓中的许多独立单元面积都达280m²，它们将居住和工作空间结合起来，几乎没有固定的隔墙，这样租户就可以随心所欲对空间进行分割。

CBD中心最大花园，面积超过30,000m²，相当于两个工人体育场。

28,000m²，地下2层，地上3层，游泳池、壁球厅、网球场、健身房、蒸汽桑拿、按摩、咖啡馆、酒吧一应俱全。

建外SOHO的精神是无中心、多层次，混合就是活力。建筑设计的手法是迷宫一样的街区小城，城里到处都有小广场,两层随意上下的花园，蜿蜒的小桥、小街，抬头望去是高高的、透明的玻璃楼。

近十多年来在日本建筑舞台非常活跃的山本理显，注重的是建筑和环境的契合。在设计建外SOHO时，他逛了北京的胡同和四合院，非常欣赏胡同中可将每个院落串起来的小街，于是就将这样的小街纳入建外SOHO的建筑中，用十多条小街将每个个别建筑串联。小街中有不规则、不定向的小广场和小花园，使住在小区中的人能交流沟通。他说“直来直去的购物场所是不会吸引人的，真正有魅力的应该是迂回曲折，时刻给人以惊喜的，如此一来，就会让顾客充满期待。所以街道都是弯曲的，让人看不见尽头，增添发掘的趣味，提升逛街的欲望。另外，还需要一个精致的空间，它会拉近人和建筑物的距离，就是极具包容力的立体结构。”由此就诞生了把很多3m×3m×3m的立方体组合成一个整体的方案。其中一个立方体就像一个细胞，细胞不断分裂，整体也就扩大。所以，将这个方案命名为“细胞的城市”。地下车库中又设计出下沉式的绿地，使整个小区富有立体感。由于建外SOHO处于北京CBD中心，这些富有人情味的街区，刚好弥补了CBD白天热闹晚上冷清的缺憾。

简约主义是建外SOHO最大的特点。从主干道到地面，到建筑本身的楼层，甚至地下设施都有所体现。每个区域都是商店、办公楼、住宅的结合体，极具个性。但是，各自之间又彼此联系，形成一个整体。

开放应是另一大设计特色。山本理显表示：建外SOHO的灵感来自一个名叫休达的摩洛哥城市，“我的做法就是把建外SOHO做成开放的，把一条条街穿插到建筑中去，打破了北京胡同和四合院的封闭概念”。因为建外SOHO包含住宅、商铺、办公室和花园等多种功能，各个空间又可融合互通，于是便造就了一个开放式的社区，而不是传统的单调封闭式的住宅区，这在北京是种新的概念。

在这里，高层公寓与小型商场相结合，形成了一种松散的棋盘式格局。通过把服务区入口设于商业设施内并将车行入口布置在每幢公寓的地下停车场，使整个区域的地面空间成为纯步行空间。在地面层布置了许多下沉式花园，这些花园把两个层面结合在一起，让阳光能够透射到最低的幽深处。

建外SOHO也是北京市前所未有的斜向大型建筑。在历史悠久的北京城内，几乎所有的建筑都是正南正北，斜向的大型建筑根本无处可寻。但建外SOHO的大楼方向却定为向南偏东30°，顺应了周围的街道以及从旁流过的通惠河。这种旋转和公寓塔楼的高低错落(离河水越远的建筑越高)充分利用了基地，并使每幢公寓都能享受到充足的阳光，消除了楼与楼之间的阴影影响，进一步维护了住户的隐私。

+hp
¥1499!
OFFICEJET 4110

惠普个人激光打印机系列
黑白本色 超越自我
東田造型
TONY STUDIO

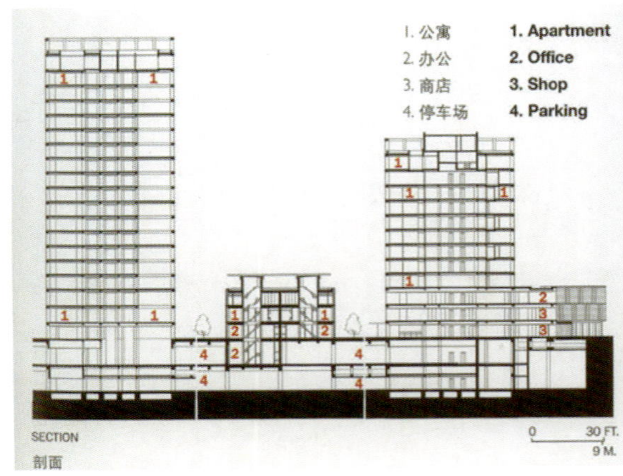
1. 公寓 1. Apartment
2. 办公 2. Office
3. 商店 3. Shop
4. 停车场 4. Parking
SECTION
剖面
0 30 FT.
9 M.

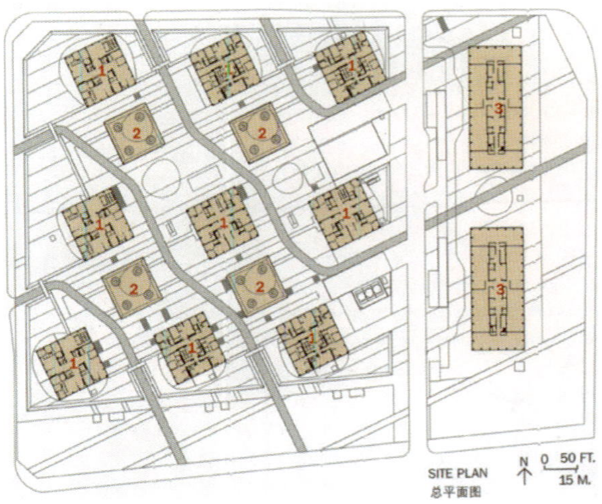
SITE PLAN
总平面图
N
0 50 FT.
15 M.

Fulton center

富顿中心

Developer: Beijing Kun Taijia Real Estate Development Co., Ltd.
Project address: southeast of Jinsong Bridge, Chaoyang District, Beijing
Flooring size: 13,900m²
Building size: 105,400m²
Landscaping radio: 30%

开发商：北京昆泰嘉业房地产开发有限公司
项目地址：北京市朝阳区劲松桥东北角
占地面积：13,900m²
建筑面积：105,400m²
绿化率：30%

Fulton center is located in northeast of Jinsong Bridge of Chaoyang District in Beijing. It closes to the East Third Ring Road, steps away from the ITC, with excellent supporting unique CBD area. The traffic is convenient here. It is large supporting integrated property setting with offices, apartments, matching business. The offices of the Project and apartments are an organic whole to support each other. Office functions structured reasonably. the smallest area is to 65~180m². 500m² sunny lobbies, Mitsubishi Elevator for entrances and exits, exclusive elevators, and corridors, are very suitable to buy the whole floor.

Fulton made the concept of "Fifth Generation Apartments", to learn the advantages of apartments in European and American countries. Creative thinking was from the small size, but higher than the small units. Through the optimal design of the units, and according to the normal living comportment, area distribution of life function is accurate calculated. In particular reinforce the luxury living space, in the conditions of area still restricted, the substantial increase housing using rate, and stroke up to 80% of the high level. The building

富顿中心位于北京市朝阳区劲松桥东北角，紧临东三环，距国贸咫尺之遥，拥有CBD区域得天独厚的优秀配套，交通畅通，出行便捷，是集办公、公寓、配套商业为一体的大型综合物业。项目写字楼与酒店式公寓相互支撑形成一有机整体。写字楼功能结构合理，最小开间面积达65~180m²，500m²阳光大堂，进口三菱电梯，独有的电梯间、走廊，非常适合整层购买。富顿提出"第五代酒店式公寓"的概念，吸取欧美等国家酒店公寓优势，创意发想源于小户型，又高于小户型，通过户型优化设计，根据正常的居住舒适度，精确计算生活功能的面积分配，特别强化了人居豪华空间，在面积仍然受限制的条件下，大幅度提高房屋使用率，并一举达到80%的超高水平。内设欧式、现代式、中式、日式装修风格，不同楼层不同风格，将鲜明的世界风情引入居住中，再一次提升小户型的人文高度。

富顿中心位于北京市朝阳区劲松桥东北角，北距国贸中心2.5km，地处CBD南缘的商务辐射区，临城市主干道东三环路，具备优良的区位优势。中心由两栋写字楼和两栋服务式公寓组合而成，其中A、B座为富顿中心国

富顿公寓

is equipped with European, modern, Chinese, Japanese style decoration, different floors with different decoration styles. The distinctive world styles are introduced into residence. The humanity of the small size is enhanced once again.

Fulton center is located in northeast of Jinsong Bridge of Chaoyang District in Beijing. 2.5km is to Trade Center in the north. It is located in the southern edge of CBD commercial radiation zone, with excellent geographical advantages.

The center is combination with the two offices and two service apartments, of which building A and building B are Fuilton International Center. Building C and D are Fuilton International Apartments. Business, commercial service facilities, functions as the city's comprehensive body, are matched. The height of the building is not exceeding 80m, 8th floor to 25th floor on the ground, 3 basements. the center is built to be a new landmark of the city, as well as a new light Point of high-grade commercial real estate, having been certified by authority as 50 service apartments in China in 2004.

Building base is a square, surrounded by roads. The west and south are Urban Freeway (East Third Ring Road) and the main road.

For this, planning will be two office buildings as a combination of commercial and residential apartments conjoined, with "L" shape along the southwest side. It will be independent in the apartments, arranged in sections to form good internal and external partition function for office and residential offer a relatively suitable land environment. With the architectural layout features, both inside and outside of the ring road settings are also formed, the outer office and business portal access, forming the business flow of traffic and parking systems. The inner ring with central garden design, twists and turns, and apartment entrance phase, constitute a good casual stroll and fire transport system. But also seek to meet the dynamic properties of static, closed management requirements. Arranged in the underground garage entrances and exits the outer edge of the lot to avoid the interference of the center environment.

际大厦，C、D座为富顿中心国际公寓，配套有商务、商业服务设施，功能为城市综合体，建筑高度不超过80m，地上8至25层，地下3层。该中心的落成为都市构筑了新的地标性建筑，同时为中高档商务楼盘添置了新的亮点，2004年已被权威机构认证为中国服务式公寓50强。

建筑基地为正方形，四周环路，其中西、南两侧分别为城市快速路（东三环路）和主干道。为此，规划将两座写字楼与一座商住式公寓联体组合，呈“L”形沿西南侧布置，将独立式公寓布置在地段内侧，从而形成良好的内外分区，为办公与居住功能均提供了相对适宜的用地环境。配合建筑布局特点，道路设置也形成内外两环，外环接写字楼及商业入口，组成商务流交通和停车系统，内环结合中心花园设计，曲折变化，与公寓入口相接，构成良好的休闲漫步兼消防交通系统，同时也满足了动中求静的封闭式物业管理要求。地下车库出入口布置在地段外缘，避免了对中心环境的干扰。

建筑设计首先从打造地标性建筑着眼，结合CBD区域特征采用国际式风格，通过简约的线面处理，实现了立面形式的完整统一。同时，利用细腻的虚实对比、肌理组构、色差变化和材质对比等手段，加之内外流通的现代建筑空间处理手法，采用顶部及入口等细部设计，充分体现了白领目标客户群的文化品位，并且在与周边建筑相互协调的同时，创造出具有个性化和感召力的现代地标建筑形象。

体量整合是富顿中心设计解决的首要问题。基于该建筑对地段北侧现状住宅的日照影响，以及地段内两栋公寓之间自身遮挡因素，呈“L”形布置的办公公寓综合楼需采用中间升起、两侧跌落的体量处理方案。为避免因此而产生的建筑形象散乱不整的问题，立面采用以竖线条为主的设计方法，同时通过外饰构件的肌理变化组构，配合顶部造型设计，创造了既统一又富有变化的效果，增强了视觉冲击力。为体现建筑的高尚品质和商务气质，外装饰以深灰色系的金属材料与玻璃相结合，同时与周边现有建筑也构成了完美和谐的整体街区景观环境。

First, from architectural design to create landmark buildings in mind, combine features of CBD area of international-style. Through the simplicity of line and surface processing, unity of the elevation is formed. At the same time, the using of the actual situation delicate contrast fabric texture, color change and material means of comparison, combined with internal and external circulation of the handling of the modern architectural space, using the top and detailed design of the entrance so fully embodies the culture of white-collar target audience taste. The surrounding buildings are coordinated with, while creating a personalized and inspiring image of the modern landmarks.

Volume integration is the first problem to solve in design. Based on the status of lots of sunshine construction on the north side of residences, and within lots of self-shielding factors between the two apartment buildings, "L" shaped layout of the apartment of complex office building adopts to the treatment program of being raised in the middle, and both sides of the drop. In order to avoid the result of image of building is not the whole, design method based on the facade with the vertical lines. At the same time with the texture changing structure of the exterior components of in group, the design of the roof is matched, creating a unified and full of changes both the effect and enhance the visual impact. To embody the high quality and business feature, the outer decoration adopts the combination of dark grey metal materials and glasses. It forms the perfect entire street view environment with the surrounding buildings.

With a specific functional orientation, the use of appropriate space expression means, and technical means is another design elements of Fulton. Location of office apartment with high-volume ratio and small size, and pursuit of cost-effective, the conflict between these three, presented for the architectural design many challenges. According to the exact planning of size, the clever organization of traffic space, the "tailored" of equipment system, and a special selection of "comb" structure, specialized tools as such using, the final design provides unit size with excellent conditions for natural ventilation and lighting, flat flexible, achieves a rare high-rate (75% office, apartment 80%), and creates a unique space for the performance characteristics of tension and public areas. Meanwhile rely on economical design of only 31 square meters per parking garage, reduces the investment cost of the basement greatly. Cost-effective products with unique design eventually won wide acceptance in the market.

Based on size, structure and other specific requirements, the design explores a set of technical means with the original meaning. On the one hand, the using of business unit programs with short-depth, improves the physical environment of office space with the small size; on the other hand, the way by reducing the height of floor to ensure the total building size and floor area ratio are not damaged. So the contradiction between building structure and equipment are appearance. For this, with the structure of the encrypted column grid layout requirements, the original design means of \ a "film column" and "comb-like structure" is used bravely in this design. It not only meets the requirement of clarity sense for large-size office block, but also avoids the adverse effects causing by beams to the room organization in small-size office block. The corresponding consideration of the upper and lower floors determination makes every functional areas can satisfy the requirements, such as underground garage, first floor office lobby, the upper unit and other of office space. It not only create a unique office space for artistic effect of the lobby, but the utilization of the building has also been effectively improved.

According to market planning, building size of the apartment to sale is determines in 40 square meters to the 80 square meters, with fine repair, service apartment type as the basis of positioning. Since the target customers are mainly single and simple family, family life is characterized by individual behavior basically. The integration of the two dominant functions of the case, living and sleeping is the basic design concept. According to handle means of two kinds of room relatively completive and flowing to each other, the requirements in terms of living space flexibility for the specific family patterns is met. Aesthetic fashion of the younger generation is also met. On this basis, the design proposed two basic units model of "simple type" and "compound" to achieve the optimal combination of home features.

The building size of "simple type" units is generally about 30 square meters to 40 square meters. Because of small size, single space is the main expression form. Living area and sleeping area is in an open space, which is hotel standard room type of space. But there is an essentially different: generally the beds in hotel rooms are the center, which reflects its characteristics as a guest temporary accommodation, but single room apartments housing emphasis on the characteristics of stability more, so the living area is the center, the bed and furniture are arranged on the inside, which is more privacy.

"Compound" units building size is about 40 square meters to 60 square meters generally. Due to more relaxed area, independent living conditions of the region can be expended. But this design did not use the rooms, separated by conventional fixed program, and emphasis on the rooms, between the half-closed spaces of liquidity as the primary means of expression, a feature that better adapt to the life characteristics of the customer, but also reflects the Apartments and common patterns of residential dwelling significant differences between each other.

Common feature of both models is fully functional and the requirements of the basic function of living can be met. Facilities content specifically includes the independent health facilities, kitchen operating area, dining area, writing the work area, dressing Area, storage area, clothing and so on. An open kitchen

结合特定的功能定位、采用适宜的空间表现手法和技术手段，是富顿中心设计的又一要点。高容积率、小户型办公公寓的定位与高性价比的追求，这三者之间的矛盾，为建筑设计提出了许多难题。通过平面尺寸的精确推敲、交通空间的巧妙组织、设备系统的“量身定做”以及“梳状”结构形式的特殊选型等专业化手段运用，设计最终提供了自然通风采光条件优越、平面灵活多样的单元户型，实现了难得的高使用率（办公75%、公寓80%），创造了独具空间张力和表现特色的公共区。同时，依靠每车位仅31m²的经济型车库设计，大大降低了地下室的投资成本。高性价比与独具特色的设计产品最终也赢得了市场的广泛认同。

基于户型、结构形式等特定需求，设计探索出一套具有原创意义的技术手段。一方面，采用短进深办公单元方案，有效提高了小户型办公空间的物理环境；另一方面，通过降低层高的方式保证了建筑总规模和容积率不受损失，于是建筑结构与设备之间的矛盾也凸现出来。为此，在结构柱网布局要求加密的前提下，设计大胆采用了“片柱”与“梳状结构”的原创性设计手段，既满足了大户型办公区对于空间通透感的要求，又使小户型办公间避免了梁柱对室内布局的不利影响。同时，开间尺寸的确定将上下层对应考虑，使得地下车库、首层办公大堂以及上层办公单元等各功能区空间均能很好地满足其使用要求，这样不但创造了办公大堂独特的空间艺术效果，而且使建筑的使用率也得到有效提高。

根据市场策划，公寓销售建筑面积确定在40m²~80m²左右，并且以精装修、服务式公寓类型作为定位基础。由于目标客户主要为单身及简单家庭，家庭生活基本以个体行为为特征。本案以起居和就寝两项主导功能一体化作为基本设计理念，通过两种空间既相对完整又互相流通的处理手法，满足了特定家庭形态对生活空间弹性方面的要求，也符合了年轻一代的审美时尚。在此基础上，设计提出“简单型”和“复合型”两种基本户型模式，实现了居家功能的优化组合。

“简单型”户型套内建筑面积一般在30m²~40m²左右。由于面积较小，以单一空间为主要表现形式，起居区与就寝区在一个开放空间内划区，基本属于酒店标准客房的空间类型。但两者又有本质不同：酒店客房一般以床为中心，体现其作为客人临时居所的特点，而单一空间的酒店式公寓更强调稳定居所的特征，因而以起居区为中心，床组家私则布置于内侧，更具私密性。

“复合型”户型套内建筑面积一般在40m²~60m²左右，因面积较为宽松而具备了扩展相对独立生活区域的条件，也使套型平面具备了更多住宅的特点。但本设计没有采用厅室固定分隔的常规方案，而着重以厅室之间半开半合的流动性作为空间表达的主要手段，这一特点更好地顺应了客户的生活特征，也体现了酒店式公寓与普通住宅的套型格局之间的显著区别。

两种模式的共同特点是功能齐全、能满足基本居住功能的要求，设施内容具体包括独立卫生设施、厨房操作区、就餐区、书写工作区、梳妆区、衣储区等等。其中厨房在本案采用了开放布局的形式，适应了简单家庭厨房低使用率的特点。

基于小户型特点，办公、公寓产品选择了可独立控制的户式空调系统。根据建筑综合体市场运营的需要，在裙房和地下室提供了综合的商务和生活配套设施，包括商务中心、银行保险、快餐、餐吧、超市以及物业管理等内容。

layout in the form of opening in this case is adopted, suiting with the characteristics of kitchen low usage for a simple home.

Based on characteristics of small units, office, apartment selection of products that can be individually controlled air-conditioning systems account. According to the building complex needs of market operators, in the podium and the basement comprehensive facilities for business and life are offered, including the business center, banking and insurance, fast food, bar, supermarket and property management.

富顿中心
FREETOWN

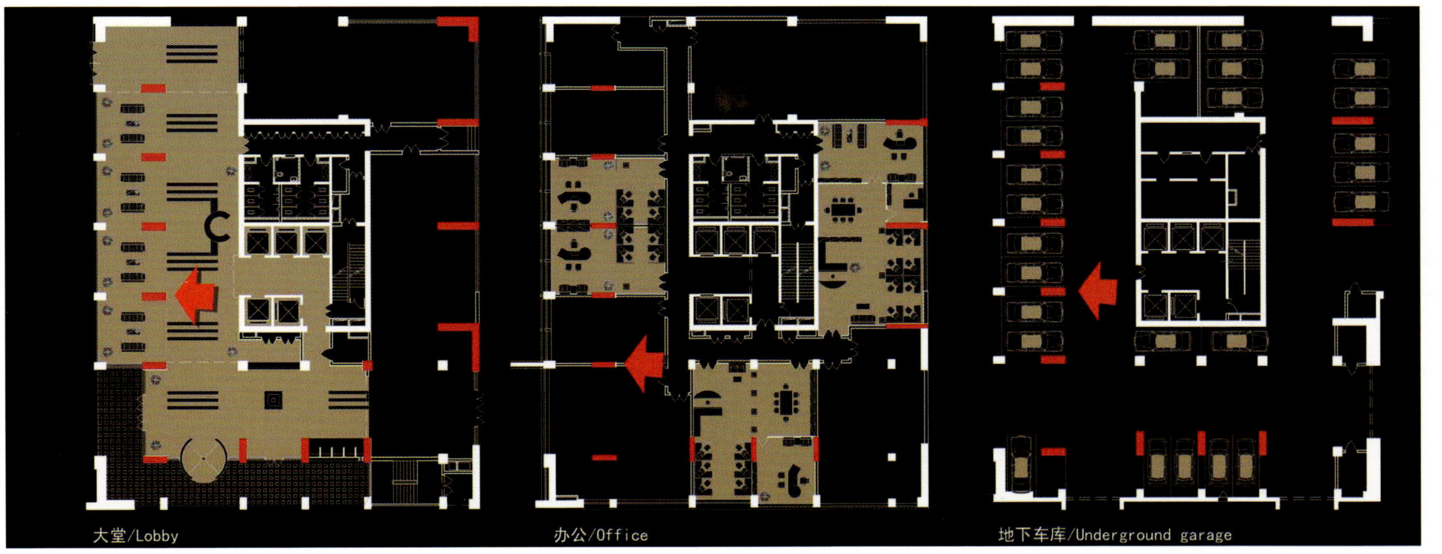

大堂/Lobby　　办公/Office　　地下车库/Underground garage

FREE TOWN

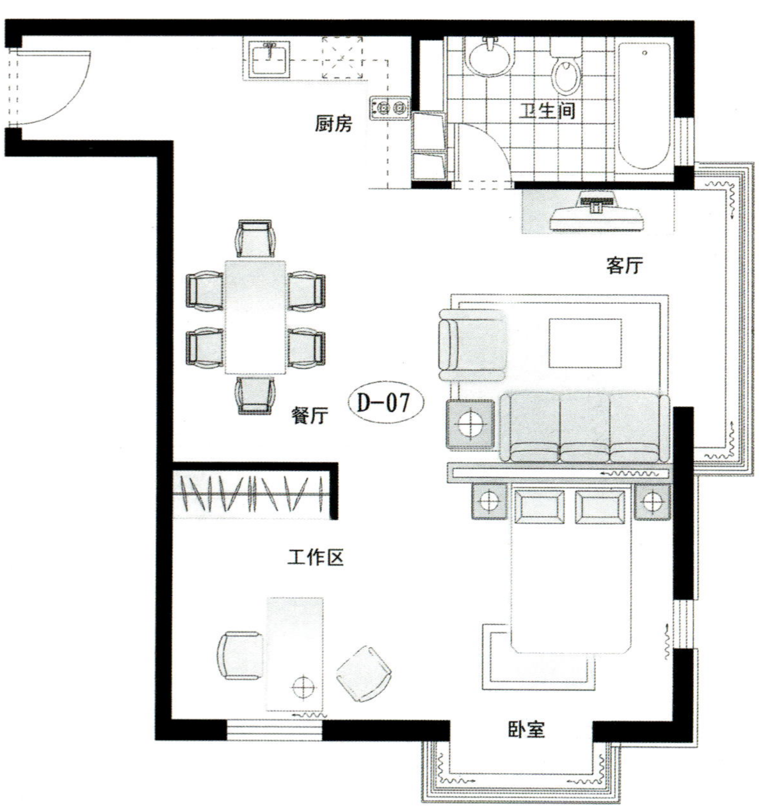
厨房
卫生间
客厅
餐厅
D-07
工作区
卧室

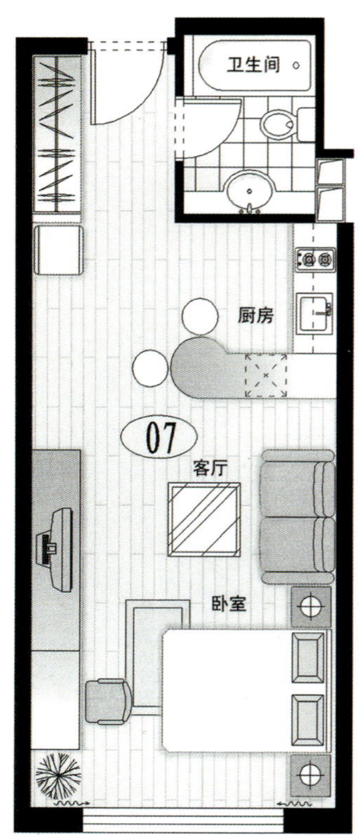
卫生间
厨房
07
客厅
卧室

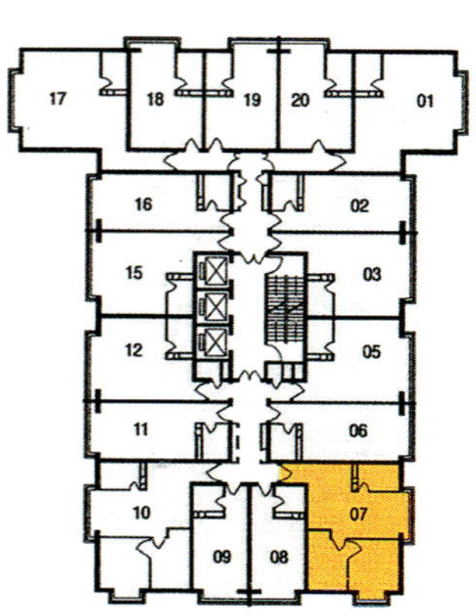
17
18
19
20
01
16
02
15
03
12
05
11
06
10
07
09
08

复合型

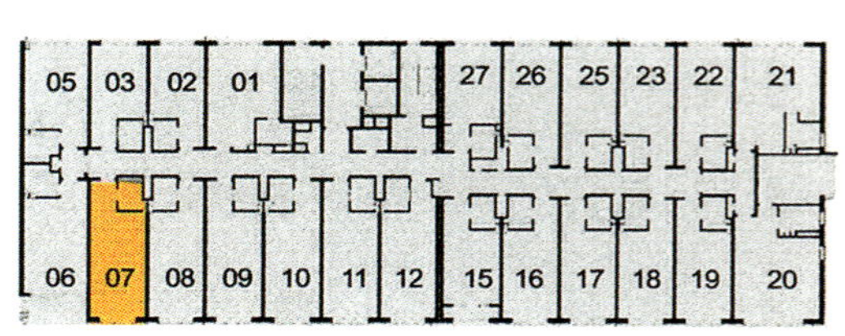
05
03
02
01
27
26
25
23
22
21
06
07
08
09
10
11
12
15
16
17
18
19
20

简单型

Xingfuli, Shenzhen

深圳幸福里

Developer: Huarun (Shenzhen) LTD.CO,
Building design: Chinese international (Shenzhen) design consultant LTD.CO,
Project address: Luohu distract, Shenzhen
Flooring size: 41,069.99m²
Building size: 293,941.9m²

开发商：华润（深圳）有限公司
建筑设计：中建国际（深圳）设计顾问有限公司
项目地址：深圳市罗湖区
总用地面积: 41,069.99m²
总建筑面积: 293,941.9m²

Huarun center second is located in the west of Luohu Distract Bao'an Road of Shenzhen, the south of Jiabin Road, and the east of Kapok Street. The center is in the opposite side of Huarun center first. The northwest of this area is high and southeast is low. Height difference is up to 1.8m or so, and restriction of this area is very much. The program is a multi-functional integrated projects, leisure shopping, cultural entertainment, specialty restaurants, five-star hotels and high-end residential and other functions into one.

The project with a new design concept, aimed at creating a new business center, thus enhancing the urban vitality and competitiveness. With a new design ideas for the Shenzhen public , it is to create a new business, entertainment, leisure environment. Continuing an architectural language of First, it will add a new element. The image of its construction not only meets the spirit and characteristics of Shenzhen, but also enriches the level of urban space for a wide range of Shenzhen Special Economic Zone to add brilliant color.

华润中心二期位于深圳市罗湖区宝安南路西侧，嘉宾路南侧，木棉花街东侧。与华润中心一期隔街相对。地形西北高、东南低，高差达1.8m左右，地形复杂，场地限制较多。该方案是一个多功能的综合性项目，集休闲购物、文化娱乐、特色餐饮、五星级酒店及高档住宅等多功能于一体。

该项目以崭新的设计理念，旨在创造一个新的商务活动中心区，从而增强城市活力及竞争力。以全新的设计思路为深圳市民创造一个全新的商业、娱乐、休闲环境。延续一期的建筑语言，并添加了新的元素。其建筑形象不但符合深圳精神与特色，也丰富了城市空间层次，为多元化的深圳特区增添绚丽色彩。

二期建筑体块是一期空间效果的延续。一期的华润大厦屹立在基地北端，带绿化及水量的下沉广场呈弧形，以景观流水墙从地下一层开始，呈螺旋状上升，然后变成玻璃“围巾”，轻柔地围绕着“金三角”旗舰店，再以色彩活泼的红色金属墙壁将弧线继续从北往南延继下去。二期，弧线继续从嘉宾路和宝安南路交叉口，形成浮起的桥梁。在基地南端，弧线牵引上升至酒店塔楼，呈螺旋状向天空延伸。

君悦酒店
82666666
幸福里
LOUIS VUITTON

BVLGARI
HERMÈS

Body mass of building second is a continuity of building First in spatial effect. Building first is standing at the northern of the base, with green and the water curved sunken plaza. Water wall of the landscape started from the ground level, rising spiral, and then into the glass "Scarf", gently around the "Golden Triangle" flagship store, and lively colors red metal wall will extend from north to south arc to continue. Building second, the arc continues from Bao'an South Road and the intersection of guests to form a bridge float. The southern end of the base, arc is pulled up to the hotel tower, spiral extension to the sky.

The commercial project is the perfect continuing of the building first. Southwest side of the base set three high-rise residential with 49 floors. To 3 layers underground can accommodate 1,382 parking spaces for the large underground parking space and equipment room. Four multi-storey commercial buildings are in the north side, including commercial retail, restaurant, and multi-hall cinema. A 38-floor high-rise hotels and annex with two floors in Southeast.

High-rise residential building is located in the southwestern part of the project, cleverly drawing a block south skyline. In addition to full vision, direction, and the advantages of the layout, the shape of the arc also the same hotel tower echoes. Simple and elegant residential building volume and outdoor shopping centers expressed a similar affinity, human scale.

In order to create a quiet and comfortable environment for residential areas, the entire residential and residential roads are treated for elevation, with an average elevation of 2 meters.

Two entrances for cars and two entrances for people are set. For shortage of land for the project, an open courtyard space for more tenants, housing for low-level elevated green are designed with the clubhouse and residents interspersed lobby.

The lobby of each household is set in the side rest area, clubs are in one, two, three parts of dispersed layer of overhead to make room for circulation. Three residential towers, a five to four ladder, and the other two to four ladder six, all 49 layers. 3 Penthouse units are top of the tower overlooking the city, take advantage of the surrounding landscape resources.

A total of three underground can accommodate 1382 car parking spaces. Third underground garage is parking area for residential, second floor in underground is parking area for commercial and residential, ground floor is parking area for commercial, hotel. Residential tower elevator can reach the underground garage, household use conveniently.. Automatic escalator is located in Ground floor and basement garage; customers can go directly to the ground taking advantage of the 2 dispersed groups of buildings with the escalators. Residential garage entrances located at the base of the south and west, business hotel located at the base of the garage entrances and exits to the east, commercial freight garage entrances located at the base of the north.

Office and hotel towers are both standing at the end of the base. The former is a solid foundation, and the other extend gracefully to the sky, and embrace a base full of outdoor space.

Downtown is in the North of the base, reflecting the modern commercial building design concepts. Building second is complement and extending of Building first. It will give people a whole new experience. Extensive commercial space provides customers with a clear sense of direction. And exciting "OUTDOOR space" is the favorite place for all in Shenzhen, and the best place of international

项目的商业区将巧妙地作为华润中心一期商业区恰到好处的补充。基地西南侧设3栋49层的超高层住宅，地下设3层为能容纳1382个车位的大型地下停车场及设备用房；北侧设4栋多层商业建筑，包括商业零售、餐饮、多厅电影院；东南侧设1栋38层超高层酒店及2层裙房。

高层住宅楼位于项目西南部，巧妙地点出地块的南天际线。造型除充分发挥视野、方位及布局的优势外，亦能同酒店塔楼的弧线相呼应。住宅楼简洁和典雅的体量表达出与室外购物中心相似的亲和、人性的尺度。

为了营造住宅区安静舒适的环境，整个住宅部分与小区路作抬高处理，平均抬高2m。设置两个车行出入口和两个人行出入口。因项目用地紧张，为开放更多的庭院空间给住户，住宅底层架空作绿化，其中穿插设计了会所和住户大堂。每个住户大堂均在侧边设置休息区，会所均在1栋、2栋、3栋的架空层部分分散设置，使之空间流通。3栋住宅塔楼，一栋为四梯五户，另两栋为四梯六户，均为49层。3栋塔楼顶层均为Penthouse户型，俯瞰全城，充分利用了周边的景观资源。

地下共3层，能容纳1382辆车位的停车库。地下三层车库为住宅停车区，地下二层为商业和住宅停车区，地下一层为商业、酒店停车区。住宅塔楼客梯可以直达地下车库，方便住户使用。地下二层和地下一层车库设有自动扶梯，顾客可以利用分散于建筑内的2组自动扶梯直达地面。住宅车库出入口设于基地的南侧和西侧，商业酒店车库出入口设于基地的东侧，商业货运车库出入口设于基地的北侧。

办公楼和酒店塔楼屹立在基地的两端，前者是坚实的地基，后者优雅地向天空延伸，并拥抱充满户外空间的基地。酒店裙房延伸到嘉宾路与宝安路交界处，引人瞩目，主要入口在宝安路设有一个专用道，形成到五星级酒店的超豪华的入口空间序列。

商业区在基地北侧，体现了现代商业建筑的设计概念。二期商业作为一期的补充和延续，它将给人们带来全新体验。丰富的商业空间，为顾客提供了明确的方向感。富有情趣的“OUTDOOR空间”是深圳各界人士及国际旅游者的最佳工作、餐饮和娱乐场所。沿嘉宾路设置两个商业广场，将人流由两个商业广场引入，利用自然地形，形成富有城市商业氛围的酒吧一条街。它是一个开放的公共空间，利用丰富的景观元素和设计手法，作为城市和建筑的软性过渡面。跨越嘉宾路的四层廊桥，是一期和二期的自然过渡。一层架空5.4m，二层为带天棚的公共人行通道，三、四层为多厅电影院。在廊桥底部设发光天棚，保证立面采光照明要求。

酒店在基地东侧，1~3层是酒店配套设施，包括酒店过渡大堂、宴会厅、特色餐饮、高级时装、会议室、SPA、健身及泳池。4~23层为酒店客房，俱乐部休息室位于23层。25~29层为服务公寓，酒店后勤设于地下一层。24层、30层为设备层，会议室、中餐厅位于31~32层，33层为酒店空中大堂，由主入口进入酒店后，乘坐四台高速电梯可直达33层的空中大堂，办理好入住手续后转乘专用客梯到达客房层，顶部餐饮设有专用客梯。34层为酒店办公、管理，35~38层为各式餐厅及穹顶酒吧。共有客房400套，公寓58套，设有客梯15部，自动扶梯2部，货梯、服务梯6部。从地下停车场经两部客梯到达首层过渡大堂。

住宅区、商业区、酒店区完全分离，相互独立，形成了互不干扰的社区环境。

住宅立面采用两层的横向分割与竖向的短支柱形成简洁的几何网格，线条流畅、棱角分明，体现高品质的居住氛围，同时与酒店造型呼应，上演一幕在城市空间大舞台中的造型表演。局部结合出挑阳台的设计，并点缀色彩，缩小了整个建筑的视觉尺度。顶部Penthouse结合内部平面设计成一倾斜的几何玻璃形体，一是呼应了酒店顶部退台的形象，再是运用双层玻璃的立面外墙为住宅创造顶级的居住品质。同时，这一设计既能满足遮阳、节能，又能调节气流。

酒店和商业的造型和立面处理，是一期“缤纷万象”的延续，以韵律曲线呈螺旋状向天空延伸，并由一期室内空间延展为二期的outdoor都市街区气氛。一、二期的重要联系纽带—四层廊桥更被赋于了戏剧性效果，犹如当代的佛罗伦萨“Pontevecchio”，跨越嘉宾路上空。运用与一期相协调的材料、色彩、抛光及细节设计，为两期工程之间提供一个巧如天成的自然过渡。闪烁的金属、优雅起伏的玻璃天棚、扶手、充满活力的米色石墩及托架，将人们引向小型都市广场及带佛罗伦萨风情的露台建筑，仿佛一期“伸手与亲朋招呼联络、把臂同欢”。最后的高潮为酒店主体塔楼，呈步步升高之势，简洁典雅的金属加玻璃幕墙与蓝天、白云交相辉映，一系列如飞羽般的楼顶玻璃体，将塔楼的各体节天衣无缝地接成一个连续的、完整的流线弧体。北立面两侧为旅馆客户的窗户，中心是一整片直通塔楼顶端的玻璃电梯厅，与塔楼服务核心的固体外墙形成刚柔相济的平衡美。

三个主要功能区块都有自己的交通系统。

住宅区的交通组织自成一体。沿基地西侧和南侧设有住宅区的车行出入口。住宅车辆由西侧双车道直接进入地下车库，住户通过西侧小区的主

tourists, restaurants and entertainment venues. Guest Road, along the set of two commercial plazas, will flow from the introduction of two commercial plazas, use of natural terrain, the emergence of a commercial atmosphere of the city's bar street. It is an open public space, the use of the rich landscape elements and design practices, as the city and construction of the soft transition surface. Four bridges across the road guests, is a natural transition to Phase I and II. 5.4 m layer of overhead, floor to ceiling with a public sidewalk, three, four multi-hall cinema. Luminous ceiling set in the bottom of bridges to ensure the lighting facade lighting requirements. The hotel is at the east of base.

1 to 3 layers are mating facilities of the hotel, including hotels transitional lobby, banquet rooms, specialty restaurants, high fashion, meeting rooms, SPA, fitness and swimming pool. 4 to 23 layers for the hotel rooms, club lounge located in the 23 layer. 25 to 29 layers for the service apartments, a hotel located in the basement back. 24 floors, 30 floors for the equipment level, meeting rooms, the restaurant located in the 31 ~ 32 layers, 33 layers of air for the hotel lobby, from the main entrance into the hotel, take four 33-story high-speed elevators can reach the sky lobby, for a good stay After the transfer procedure room floor dedicated passenger elevators to reach the top restaurants with private passenger elevators. 34 level for the hotel business, management, and 35 to 38-story dome for the variety of restaurants and bars.

There are 400 sets of rooms in total. 58 sets of apartments, with 15 passenger elevators, 2 escalators, freight elevators, 6service elevator are in there. Two passenger elevators from the underground car park are to reach the first floor by the transition to the lobby.

Residential, commercial, hotel area are completely separated, independent of each other, forming a non-interfering environment of the community.

Residential façade is divided by two horizontal and vertical pillars of the formation of short simple geometric grid, smooth lines, angular, reflecting the high quality living environment, while the hotel echoes style, staged a scene in the urban space in the big stage modeling performance. Designed with balcony overhangs, and decorative color, reduce the visual scale of the building. Top Penthouse with internal graphic design into a tilted glass geometric shapes, one echoed back to the hotel at the top of the image of Taiwan, and then is to use double-wall glass facades for residential housing to create top quality. Also, this design can meet shade, energy conservation, but also adjust the airflow.

Hotel and commercial modeling and elevation treatment, is a "Fun in Vientiane," the continuation of the rhythm of the spiral curve extending to the sky by an extension of interior space for the second phase of the outdoor atmosphere of urban neighborhoods. Metal flashing, elegant undulating glass ceiling, handrails, full of vitality and bracket beige pier will lead a small city square and people with a terrace-style buildings in Florence, as if a " call to reach out and connect with friends and family with the arms. " Finally, the climax of the main tower for the hotel, the trend was increased step by step, simple and elegant metal and glass curtain wall with blue sky, white clouds off each other, such as the flight feathers like the roof of a vitreous, the tower section of the body are connected to form seamless a continuous, full arc of the body flow lines. Northern elevation windows on both sides of the hotel customers, the center is a whole slice through the tower to the top of the glass elevator hall, and the tower wall to form a solid core of services, the balance of hardness and softness.

Three main functional blocks have transport system. West and south along the base are entrances and exits for residential. Two-lane residential traffic from the west directly into the underground garage area residents through the west side of the main entrance and lobby to reach their home, and then take the elevator home of the tower. Owners can also passengers in the underground parking garage elevator to go home.

Four separate buildings downtown commercial pedestrian street ring formation. Freight Entrance Road at guests, underground car parking ramp is located in Bao'an South Road. Another car under the passenger area is located in west side of Bao'an South Road.

the main hotel entrance of the hotel is located in the southern tip of the Base, accessing to Bao'an South Road, forest uphill to the hotel main entrance of the vagina.

入口进入并到达各自住宅入口大堂，然后再乘各塔楼的电梯入户。业主也可在地下停车库乘客梯回家。

商业区的4栋独立建筑形成环形的商业步行街。货运入口设于嘉宾路，地下停车场的汽车坡道位于宝安南路。另有汽车下客区设在宝安南路西侧。

酒店主入口位于基地的南端，由宝安南路进入，林阴道上坡到酒店主入口。酒店宴会厅从酒店南侧主入口进入，在酒店主入口的西侧设置进入地下停车场的客用汽车坡道，沿嘉宾路与商业区共享货运地下通道。

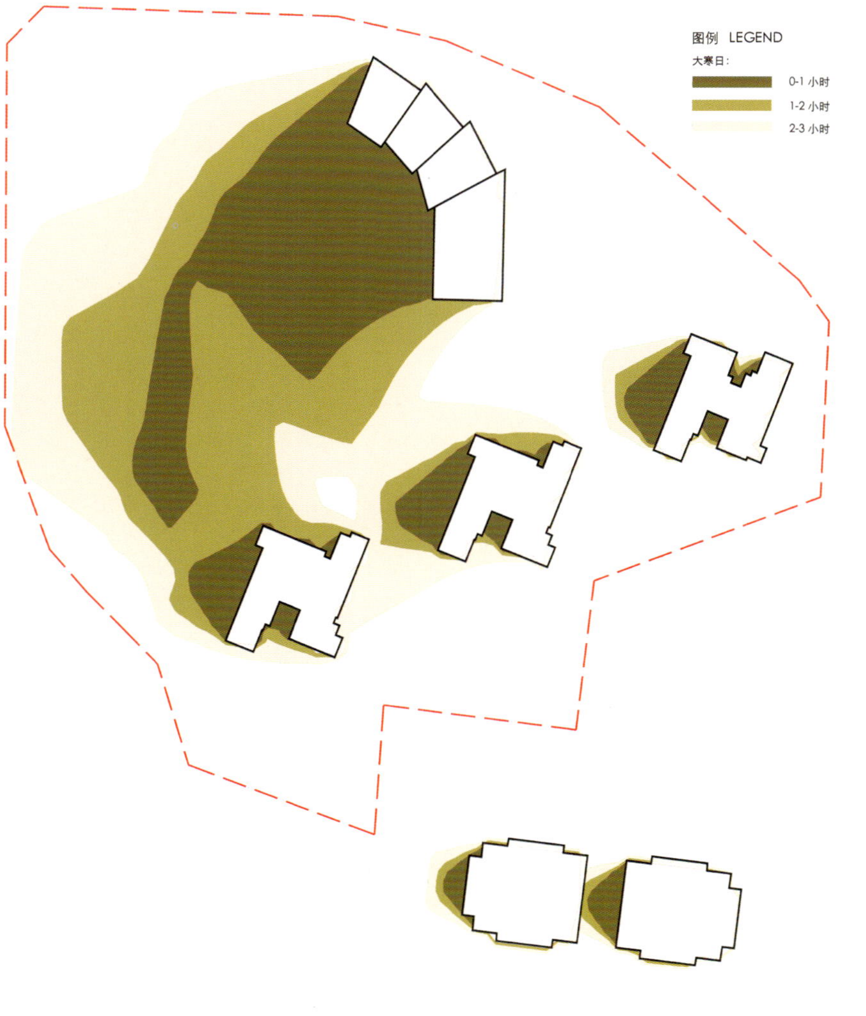
图例 LEGEND
大寒日:
0-1 小时
1-2 小时
2-3 小时

华润中心二期
工作范围
CRC PHASE II
LIMIT OF WORK
9G/31.01
S. BAO 'AN ROAD
宝 安 南 路
过街天桥
BRIDGE ABOVE
嘉宾路（双向六车道）
JIA BIN ROAD
POWER PLANT
KAPOK STREET
木 棉 花 街

华润中心二期
工作范围
CRC PHASE II
LIMIT OF WORK
9G/31.01
S. BAO 'AN ROAD
宝 安 南 路
过街天桥
4层
底层架空
BRIDGE OVER
STREET
4 STORIES
4层
4 STORIES
2层
2 STORIES
3层
3 STORIES
4层
4 STORIES
49层
RESIDENTIAL TOWER 1
49层
RESIDENTIAL TOWER 2
49层
RESIDENTIAL TOWER 3
POWER PLANT
KAPOK STREET
木 棉 花 街

图例 LEGEND
市政道路
出租车
商业货运车流线
住宅区车行流线
酒店车行流线
商业街人流线
商业主入口
住宅主入口
货运主入口
酒店主入口
9G/31.01
华润中心二期
工作范围
CRC PHASE II
LIMIT OF WORK
S. BAO 'AN ROAD
过街天桥
BRIDGE ABOVE
POWER PLANT
KAPOK STREET

图例 LEGEND
消防登高面
消防车道
9G/31.01
华润中心二期
工作范围
CRC PHASE II
LIMIT OF WORK
S. BAO 'AN ROAD
过街天桥
4层
底层架空
BRIDGE OVER STREET
4 STORIES
4层
4 STORIES
36层
2层
2 STORIES
RESIDENTIAL TOWER 1
POWER PLANT
KAPOK STREET

Jinqiuzhichun International PLAZA

锦秋知春国际大厦

Developer: Beijing Jinqiuzhichun real estate development LTD.CO,.
Building design: Zhongyuan international engineering design research commity COX design group
Building size: 74,824m2
Writer: Zhongyuan international engineering design research commity COX design group

开发商：北京锦秋知春房地产开发有限公司
建筑设计：中元国际工程设计研究院 COX设计集团
建筑面积：74,824m²
供稿：中元国际工程设计研究院 COX设计集团

Jinqiuzhichun International Building situated in the high-tech development zone and the "Silicon Valley" surrounded by the college town. It's mainly leased (on sale) to small and medium sized companies in scriptorium and in the form of modular office. To comply with the design concept

"Return to City Building", the building determines its own image and the general layout cloth board based on the geographical location and implement all subsequent design phase, trying to create a landmark high-tech image.

There are two tower residential buildings of 16 storeys in the north side of the project base. In order to avoid the sunlight occlusion, a "prevailing circumstances" approach is applied with appropriate architectural vocabulary to create two independent but joint constructions. On the premise of a complete image of the north façade forms the "One one board" combination, and solves the "solar shield block" problem and builds up its own unique elevation image, which help achieve the respect for the natural environment and establish their own city landmark image of the fold. The construction site leans the east and

锦秋知春国际大厦基地位于由高科技发展区及大学区所围绕的北京“硅谷”，主要以写字间及单元式办公室的形式向中小型公司出租(售)。工程遵从“建筑回归城市”的设计理念，根据所在地域来确定自身形象特征和总图布局，并贯彻于随后的各个设计阶段，力图创造出高技术的地标意象。

本工程基地北侧有两座16层高的塔式住宅楼，体形设计上应避免对其造成日光遮挡，因此设计采用“因势利导”的手法，借助适宜的建筑语汇创造两栋既独立又联合的建筑，在保证北立面形象完整的前提下，形成“一点一板”的组合，解决“日光蔽挡”问题的同时，形成了自己独特的立面形象，实现了尊重城市固有环境和树立自身地标形象的双重目标。工程的东、西两侧分别靠临锦秋东路和锦秋西路，在总平面布局上沿知春路和锦秋西路布置成一个L形，这样与南侧已建设中的锦秋知春住宅小区的9、10号楼自然围合成一个小的庭院，延续了锦秋知春住宅小区已有的以建筑实体围合庭院的空间组织手法，顺应了原有的空间环境特征。此庭院的东侧则布置大厦的地面停车场。

west sides of Jinqiu Road respectively, forming an L-shaped in the overall layout by Zhichun Road and Jinqiu West. In this case, it forms a small natural garden with the 9th, 10th buildings in southern Jinqiu Zhichun residential area which is under construction, continuing the spatial organizational practices of entity enclosed courtyard building method, and conforms to the original characteristics of the space environment. The courtyard east is dedicated to be the building parking lot.

Underground Layer 1 is for the internal staff dining room, kitchen, partial equipment room and car garage; Underground Layer 2 and 3 are also for small car garage.1st floor is the lobby, the business office, the lounge and so on. F2 is all for business office. A F3 and up above layers are leased (on sale) to be scriptoriums, 25th, 26th floor being the equipment storage room. B F3-19 are the layer modular offices with the 20th being the technical floor and outdoor evacuation platform. C F3-12 are the modular offices while F6, 10, 14 being 2-storey lounge.

A pedestrian flow entrance is set up individually at the gates leading to Zhichun Road and Jinqiu Road. An outdoor plaza is installed in the front in order to form a natural transition between the building and the arterial street, which? Suitable furniture design is placed at the ground, which performs as both outdoor self-open-space and urban nodes. Thus it forms a building - square (intermediate space) -three levels of hierarchy of the city and eventually the buildings and urban space functionally overlap each other and infiltrate mutually.

The periphery of the building is made of aluminum and glass curtain wall, trying to create a hi-tech vertical surface form, which is also in conformity with the design idea of ecology. The eastern and western vertical service adopts the crosswise glass overshadow, which is used for shielding hotness on the premise of which the view contacts between indoor and outdoor are not affected and it can also admit the natural light, can avoid being sunburn from the east and west and reduce the energy consumption of refrigeration, thus is given a distinct personality magnetism. Southern vertical surface adopts a horizontal spindle surface, which can keep out the sun, also guide air. The entire vertical composition is combined with the horizontal and upright, make the building simple and erect, especially the perpendicular "tough wall" in the northern vertical surface is amply represented under the upright.

The entire huge construction dimension in the northern and southern vertical surface can be attenuated through the balcony convex-concave change and the vivid shadow in the middle intersection. The verdant roof terrace on the top of Building B clearly reflects the function its internal department office. Its beautiful exquisite house morphology creates the unique sculpture feeling for the construction. Overlooking from Building A located in the higher position, it likes an elegant hanging garden. The light and lively vertical metal purlin, backlash's-like metal node and the delicate aluminum 100-page system and high quality glass curtain wall in grid shape together create a hi-tech architectural image appropriate to the hi-tech environment of Zhongguan Village.

Project seeks to provide owners to maximize rentable space, the rational allocation of A, B Block, the core tube, reducing the room number of load-bearing frame columns, in order to increase the use of space and rental rates, and to minimize the length of the corridor to increase the utilization rate . Integrated logistics services and facilities layout, equipment layout in the tube well outside the core tube, so that the equipment on each floor of the vertical lines are convenient interface, reducing cross each other, but also reduced the height of the ceiling layer and improve indoor clearance.

The forms of gross floor partition are flexible. Just one or more households on one floor is flexible. wall between households and households is used with light materials, to facilitate flexible use of interior space tenants, to correspond to the diverse market needs. Building B, standard layer in the core tube surrounded by four sets of kitchen, health facilities, to ensure that, when conducting separate floors, each business unit has a separate kitchen, bathroom.

Standard layer profile is designed with 3.7m, in line with people on the office building vertical spatial scale of psychological requirements.I n the main entrance of building C, the first three regions of the 2 layer (Layer 3 local) high atrium space, and in its internal layout of the pedestrian bridges, staircases and other landscape. In the contact A, B Block, at the same time, create a human dynamic space, breaking the dull atmosphere inherent in the office building. Building C, along the vertical is set with the three two-storey leisure room, removing the 3-layer slab for the lounge to introduce more natural light, and through appropriate interior decoration, such as green landscape, so that inherent in the building leisure complement the functions of office space.

地下1层为内部职工餐厅、厨房、部分设备用房及小汽车库；地下2、3层为小汽车库。地上1层为大堂、商务办公、休息厅等，2层为商务办公。A座3层以上为出租(出售)写字间，25、26层为设备间。B座3-19层为单元式办公，20层为设备层兼室外疏散平台。C座3-12层为单元式办公室，其中6、10、14层为2层高的休闲厅。

大厦沿知春路和锦秋西路分别设置了人流入口，为使城市干道与大厦之间形成自然过渡，入口处设置了室外前庭广场，对来自大厦外围的各种交通进行“接驳”。广场上进行了适宜的家具设计，使之扮演大厦自身室外开放空间和城市节点的双重角色，这样就形成了建筑—广场(中介空间)—城市三个层次的阶层关系，最终使建筑和城市空间在功能上互相交叠，互相渗透。

大厦的外围护材料采用铝板和玻璃幕墙，力图创造高技术的立面形象，这也与生态设计理念相一致。东、西立面采用横向的玻璃遮阳板，在保证室内外视线交流不受影响且能引入自然光的前提下，起到屏蔽暑热的作用，防止东、西晒，降低了制冷的能耗，并赋予了鲜明的个性形象。南立面采用水平梭形铝面页，既可遮阳，又可导风。大厦的整体立面构图为水平与竖向相结合，使建筑的体形简洁挺拔，尤其是北立面垂直的一片“刃墙”在竖向得到了充分的表达。

整个庞大的建筑体量，在南北立面上通过中间结合部的阳台凸凹变化及生动的阴影得以弱化。B座顶层葱绿的屋顶平台清晰地反映出其内部公寓式办公的功能，其优美的斜屋面形态，为建筑创造出别致的雕塑感，从较高的A座俯视，宛若一座优美的空中花园。轻灵垂直的竖向金属檩条、后退式的金属节点、优美的铝百页系统和网格状的高质量玻璃幕墙共同创造出了与中关村高科技氛围相适宜的高技术建筑形象。

项目力图为业主提供最大化的可出租空间，合理配置A、B座的核心筒，减少房间内承重框架柱的数量，以此来增加使用面积和出租率，并尽量减少走廊长度以增加使用率。综合布置后勤服务设施，设备管井的布局位于核心筒外围，这样各楼层的设备管线均有便捷的竖向接口，减少了相互间的交叉，也缩小了吊顶层的高度，提高了室内净空。

项目楼面分割形式灵活机动，既可一层一户又可一层多户，户与户之间的隔墙采用轻质材料，便于租户灵活使用室内空间，以对应变化多样的市场需求。如B座标准层在核心筒外围有四组厨、卫设施，可以确保在进行楼层分割的时候，各个办公单元均有独立的厨房、卫生间。

标准层剖面设计层高为3.7m，符合人们对办公建筑竖向空间尺度的心理要求。在主入口C座的首三层区域提供了2层(局部3层)高的中庭空间，并在其内布置了人行天桥、景观楼梯等，在联系A、B座的同时，创造出人性化的动态空间，打破了办公建筑固有的沉闷氛围。C座沿竖向布置了3个2层高的休闲厅，抽掉3层楼板为休闲厅引入更多的自然光线，并通过适宜的室内布置，如绿化景观等，使休闲性与大厦固有的办公空间职能相辅相承。

锦秋·国际
Tower

锦秋·国际
Horizon
International Tower

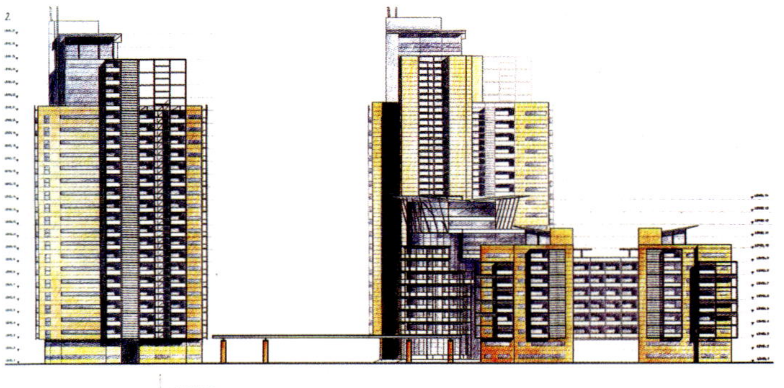

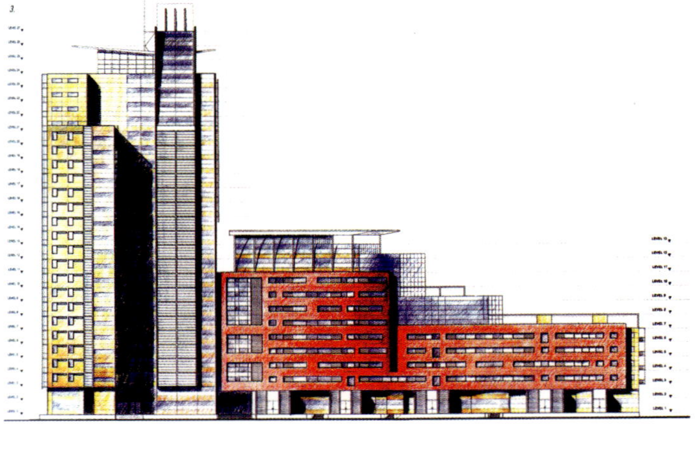

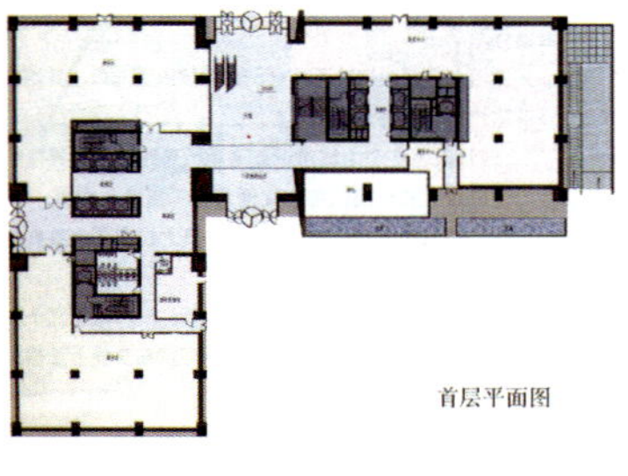

首层平面图

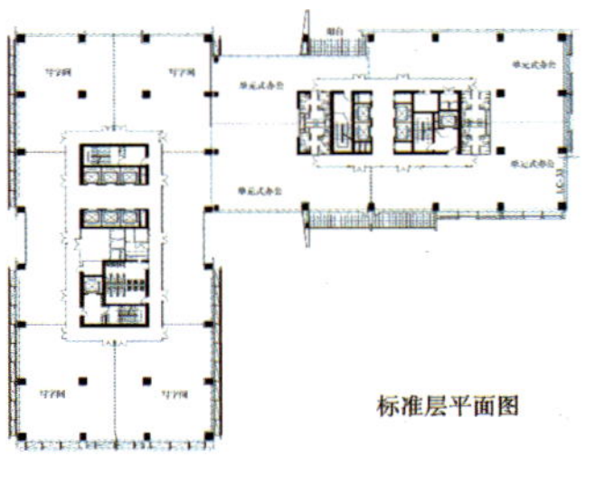

标准层平面图

Golden Corridor No.1, Shenyang

沈阳金廊一号一期

Developer: Shanghai Shenhu Group
Building design: Antu(International) design LTD.CO, Shanghai Shenhu Group
Project address: Liaoning Province Shenyang Heping Distract
Flooring Size: 49,800m^2
Building size: 240.000m^2
FAR: 4.8
Landscaping radio: 32%
Writer: Antu(International) design LTD.CO, Shanghai Shenhu Group

开发商：上海申湖集团
建筑设计：安图（国际）设计有限公司上海安图建筑设计咨询有限公司
项目地址：辽宁省沈阳市和平区
占地面积：49,800m^2
建筑面积：240,000m^2
容积率：4.8
绿化率：32%
供稿：安图（国际）设计有限公司上海安图建筑设计咨询有限公司

"Golden Corridor" is the corn concept of the development strategy of Shenyang city which runs through the main functional area of Shenyang and the length of it is 17km from south to north. It crosses one and two ring roads and the east and west high ways. Through exploration and development, it will be the focus area to build reprehensive, environmental friendly, road landscape of the modern service industry, the image display area of international city and the core control areas of northeast central city. It forms the golden corridor of Shenyang Axis and the creation of the new century. "Golden Corridor" North from the famous historical and cultural heritage - Zhao Ling, south to the Hun River, which means blending into the future from history.

"Golden Corridor No.1" precisely settled in the endpoint of this " future", that is, the southern of "Golden Corridor", is the first piece of base to enter the "Golden Corridor". It's also the landmark level project of "Golden Corridor". It will hold in northeast business center and face the world by its complex industry citymall.

"金廊"是沈阳市发展战略中的核心概念，南北绵延17km并贯穿目前沈城的主要功能区域，横跨一、二环线和东西快速干道。通过逐步开发和改造，将成为建筑标志化、环境生态化、道路景观化的现代服务业产业聚集区、国际大都市的形象展示区、东北中心城市的核心控制区，形成沈阳中轴和创造新世纪的黄金走廊。"金廊"北起著名的历史文化遗产——昭陵，南至浑河，意寓从历史走向未来。"金廊一号"恰恰位于这一"未来"的端点，即金廊的南部，是进入"金廊"的第一块基地，也是"金廊"的地标级项目，将以其复合型citymall的业态，立足于东北商业中心而面向世界。

"金廊一号"位于沈阳市中央都市走廊（又称金廊）的最南端，紧邻沈阳五里河体育中心，在"金廊"总体规划中属于"世界文化产业核"的一部分，由国际国务中心、超甲级写字楼、国际酒店公寓和高级公寓组成。设计创意通过独特的造型手法取得鲜明的、具有吸引力的主题，强化城市的精神特质，体现坚实、稳固、积极发展的立意。设计则围绕主题展

SPORTBOX
KFC

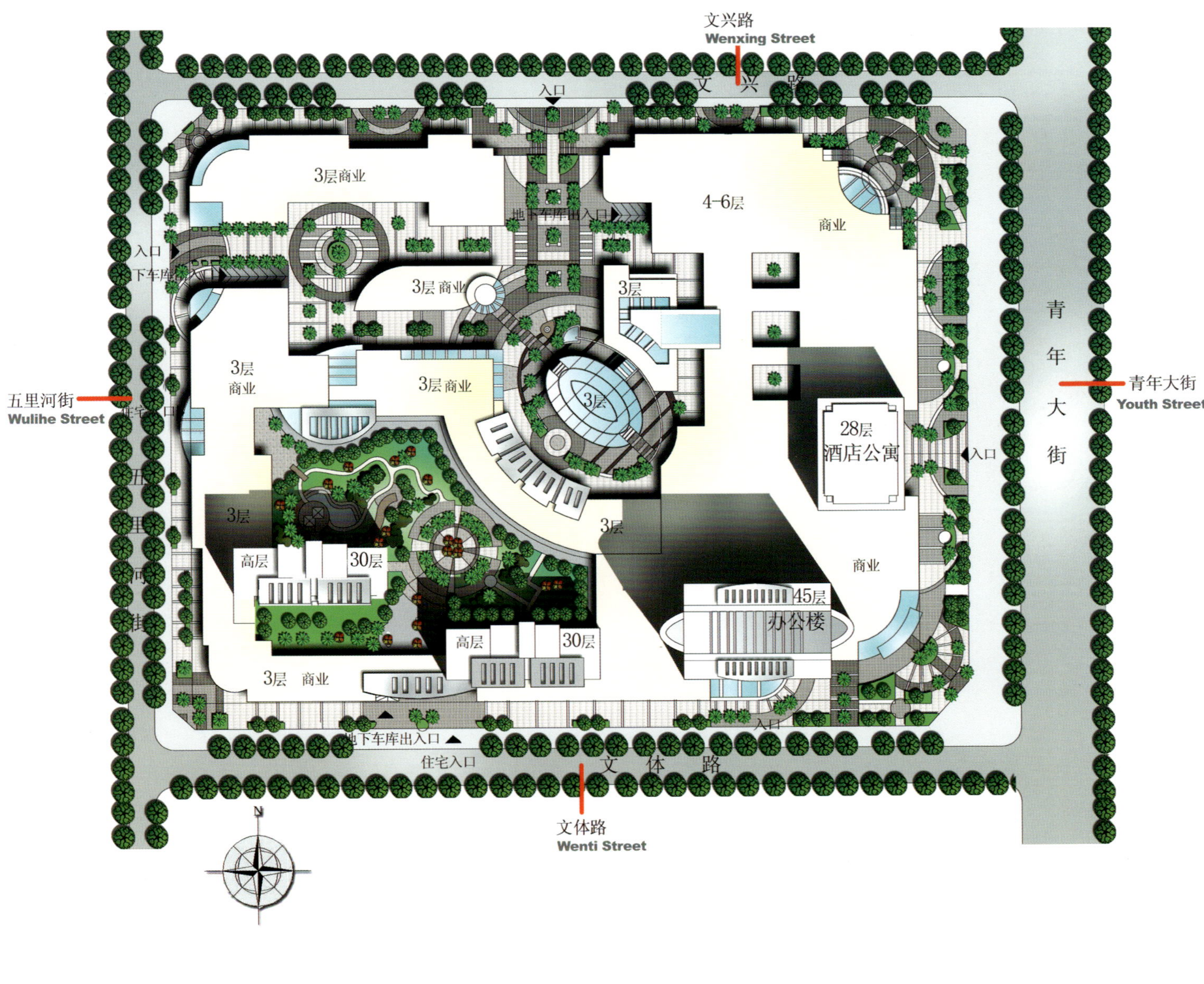
文兴路
Wenxing Street
入口
文 兴 路
3层商业
4-6层
商业
地下车库出入口
入口
下车库入口
3层 商业
3层
3层
商业
3层 商业
3层
五里河街
Wulihe Street
28层
酒店公寓
入口
青
年
大
街
青年大街
Youth Street
3层
3层
高层
30层
商业
45层
办公楼
高层
30层
3层 商业
入口
地下车库出入口
住宅入口
文 体 路
文体路
Wenti Street
N

"Golden Corridor No.1" located is the southernmost of the central city corridor of Shenyang (also named golden corridor), it near to the Shenyang wuheli sports center. In the whole plan of "Golden Corridor", it belongs to one part of the worldwide cultural industry core which was formed by international state center, Grade A office space, international hotel apartments and senior apartments. Design creativity through a unique modeling technique to obtain clear and attractive theme, and strengthen the city's ethos, reflecting a solid, stable, active development conception. Design is around the themes, according to different functions and format on the basis of the rationality highlight its uniqueness, in order to avoid duplication with similar projects. Architectural form stable contains luxurious, taking into account the future construction of the light, dream fusion of the two interludes. Design follows the principles of sustainable development, in the recognition, construction and economic aspects to achieve a harmonious whole, the architecture and environment to achieve the best results and become a classic model of urban architecture, to ensure the quality of the overall program to meet the highest in northeast standard construction requirements. "Golden Corridor No.1" occupied a very important site, its shape is quite founder. Considering it in the north of wuheli sports center with a widen vision. And Youth Street is the main rode of axle wire, therefore taking the high buildings into its south and east. The southeast of the base is the first vision focus to enter "Golden Corridor" which has the reputation of "Golden Angle". It's the best location to set high offices. From here, you can see the whole view of "Golden Corridor", at the same time you also can enjoy the whole northeast area. For the city space, it drew a perfect period to the "Golden Corridor" as a commanding height. Arrange two high-rise residential in the south side, from the vehicle entrance and pedestrian flow line with other separate business office, home range around the independence of the central garden, enjoying the quiet.

On the north side of base is the Zijinyuan of Vanke which is high-side residential occupancy, so the plan has to consider the impact on the residential buildings of sunshine. Hotel styled apartment located in the north of office building, becoming a nature transition between business offices and residential area.

开，根据不同功能及业态在合理性的基础上突出其独特性，以避免与同类项目雷同。建筑造型稳重、华贵，同时兼顾未来建筑的轻巧、梦幻，将两者融合穿插。设计中遵循可持续发展的原则，在知名度、建筑、经济三方面达到整体上的和谐，使建筑与环境达到最佳效果，并成为城市建筑的经典样板，确保总体方案的质量能够满足东北地区最高标准的建筑要求。

"金廊一号"在"金廊"中的位置极为显要，基地形状较为方正，考虑到其基地南侧紧邻五里河体育中心，视野较为开阔，而青年大街又是中轴线的主干道，因此将高层建筑布置在基地南侧和东侧。基地的东南角是进入"金廊"的第一个视觉焦点，有"金角"之誉，是布置超高层办公楼的绝佳位置。从这里可以纵观"金廊"全景，同时还可以俯瞰浑南中心区全区。从城市空间上，以制高点的姿态为"南北金廊"画上一个完美的句号。南侧布置两栋高层住宅，从车辆出入口和人流流线上同其他商务办公区分开，居者活动范围环绕独立的中心花园，闹中取静。

基地北侧靠万科紫金苑，为已入住的小高层居民住宅，因此规划中要考虑对住宅楼的日照影响。酒店式公寓位于办公楼北侧，成为商业办公与住宅区之间的一个自然过渡。基地周边布置商业裙房，各街面均形成完整的商业街面，在城市主要景观道路——青年大街一侧强化大型商业裙房的造型。基地内部的商业用房化整为零，通过"商业街"的形式分成若干个灵活多变的中小型商业体。总体布局中在长方形的基地内强调几个"角"部的变化，裙房的处理上采用"曲线"创造出流动的内街空间，商业、办公、住宅合理进行整合，高效利用土地，合理分配空间。

金廊一号的商业用房可分为两种类型，集中式大型商场和休闲商业区。青年大街是城市的主要交通干道，尺度大而且车流繁忙，因此沿青年大街留10m宽的市政绿化带，大型商场的主入口正好面对这条绿化带。商场的主入口立面营造了一种恢宏的气势，为青年大街街景创造了一个完整的、有合适尺度的立面。休闲商业区由小体量的商业体组成，商业体之间形成内街，内街利用植物、声音及相位，形成趣味空间，建立了亲切感，增强了对比度及流动性。同时商家将店面及其货品直接开放至人行道，以

Around the base are business annex, every street is a complete business street. In one side of the main landscape roads---- Youth Street strengthen the model of large business annexes. Commercial space within the base of piecemeal, through the "Commercial Street" in the form is divided into a number of flexible small and medium business bodies. Overall layout of the rectangular base to emphasize several "corner" the Department of changes in the handling of podium a "curve" to create a flow within the street space, commercial, office, residential and reasonable integration, efficient use of land, a reasonable allocation of space.

The business houses of"Golden Corridor No.1" can be divided into two types Centralized large shopping malls and leisure business area. Youth Street is the traffic main road with wide scale and heavy traffic, thus keep widen 10m municipal greenbelt, the main entrances of large shopping malls right faced this greenbelt. The main entrance of the mall creates a magnificent façade of the momentum, also creates a complete and right scale of façade for the streetscape of Young Street. Leisure business area was formed by small volume of commercial bodies. The small commercial bodies formed inside street which forming funny space by using plants, voice and phase to establish intimacy and enhance the contrast and fluidity. Meanwhile businesses store their shops and goods directly open to the sidewalk t create a unique and good shopping environment. According to the shopping numbers determine the shopping space. After a scale comparison, 6 ~ 10m wide sidewalk is more appropriate. Wencui Road entrance by the collection within the axis into the business, is facing an oval-shaped vitreous, vitreous by the square parcel, the square is at the intersection of the moving line, that human nature is most active place. On the one hand used to provide activities other than shopping, such as assembly, pop display, dynamic performances, on the other hand is a place of public activities and rest. Plaza lighting and use of decoration, furniture and create space for tension, making it the focus of the image of shopping.

"Golden Gallery One," a landmark decision of its landmark building facades. Located in the "Golden Corridor", "future" one end of the end, her position both as a magnificent climax of the end of movement, is also a prelude to the future unlimited dreams. Stand the "Golden Corner" of the shape of tall high-rise office building, the whole formation of the Hanging Gardens of vitreous wings, floating in the sky, with a "prospect" gesture stands in the southernmost city center axis. Apartments and office buildings echo the shape and the formation of primary and secondary contrast. Around the base are large buildings, such as sports centers, schools, space is relatively pure, the main commercial buildings at the entrance to enhanced treatment to the city with generous scale curved space, internal space echoes the shape and appearance, rational use space while creating a highly infectious and futuristic atmosphere, reflecting the city's high-speed operation and development, focusing on fashion and efficiency. Flexible division of space, changeable space, light and shadow contrast to interpretation of modern business office.

塑造独特、良好的购物环境。根据购物者的人数，空间不应过窄以至感觉太过拥塞。经过尺度对比，人行道6~10m宽较为合适。由文萃路入口轴线进入商业集合体的内部，正对着一个椭圆形玻璃体，玻璃体由广场包裹，广场处是各动线交汇点，亦即人性活动最频繁处。一方面提供购物以外的活动使用，如集会、流行展示、动态表演等，另一方面也是公共活动及休息的场所。广场内利用照明及装修、家具等塑造空间张力，使其成为购物的意象焦点。

“金廊一号”的地标性决定了其建筑外立面的标志性。位于“金廊”“未来”一端的端点，它的位置既像一个华丽乐章的结尾部分的高潮，同时又是未来无限梦想的序曲。立于“金角”的超高层办公楼造型挺拔，全玻璃体的空中花园形成两翼，漂浮在天际，以一种“展望”的姿态矗立于城市中心轴的最南端。酒店式公寓和办公楼的造型相呼应，并形成主次的对比。由于基地周边均为大型建筑，如体育中心、学校等，空间相对纯净，商业用房的主入口处强化处理，以大方的弧形空间配合城市的尺度，内部空间造型与外观相呼应，合理利用空间的同时创造了极具感染力和未来感的氛围，体现城市的高速运行和发展，着重时尚、效率。空间的分割灵活多变，利用变幻的时空和光影的对比来诠释现代化的商业办公。

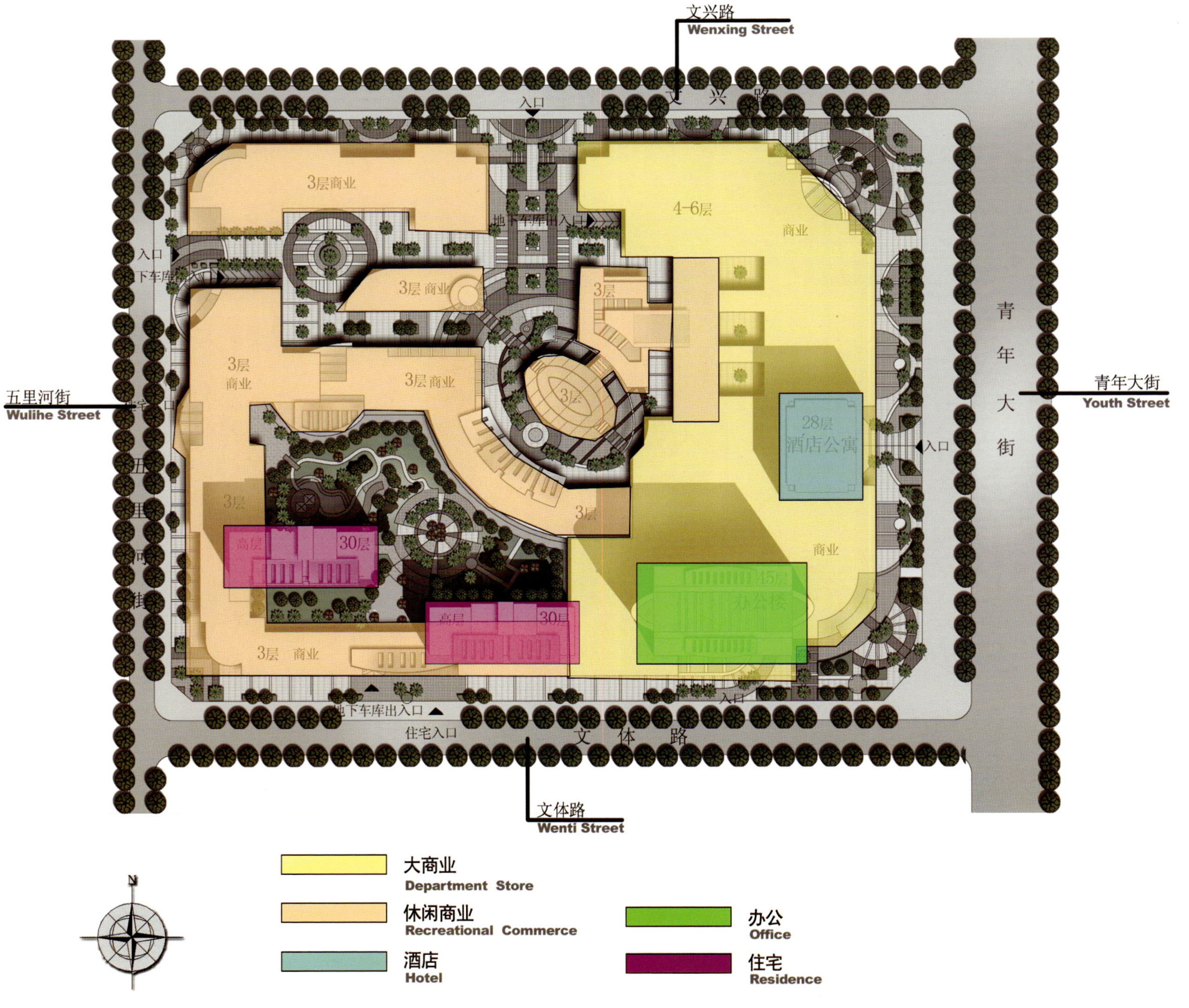
文兴路
Wenxing Street
五里河街
Wulihe Street
青年大街
Youth Street
文体路
Wenti Street
3层商业
4-6层
商业
28层
酒店公寓
15层
办公楼
30层
入口
地下车库出入口
住宅入口
大商业
Department Store
休闲商业
Recreational Commerce
酒店
Hotel
办公
Office
住宅
Residence

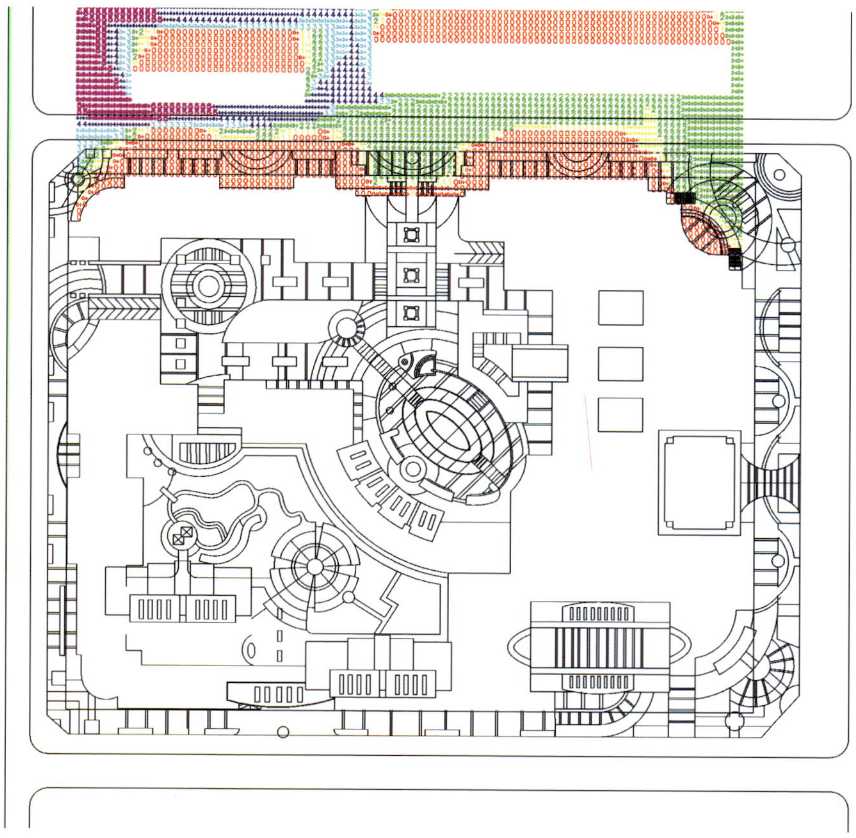

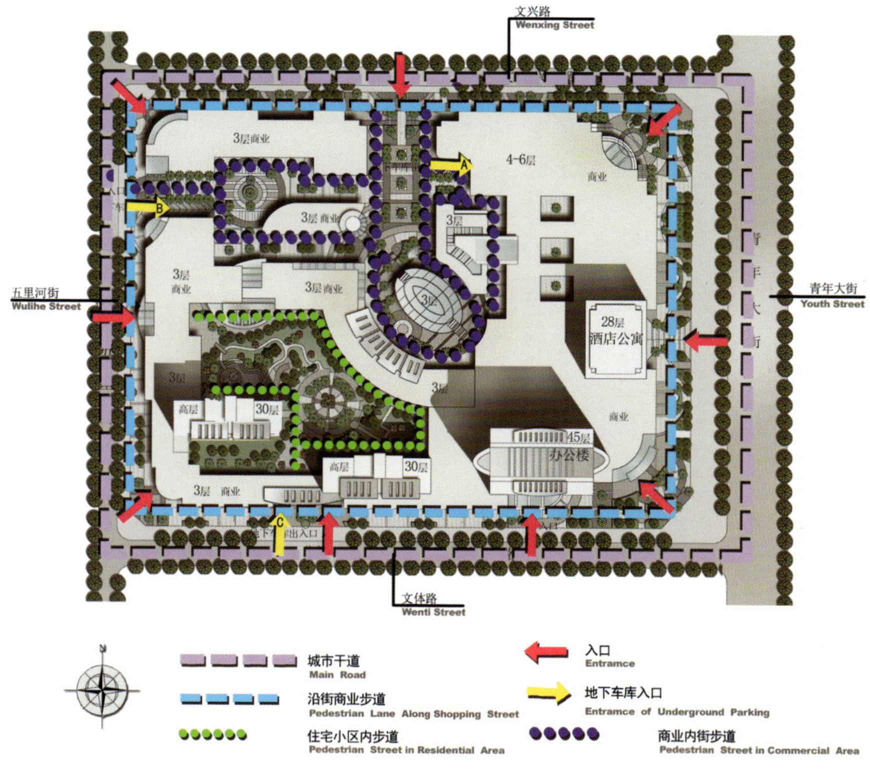

办公楼标准层平面图

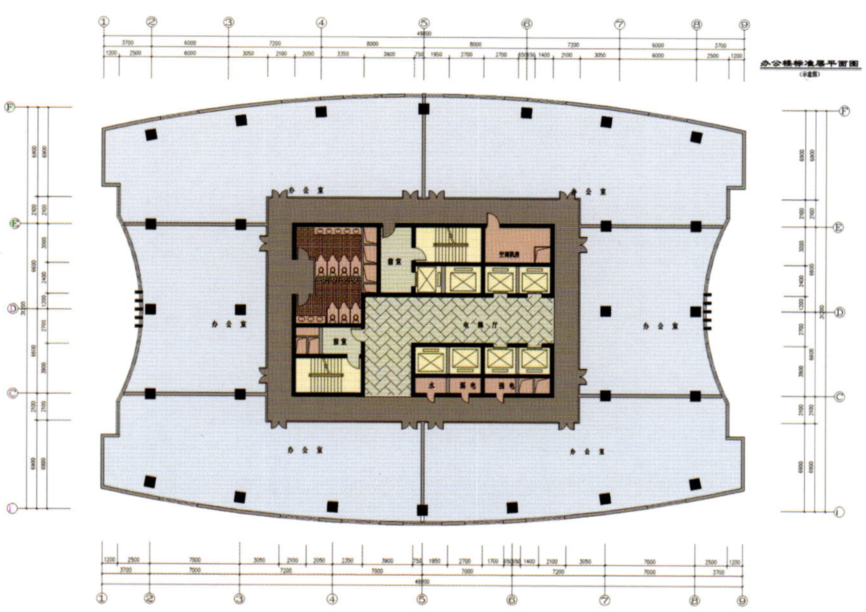

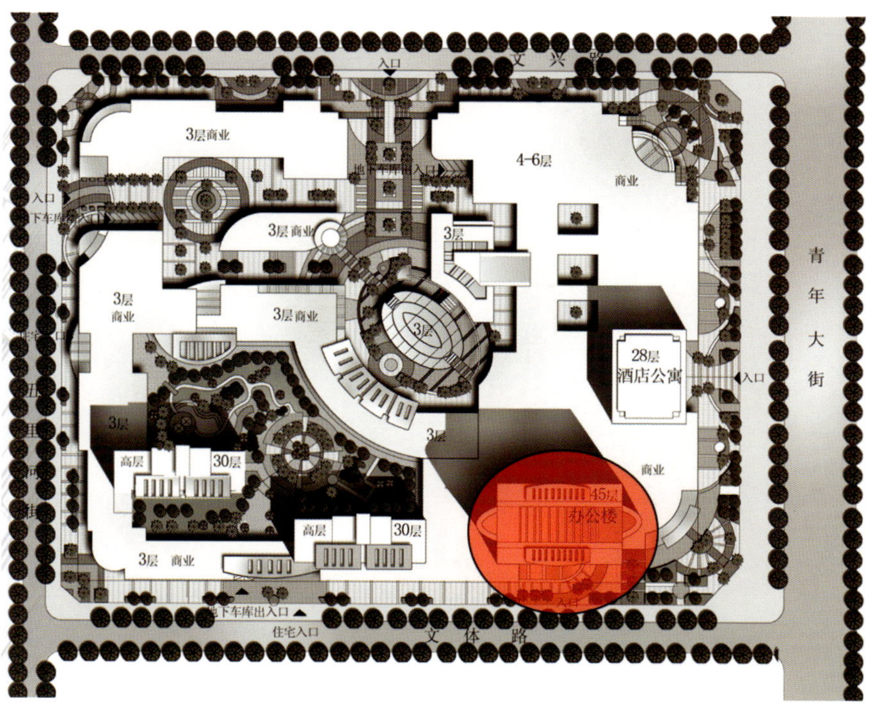

住宅楼标准层平面图

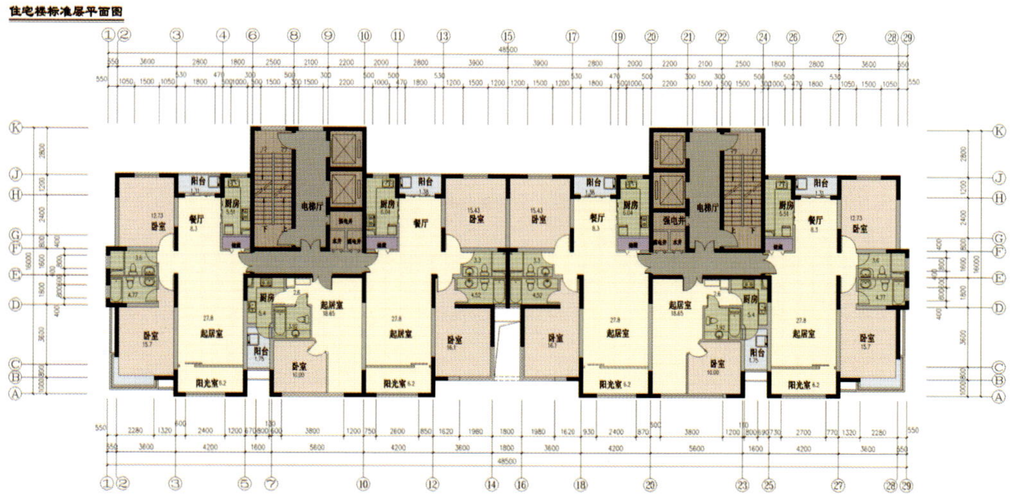

Shanghai Jinqiao software park office building

上海金桥软件园办公楼

Developer: Shanghai Jinqiao (Group) LTD.CO,.
Design company: Henglongwei design consultant LTD.CO,.
Project address: Shanghai Pudong Jinqiao Development Park
Flooring size: 40,074m²
Building size: 40,601m²

开发商：上海金桥（集团）有限公司
设计公司：恒隆威建筑设计咨询（上海）有限公司
项目地址：上海浦东金桥开发区
用地面积：40,074m²
建筑面积：40,601m²

Software Park, Shanghai Jinqiao office building located in the scenic variety of the park trees, fully reflect that emerged in the park office by transnational corporations have the demeanor and the international style.

Shape of the case based on the construction of six office buildings in different but similar in style, while maintaining the unity of architectural style to create a diversity. Software designers take advantage of the rich landscape park resources and the different respect of each office to plan layout and provide a flexible lease. Architectural interior design search for chic and simple structure and the close links between the landscape, blending the pure geometric form, rules law lines, rich materials and strong colors, with giant revolving door, creating a surface new commercial areas to the era of Shanghai style.

A commemorative painting by scaffolding and direct access to the grand reception hall, through the active pattern on the ground, in contrast to the walls and ceiling lines become very rich. The center of the visual is from the side of a two-storey wooden wall and the wall out of a great eccentric rotation.

上海金桥软件园区办公楼位于风景优美、树木繁多的园区内，充分体现出跨国公司在园区内办公所具备的大家风范和国际气派。

本案立足建造6幢外形不同但风格相似的办公楼，在保持建筑风格统一的基础上创造多样性。设计师充分利用软件园内丰富的景观资源和每幢办公楼地块不同的特定规划平面布局，提供灵活的租赁方式。建筑内部设计则寻找别致而简洁的结构与景观之间的密切联系，融合纯几何形体、有规律的线条、丰富的材料和强有力的色彩，结合巨型的旋转门，创造一种面向上海新兴商业领域的时代风格。

经由一个纪念性的烤漆钢棚架可直达宏大的接待大厅，大厅通过地面活跃的图案，强烈对比的墙面和吊顶线条变得异常丰富。而整个视觉中心源自一面两层高的木质墙面以及墙面上开出的一个巨大偏心旋转。

建筑上层的楼面布局结合了由底层及连廊相互结合而成的中庭空间风格，楼内公共走道使用了类似底层的活泼线条。上层楼面的空间布局较为

Upper floor layout of the building comes from the bottom and corridor with a combination of each style atrium space grid. The public hallway use lively line with a similar of bottom. The layout of upper floor space is more formal, design mainly around private offices and independent meeting rooms to conduct. Glass walls, horizontal window and the inner courtyard structure are beneficial to light, to maintain the permeability of space. Some of the space through the "self-by the floating body "to define but not regular walls and corners, in the indoor application of strong color and distinct materials arranged just right to bring out the interior full of natural light and the sweeping park landscape. Most of the room is not decorated ceiling. The bare wire, ventilation duct and spray system painted dark gray oil but also exposed concrete building structure itself, these rough treatments were softened by wooden handling, strong patterns and such means as local enhanced colors.

Number of unique architectural elements of this case draws on the design of Mies' Barcelona exhibition grounds concept, but also integrated into the design team's own inspiration and creativity. These elements include glass music box: The entire building is a vivid metaphor. Music box contains a number of independent "object" elements, such as electronic "board" as the great revolving door pattern, others such as "memory chip" and "DNA Analysis Figure "has also become a pattern template by the wall, ceiling and floor . "Primitive canoe "is a rough access. Be sets of high-tech and interior decoration in order to form a strong contrast. "Glass box" is a special room with drop window, located in the building and its special place, by making use of external landscape as a core.

In addition, each office buildings are equipped with platform, extending interior landscape to the park zone. It can also be used as part of the roof of parking space. Distinctive canopy roof of the office buildings strengthen the uniqueness of the construction.

Six office buildings on both sides are the two main landscape areas. East is near the lake and west by the river. The two main facades face the two landscape belt. West across the river is a food-based business zone, the base across the north with a pedestrian bridge of Xiejiabang, leading business people for the office band provides a convenient, at another base in south-west of road bridge links the same office and business zone, it also be the second entrance of park. In addition the design of a north-southern pedestrian path helps the west-southern exit to cross the two bridges.

正规，设计主要围绕私密的办公室和独立的会议室而进行。玻璃墙、横窗及内部庭院构造都有利于采光，保持了空间的通透性。某些空间通过“自由浮体”的布局来定义而非常规的墙和转角，在室内应用强烈色彩，鲜明材料布置恰到好处地衬托出室内充足的自然光线及一览无余的园区景观。大部分室内不布置吊顶，而将裸露的电线、风管及喷淋系统涂上深灰色油漆，同时也暴露建筑结构本身的混泥土，这些粗犷处理再通过丰富的木工，强烈的图案及局部加强的色彩等手法进行了柔化。

本案一些建筑元素独具特色，借鉴了密斯的巴塞罗那展览馆的设计理念，但同时也融入了设计小组本身的灵感和创意。这些元素包括玻璃音乐盒：整个建筑是形象比喻。音乐盒中包含了一些独立的“物体”元素，如电子“主机板”为图案的巨大旋转门，其他如“记忆芯片”和“DNA解析图”也成为墙体、吊顶和地板的图案模板。“原始木舟”为一个粗犷的接待台，以此来与室内高科技装饰形成强烈对比。“玻璃盒子”是一个有落地窗的特殊房间，位于建筑及其特殊的位置，并可利用外部景观为要旨。

此外，每幢办公楼都设有平台，将室内空间延展至园区丰富的景观带，同时亦可做为停车场的部分屋顶。办公楼的屋顶雨棚各具特色，加强了建筑的独特性。

6幢办公楼两旁为两个主要景观区域。东面邻湖，西面靠河，办公楼的两个主立面正对这两个景观带。穿过小河向西是一个以餐饮为主的商业带，基地北面设有一座横跨谢家浜的人行天桥，为办公楼的人群通向商业带提供了便利，位于基地西南的另一座公路桥同样链接办公楼和商业带，也是园区的次入口。此外，设计了一条南北向的行人小道，有助于建筑西南出口通达两桥。

P 停车库

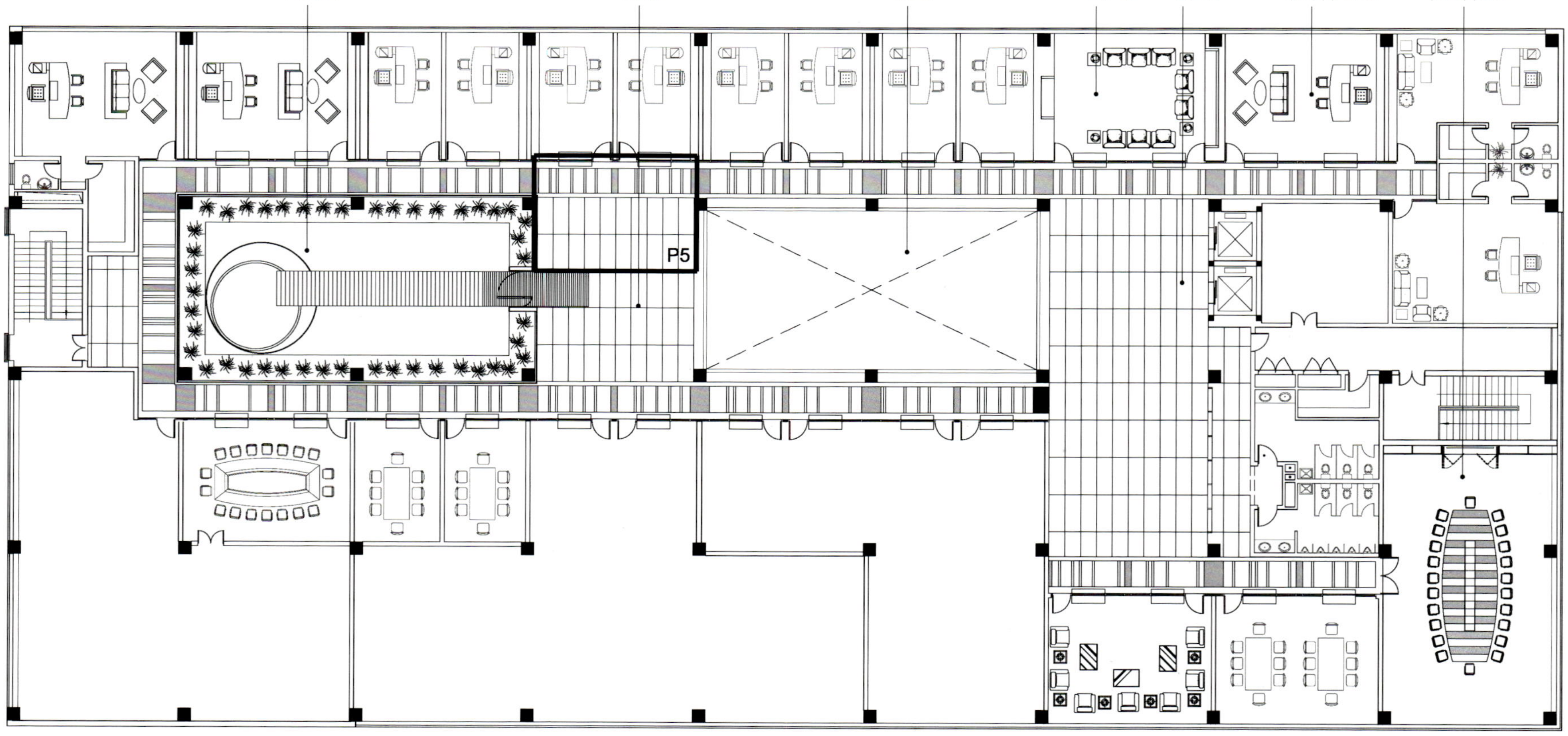
内庭园
走廊
上空
VIP室
电梯厅
领导办公室
大会议室
P5

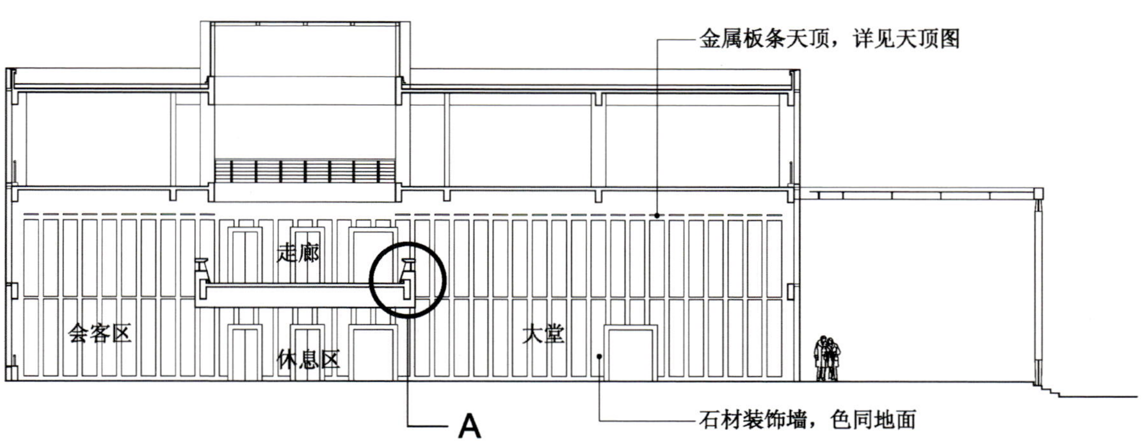

剖面 1-1

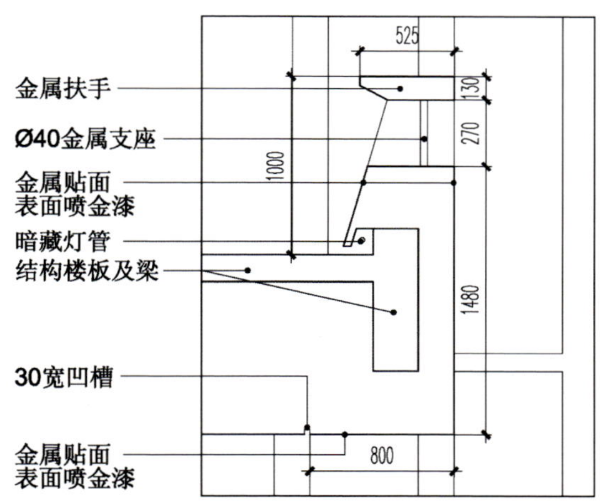

细部剖面 A

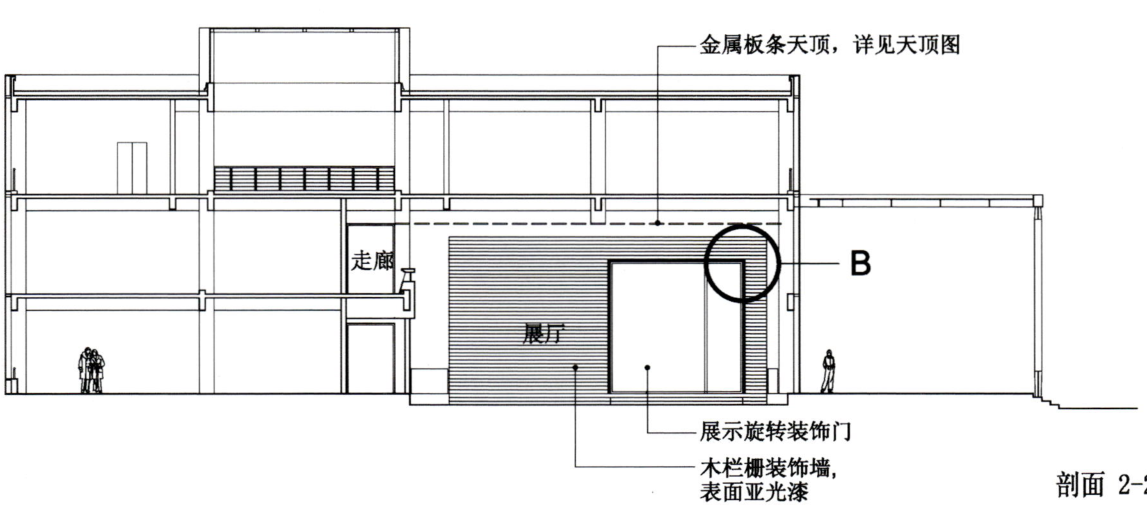

剖面 2-2

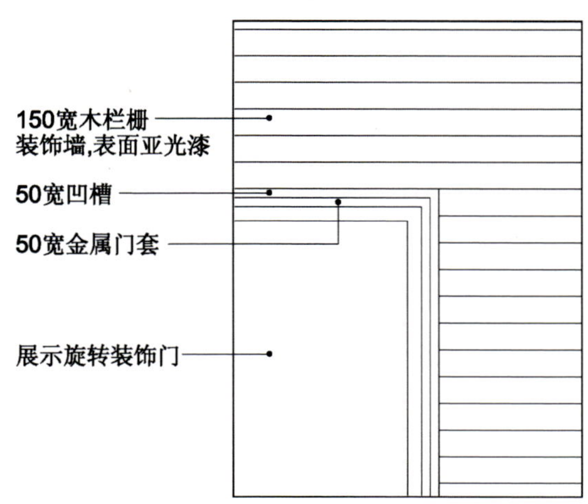

细部剖面 B

Thermo
6

7

HEALTH IMAGING
A BETTER VIEW OF LIFE

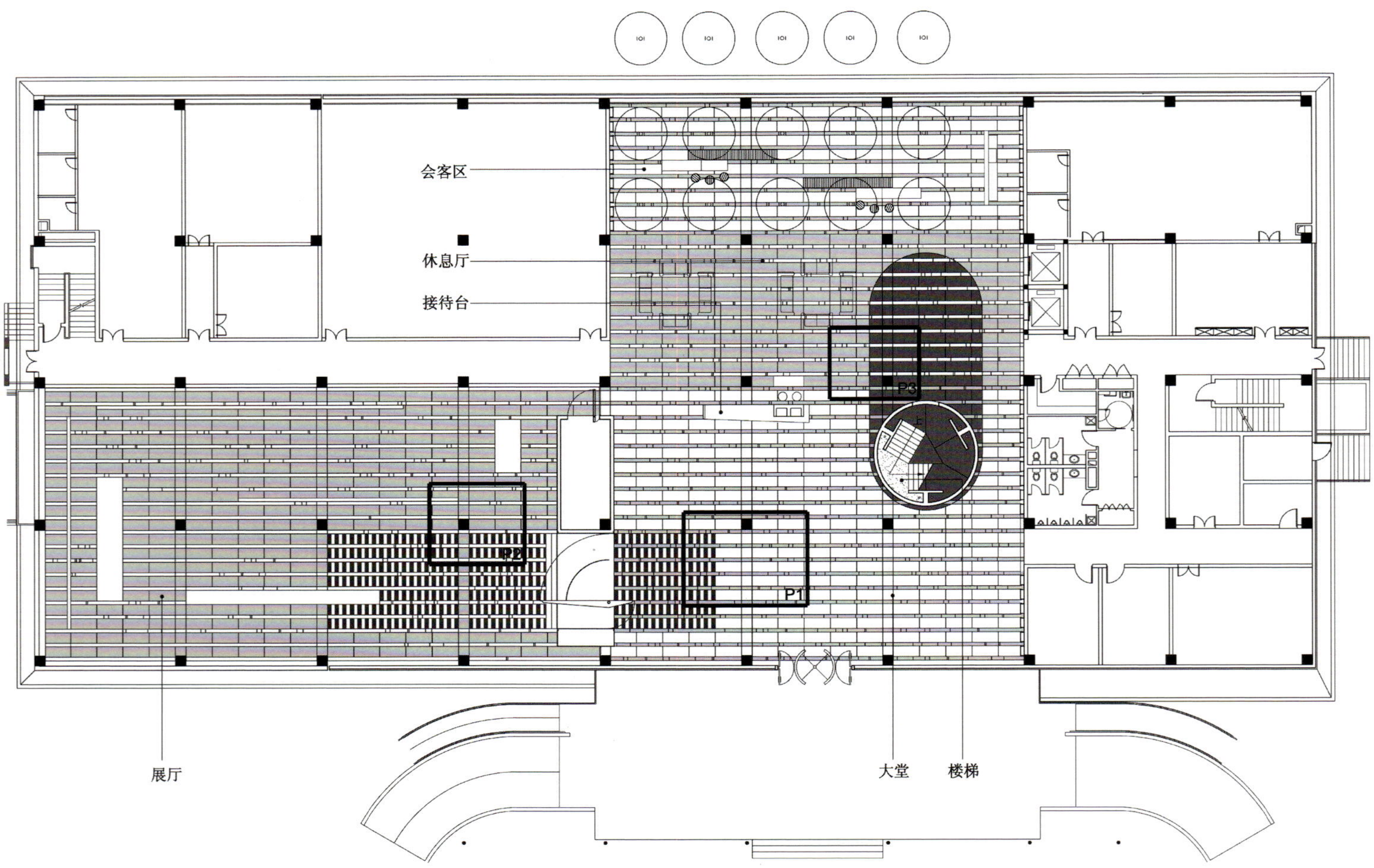
会客区
休息厅
接待台
P1
P2
P3
展厅
大堂
楼梯

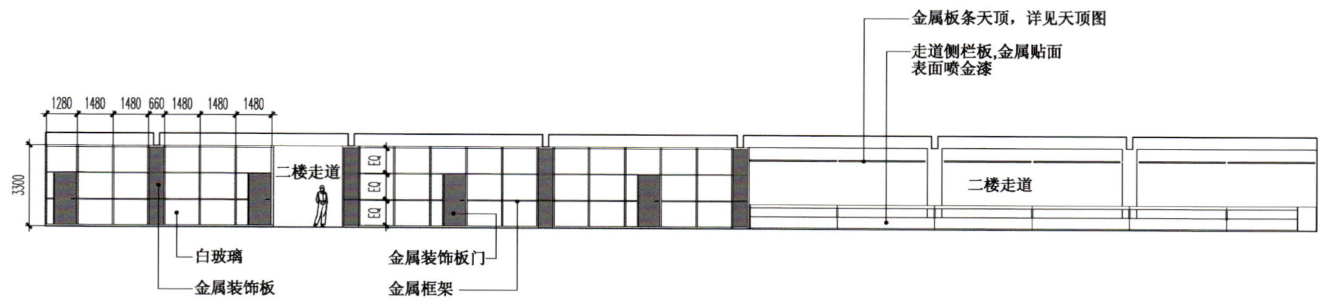

剖面 4-4

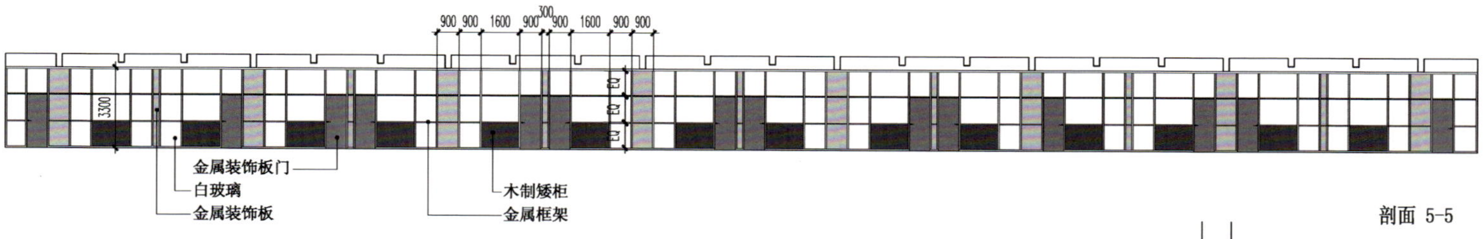

剖面 5-5

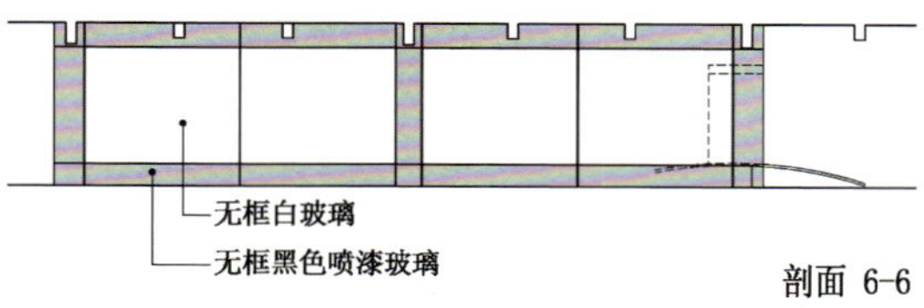

剖面 6-6

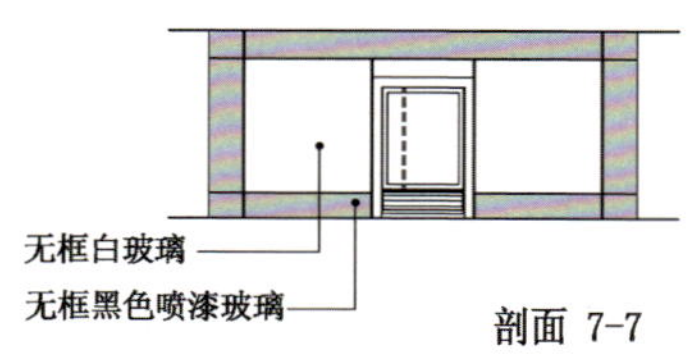

剖面 7-7

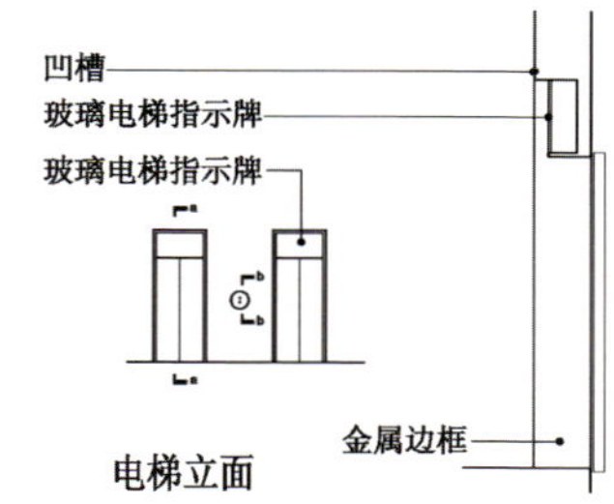

电梯立面

剖面 a-a

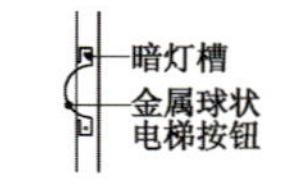

剖面 b-b

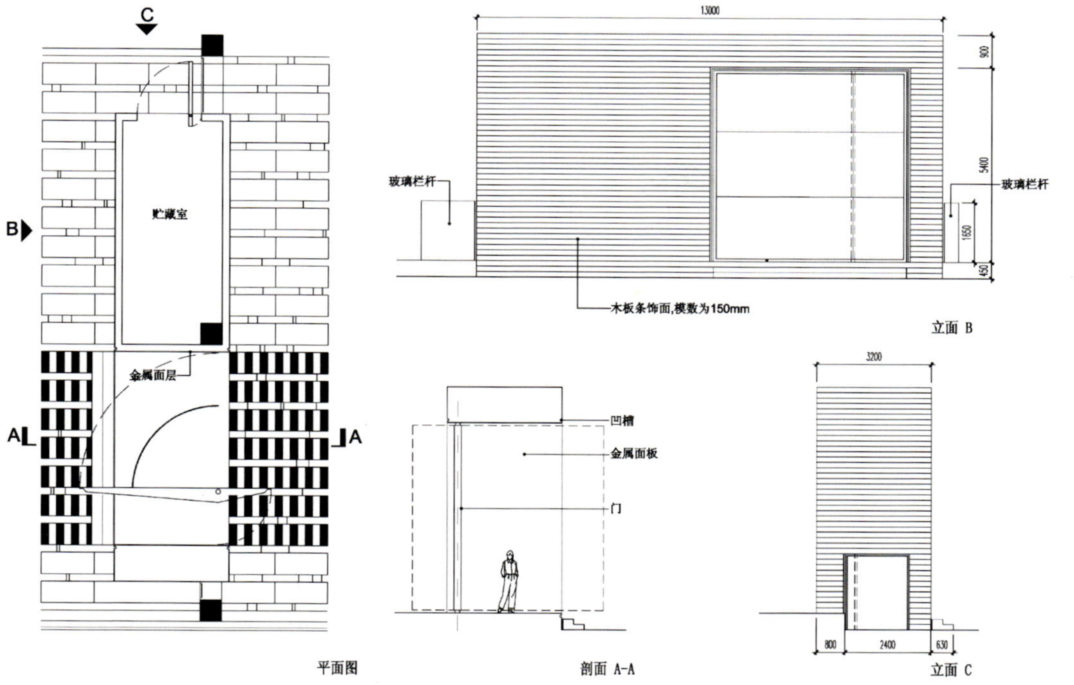
贮藏室
金属面层
平面图
木板条饰面,模数为150mm
玻璃栏杆
立面 B
凹槽
金属面板
门
剖面 A-A
立面 C

新金桥路
主入口
湖北大厦
金桥软件园区总部
红枫路
曹家浜
次入口
明月路
金桥路
马家浜

Research Center for Quality & Development of Group Danone

达能集团研发与质量中心

Client: Danone
Designer: Architecture-Studio
In Research: Choulet, Martin, Babino
Landscape Design: Franç Oise Arnaud
Acoustics: AVA
Economist: Cyprium
Area: 30,000m²
Cost: EUR 35,000,000

业主：达能集团
设计者：Architecture-Studio
参与研究室：Choulet，Martin，Babino
景观设计： Françoise Arnaud
声学： AVA
经济师： Cyprium
占地面积：30,000m²
造价：35,000,000欧元

Facilitate the exchange of space: modular design create high-quality building space to accommodate the progress of various development projects throughout the process; various informal public space for people in different research members. Between the bottom and the two-story building, between indoor and outdoor constructions, between the artificial building and the natural landscape by exchanging.

Changing the overall Natural landscape architecture: the flexibility of building structures make sure the availability of indoor building space degeneration, the extension of different core space in three directions is one of the basic concepts of architectural design, Become a unique architectural vocabulary.

A combination of architecture and nature of organisms closely: through the two broken of space inside the building and the curve treatment, the unique natural environment is naturally introduced into the interior, to arouse people's surprise and emotion; building itself also reflects the way the construction of the natural environment of respect for good.

有助交流的空间：模数化的设计创造出了高质量的建筑空间，容纳着不同研发项目的进展全过程；各种非正式的公共空间促进了在不同研究人员之间，在建筑底层和二层之间，在建筑室内与建筑室外之间，在人工建筑与自然景观之间的相互交流。

不断变换着的建筑整体：建筑结构的灵活性确保了室内建筑空间的可变性；不同的核心空间在3个方向上的延伸是建筑设计的基本理念之一，形成了独特的建筑语汇。

一个建筑与自然密切结合的有机体：通过建筑内部空间的两处断裂及其曲线化处理，周围独特的自然环境被自然而然地引入到室内，唤起了人们的惊奇与情感；建筑本身的建造方式也体现了对良好自然环境的尊重。

这个研究中心不仅仅是一座建筑，而且是若干建筑的集合体，在这里将实施一项具有试验性的开发计划，集研发机构、实验室和办公楼为一体。设计为即将在其中生活的人们创造了一幅技术与景观，热情与冷峻完美结合的景象。

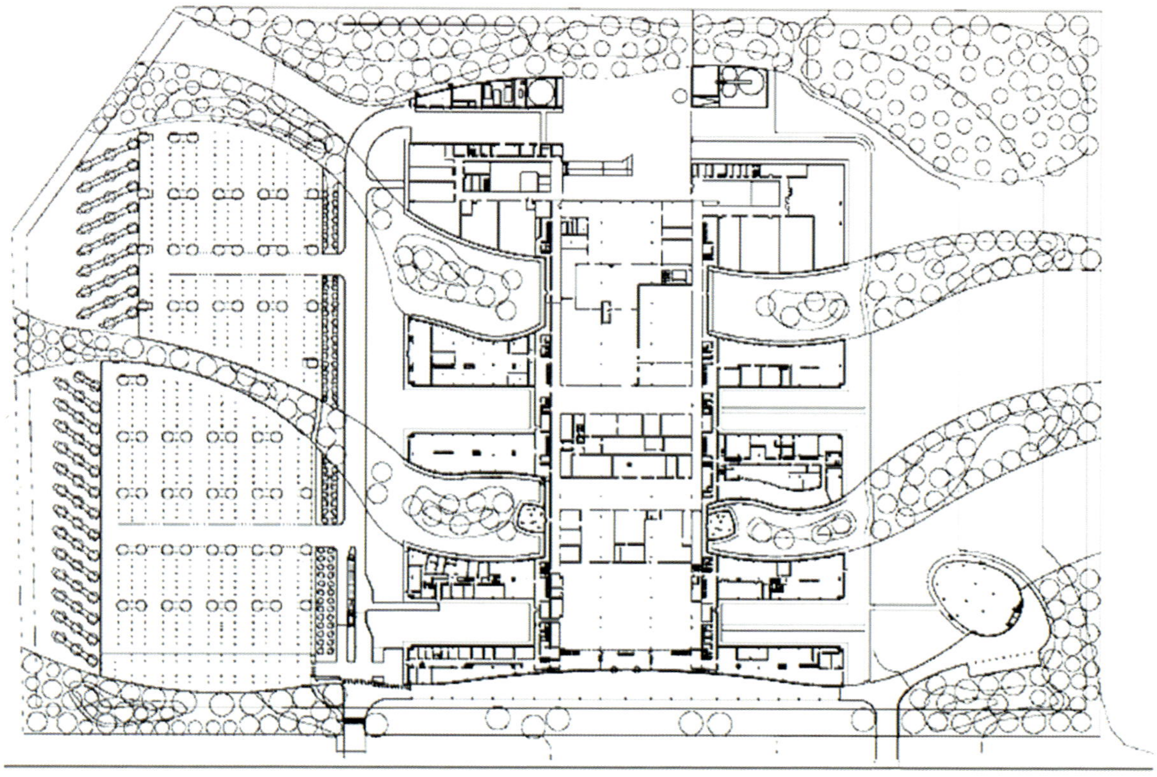

The research center is not just a building, but also a number of construction aggregates, where will implement a development plan with the pilot, R & D institutions, laboratories and offices for the body. were designed to be created in which people live has a technology and landscape, warm and Cold End The United States with the scene.

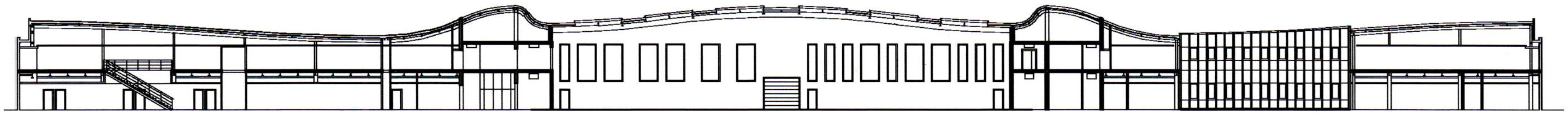

Shanghai Huisheng Chemical Group Headquarter

上海惠生化工集团总部

Building design: French AS Building Studio
Technical cooperation: Pierre Martin, Choulet
Project address: Zhangjiang science & Technology Park
Design size: 20,000m²

建筑设计：法国AS建筑工作室
技术合作：Pierre Martin, Choulet
项目地址：上海张江科技园
设计面积：20,000m²

Shanghai Huisheng Chemical Group Headquarter is located in the center of Zhangjiang science & Technology Park of Shanghai. the park is designed with a giant undulating wave-shaped roof covering up. Thus a clearly visible unified whole and visual striking effect in general is formed. The design of the roof highlights city building feature and outlines the characteristics of a representative distinctive porch. Green space is separated into different sizes of the gardens; the nature permeates every corner of the building, creating a high quality human environment and ecological environment. The design features is in the following points:

(1) The motive floor
(2) Central ecological corridor
(3) The integration of greening room and the building
(4) The integration of the water and building
(5) The adoption of nature light

The feature of this case is reflected through the motive and coordination of the roof. The case is the base for research and work. Meanwhile it is a

上海惠生生物化工园区总部位于上海张江工程园区的中心。设计方案将园区用一张巨型的波浪形起伏的屋面覆盖起来，由此形成一个明确可见的统一的整体，形成了一种视觉总体上的鲜明效果。屋顶的设计突出了城市建筑的特色并勾勒了一个有代表性、有特色的门廊。整体建筑被绿色空间东西向分割成不同规模的花园，使大自然渗透进建筑物的每个角落，创造了一个高质量的人文环境和生态环境。其设计特色主要体现在以下几点：

（1）动感的屋顶；
（2）中央生态走廊；
（3）绿化空间与建筑的渗透；
（4）水与建筑的渗透；
（5）自然光线的采用。

该方案通过屋顶的动感和协调来反映它的特征。方案是研究和办公的基地，同时也是一个交流的空间，它通过一张覆盖着工业金属板的大屋顶外壳，在立面部分透进阳光，在花园处融入天空。屋面在正立面处垂落入

WISON
惠生科技园
WISON SCIENCE & TECHNOLOGY PARK

communication room. Industrial metal plate covered with a large roof shell, in part through the facade into the sun, in the garden office into the sky. The façade of the roof get down into the water. The part is decorated with light elements, from the bottom gradually become dense, so that the façade have effect from transparent to translucent.

A clear road provides a framework of clearly visible building space. Its central axis determines its convenient location to serve the various functional spaces. The gallery space directly links offices and laboratories, biological products factories and restaurants, gardens and others. Green plants are planted along the wall to form a long Chuilian, to prevent too strong sunlight. Louvered glass curtain walls of the corridor eliminate the heavy feeling. The transparency of the two central corridors faces the water. Due to artificial drainage in the outdoor and affordable volume, the corridor became severed from the world of science "bridge." The staircase is set in the gallery side. Go with the water, so that the people are closer to nature, to strengthen the Academy of the program areas of ecological characteristics: transparent, open, full of water inspiration.

Each season, each garden has a different color. Bamboo Garden is located around restaurants, to pure the chemical air with plants. Buildings surrounded by gardens are to provide a pleasant working environment for employees. Transparent volume can be crossed to create a vision, seen through the office building gardens, trees and other construction parts.

The building is designed that water flow into the building. The ecological corridor from the north and south with the same artificial drains are throughout the buildings. Artificial lakes and the creation of artificial drains, make the water dependent and associated with the building, filled with lake water affect each buildings a vision.

Natural lighting throughout the park is the dominant ideology. It is reflected through the large transparent glass facade, ceiling and open-air atrium roof light. North-south building has plenty of sunshine and natural light. Natural light plays the theme, building a link between heaven and earth.

水中，该部分采用透光轻盈的装饰性构件，由下至上渐渐变密集，使正立面产生从透明至半透明的效果。

一个明确的路线提供一个清晰可见的建筑构架空间。它的中央主轴位置决定了它方便地服务于各个功能空间。该长廊空间使办公室与实验室、生物制品工厂及餐厅、花园等直接联系起来。沿墙面种植的绿色植物，形成长长的垂帘，防止过强的阳光。百叶窗式的玻璃幕墙消除了长廊的厚重感。两面临水的透明中央走廊，由于设在室外人工水渠及其相宜的体量，使长廊成为超脱尘世的科学之“桥”。楼梯设置在廊侧越水而上，使人群更加接近大自然，加强了该方案生态科学院区的特质：透明、开敞，充满水的灵感。

每个季节，每个庭院都有不同的颜色。竹林园设在餐厅周围，植物净化了化工厂的空气，被庭院环抱的建筑群给职工提供一个愉悦的工作环境。透明的体量创造出可穿越的视野，透过办公建筑可见花园、树木和其他建筑部分。

该建筑设计将水引入建筑，南北向的生态长廊伴同人工水渠贯穿整个建筑群，通过人工湖泊和人工水渠的创造，使水与建筑物相依而伴，湖光水影充满了建筑群的每一个视野。

整个园区以自然采光为主导思想。它通过大片透明的玻璃立面，露天的中厅天棚和透光的屋顶体现。建筑物的南北向拥有充足的日照和自然采光。自然光演奏主题，建筑成为与天地联系的纽带。

WISON
惠生科技园

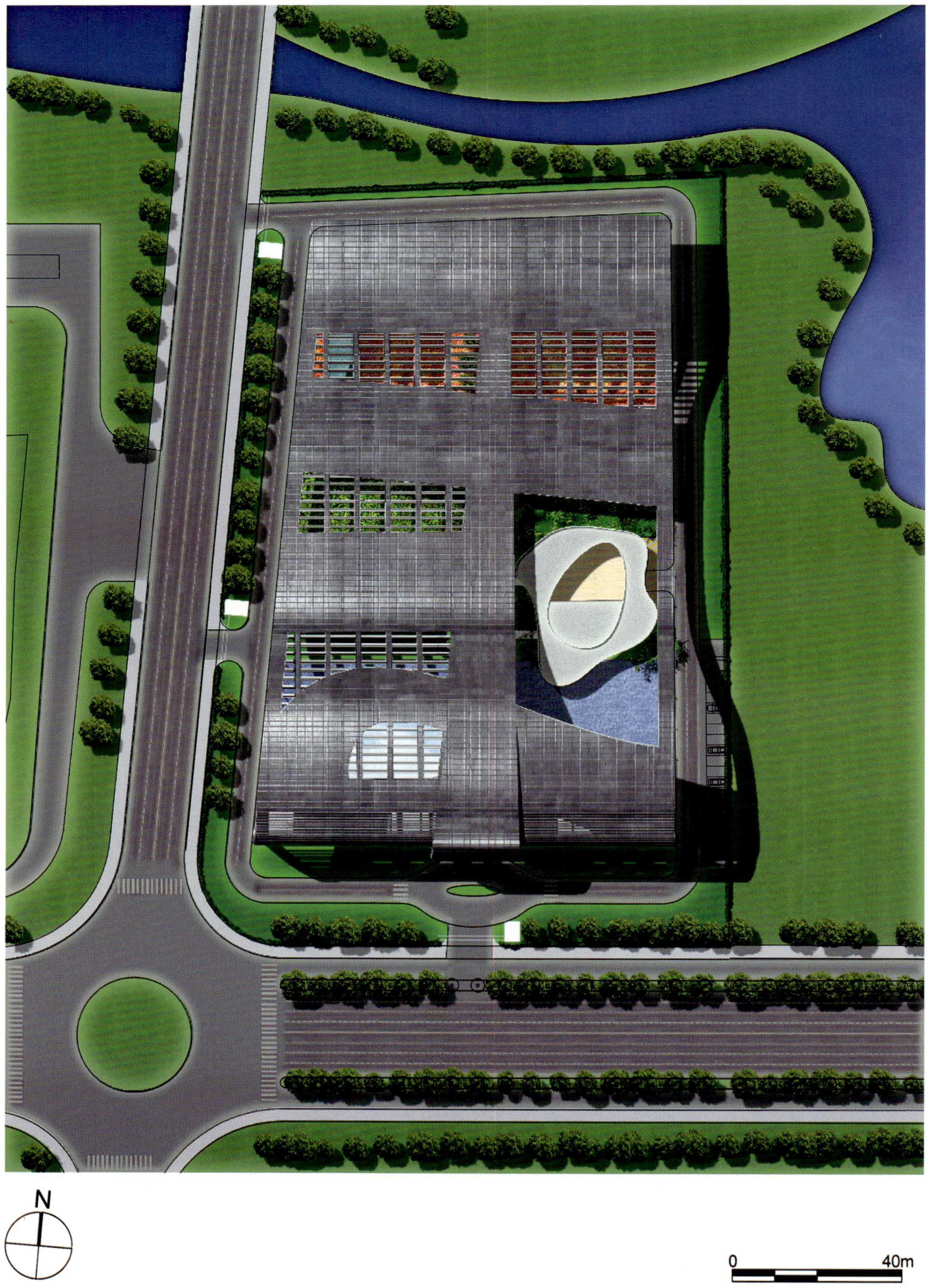
N
0
40m

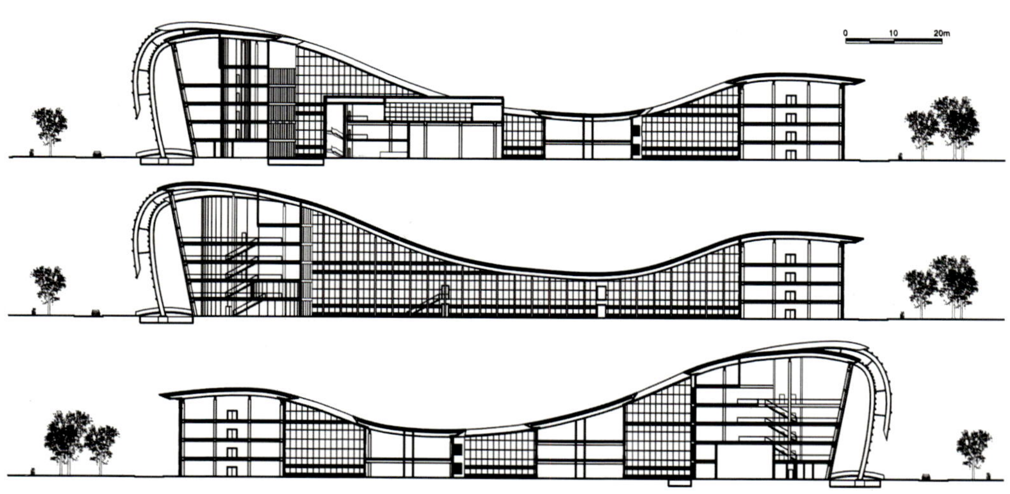
0
10
20m

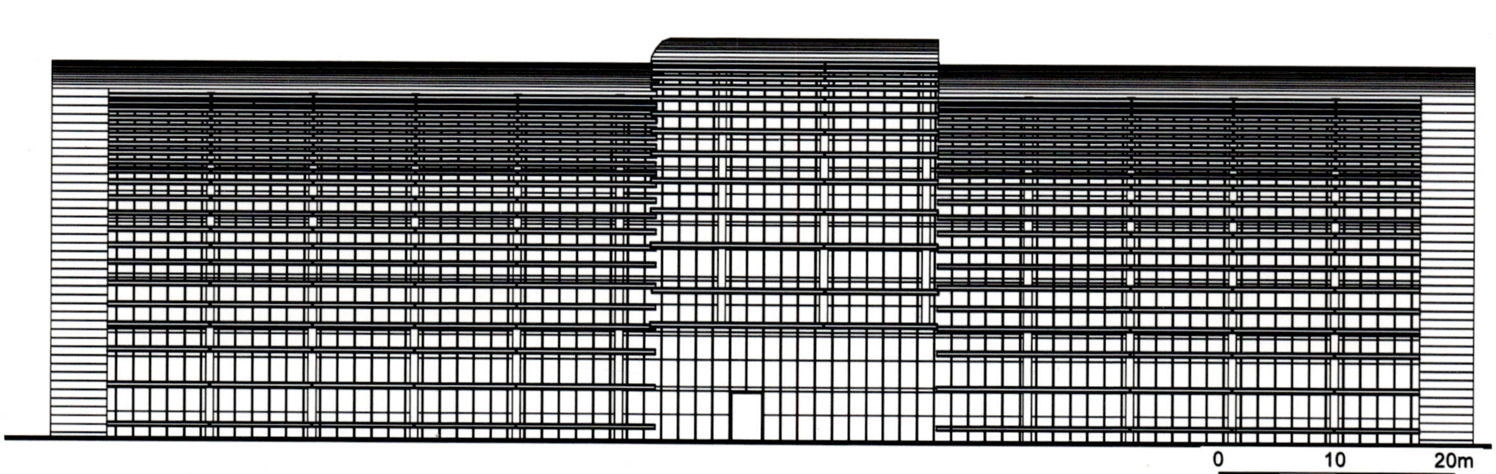
0
10
20m

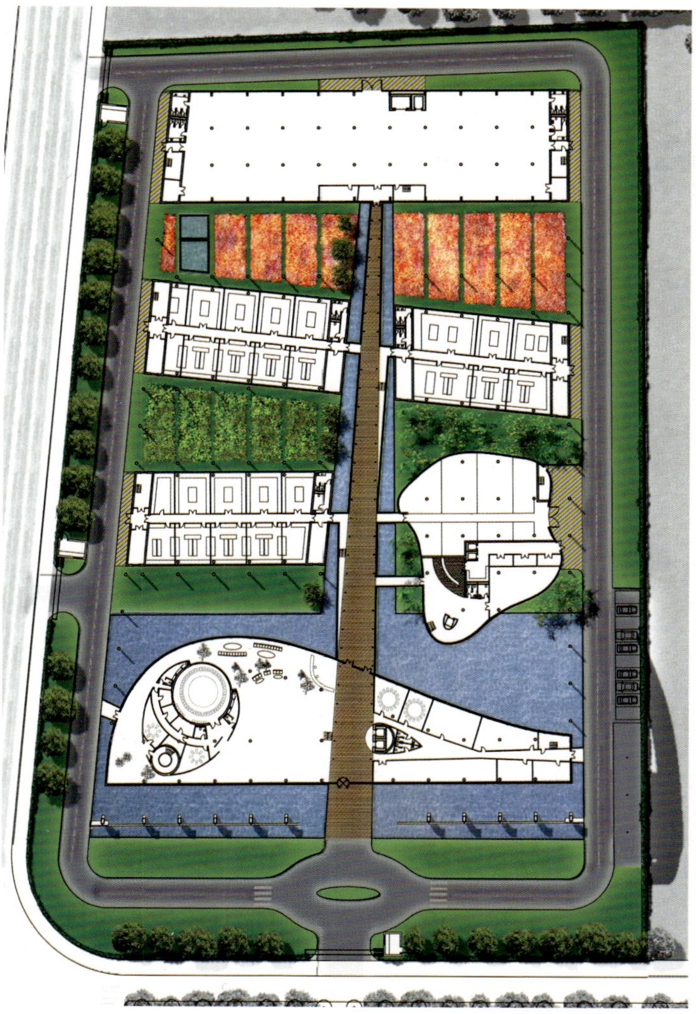

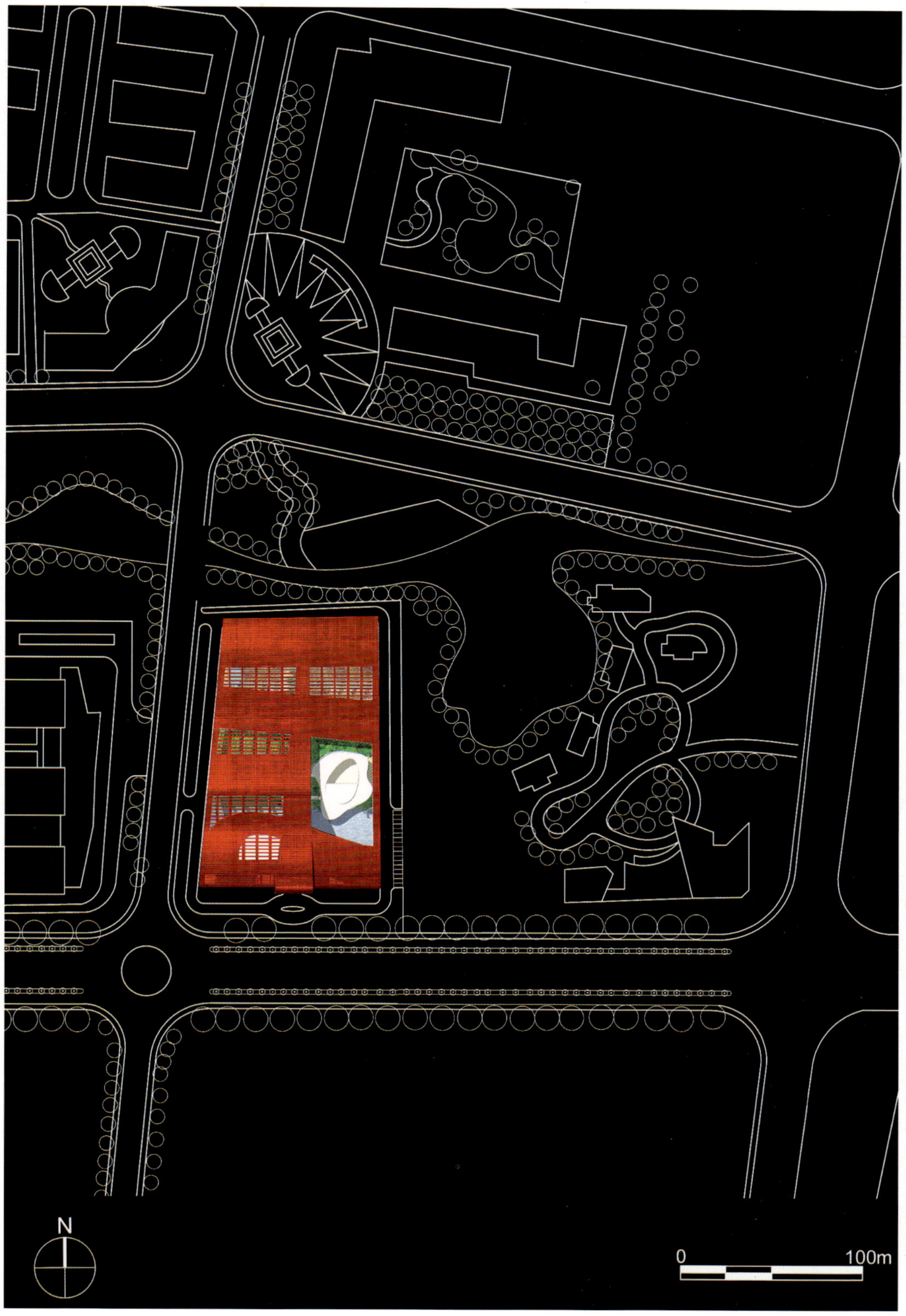
N
0
100m

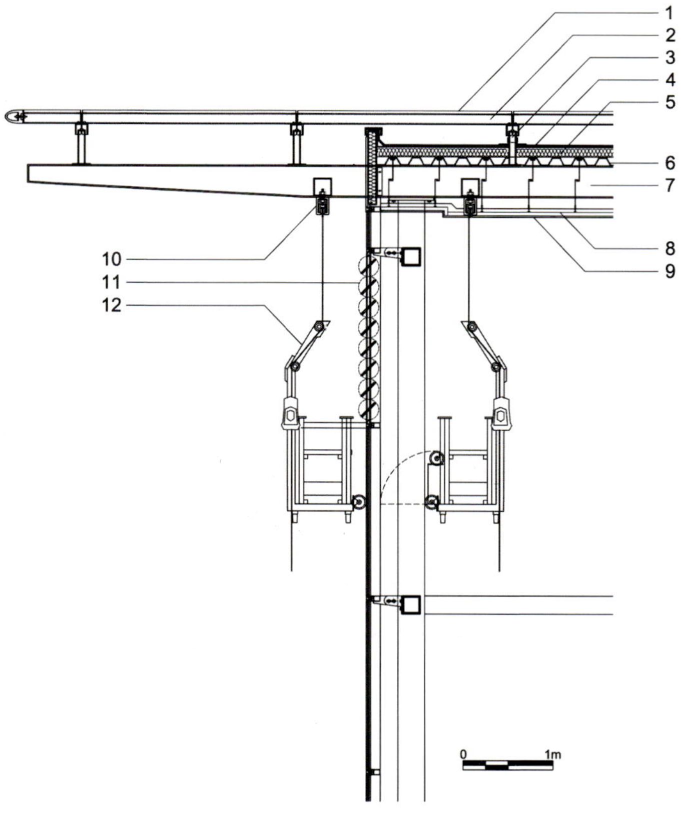
1
2
3
4
5
6
7
8
9
10
11
12
0
1m

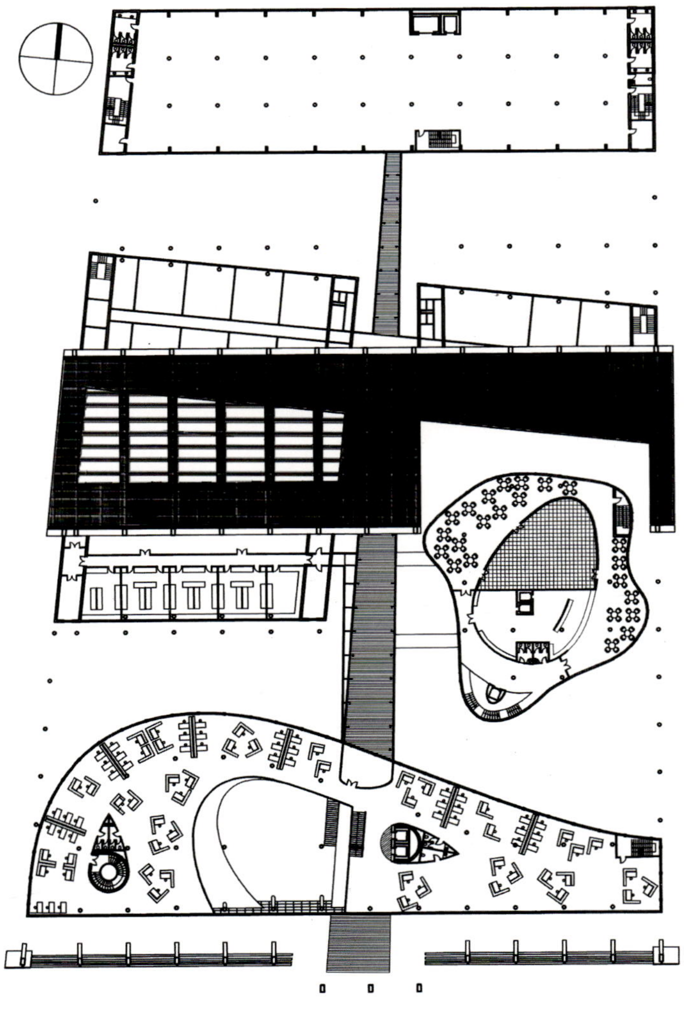